THE LIBRARY
ST. MARY'S COLLEGE OF MARYLAND
ST. MARY'S CITY, MARYLAND 20686

Carriage Terminology:
An Historical Dictionary

Carriage Terminology:
An Historical Dictionary

Don H. Berkebile

Copublished by

SMITHSONIAN INSTITUTION PRESS and LIBERTY CAP BOOKS

1978

Copyright © 1978 by the Smithsonian Institution. All rights reserved.

Library of Congress Cataloging in Publication Data
Berkebile, Donald H.
Carriage terminology.
Bibliography: p.
1. Carriages and carts—Dictionaries.
I. Title.
TS2020.B47   688.6   77-118

Smithsonian Institution Press Publication Number 6028

Cover: Buggy, see page 51.

Frontispiece: The 1848 Concord Coach in the Smithsonian collections, see page 105.

Contents

Introduction 7

Ezra M. Stratton 9

George W. W. Houghton 10

Key to Source Identification Symbols 11

CARRIAGE TYPES 13

CARRIAGE NOMENCLATURE 307

HARNESS NOMENCLATURE 411

WILLIAM FELTON'S 1796 GLOSSARY 467

Selected Bibliography 483

Introduction

A precise communication in any area of technology requires a standard terminology. Designers, manufacturers, dealers, and consumers need to communicate with the assurance that the terms used have a clear and, preferably, a single meaning. This is more than a matter of convenience; otherwise, costly and troublesome delays can result. Many industries, therefore, attempted to compile trade dictionaries at an early date. Trade or professional associations usually spearheaded these movements, but in the carriage industry this vital task was left to the trade publications, for no official effort was ever made by the manufacturers to publish a dictionary of carriage terms.

The December 1, 1888, issue of *The Hub* advised: "The need of an accurate and standard vocabulary of carriage titles has long been recognized by the trade, who, in the absence of authority, naturally used such titles in the vaguest and most confusing manner . . ." Today the intense interest in objects of the past has once more focused attention on carriages, so that the need for a standard vocabulary is again felt. The scarcity of early carriage literature has made it desirable to consolidate the information contained in these carriage trade journals into one convenient reference work.

In July 1865, *The New York Coach-Maker's Magazine*, edited by Ezra M. Stratton, began a serial entitled, *American Dictionary for Coach-Makers*, which described both carriage types and nomenclature. Stratton's magazine was absorbed in 1871 by *The Hub*, a journal that had begun publication two years earlier under the editorship of nineteen-year-old George W. W. Houghton. Later, after Mr. Stratton had completed his now well-known work, *The World on Wheels* (1878), *The Hub* purchased his carriage library, adding to it numerous other acquisitions. With the *American Dictionary for Coach-Makers* as a nucleus and this extensive library at his disposal, Houghton, aided by Adolphus Muller, J. D. Gribbon, Howard M. DuBois, F. B. Patterson, and others, prepared *The Hub Dictionary of Carriage Terms*. Serial publication began in the April 1879 number of *The Hub*, describing both carriage types and nomenclature, as had Stratton's magazine.

The editors of *The Hub* felt that the series was worthy of separate publication, and consequently prepared a revised manuscript for the purpose. Just prior to scheduled publication, the offices of *The Hub* were destroyed by fire on July 22, 1883, and the manuscript and proofs, as well as the library, were consumed. In 1891, *The Hub News*, a weekly, again published the terms in serial installments. Finally, more than a year after Houghton's death,

The Hub announced in the December 1892 number that the *Vocabulary of Vehicles* was ready for distribution. Illustrated with twenty-six plates, the volume contained only the carriage types, and sold for one dollar.

Houghton's work serves as the basis for the present volume. Other primary sources are *A Dictionary of Carriage Terms*, serial publication of which began in William N. Fitzgerald's *Carriage and Harness Journal* in 1871, and the *Carriage Trimmer's Technical Dictionary*, published in 1881 in Fitzgerald's *Carriage Trimmer's Manual and Guide Book*. Numerous secondary sources assisted in completing this dictionary. Code letters following each entry identify these various sources of information.

In compiling this dictionary, the aim has been to gather together all of the known carriage types of the Western world, as well as many of the better known vehicles of other areas, and to give the derivation of the terms, if known, and the history and characteristics of the vehicles to which the terms are applied. Obviously, such a task will result in some omissions, though it is hoped that nothing of significance has been overlooked. Many of the terms have several meanings, all of which may be entirely legitimate. In some instances, the usages given will appear to be improper, yet they are given because, in certain areas, such usages were common for one reason or another. Likewise, several different terms are at times applied to the same vehicle. Such practices often result in considerable confusion as to the precise meanings of certain terms, unless they are accompanied by qualifying statements that clarify the meanings. An effort has been made in this volume to point out the preferred usages, not to the exclusion, however, of other usages that might be encountered. Because a few of these less common and overlapping usages, some contradiction appears in this dictionary. There has not been an appreciable effort to include the many trade names that some individual manufacturers applied to vehicles that were known by a more common name, unless the trade name came into relatively common usage.

To a large extent this work has been influenced by the vernacular of the carriage trade in the larger manufacturing centers of the eastern United States. This is particularly true of the terms in the *Carriage Nomenclature* section, consequently it will not be in total agreement with the terminology of other parts of the country, such as the West and South. Hopefully, however, a better understanding of the types and construction of vehicles might be gained by a knowledge of these terms. It is also hoped that *Carriage Terminology: An Historical Dictionary* might serve as an outline history of the carriage.

Ezra M. Stratton, (*The Hub,* May 1883.)

Ezra M. Stratton

Born on May 11, 1809, on a farm in Fairfield County, Connecticut, Ezra M. Stratton received an intermittent elementary education, interspersed with periods of farm work. In April 1824 he was apprenticed for a five-year term to Charles Townsend, of the firm of Platt and Townsend in Saugatuck (now Westport), Connecticut, where he learned the carriage-building trade. Later, in serial installments in *The New York Coach-Maker's Magazine*, he published an account of these early experiences under the title, "The Autobiography of Caleb Snug." Here he described his apprenticeship in terms of hard work, sixteen-hour days, and a meager diet of pork and potatoes that, fortunately, could be supplemented on Sundays owing to the proximity of his father's home.

Following his apprenticeship, he was employed as a journeyman carriage builder for three years, until he moved to New York, where in 1836 he opened his own carriage shop. This he operated until 1858, when he began publication of *The New York Coach-Maker's Magazine* in June of that year, using the several years of experience he had gained as assistant editor of C. W. Saladee's *The American Coach-Maker's Illustrated Monthly Magazine*. Since this undertaking allowed him a surplus of time, the aggressive Mr. Stratton spent six hours a day in the study of literature, history, and modern language, which still left him sufficient time to become a prosperous coal dealer. In July 1865 he began the serial installment of his *American Dictionary for Coach-Makers* in the magazine, and, thus, might be termed the *grandfather* of this dictionary. Finally, in March 1871, his magazine was absorbed into *The Hub*, which had started as a trade newspaper in 1869, under the editorship of George W. W. Houghton. Relieved of the responsibility of the editorship, he continued his coal business, but now used his spare time to finish a carriage history begun in the mid-60s, *The World on Wheels*, which he published in 1878. Following this, he lived in semiretirement, yet kept active by writing numerous small items for publication in *The Hub*. After several months of illness, he died on November 20, 1883, at his home at 7 West 128th Street, New York City.

George W. W. Houghton. (*The Hub,* May 1891.)

George W. W. Houghton

George W. W. Houghton, born in Cambridge, Massachusetts, on August 12, 1850, graduated from Cambridge High School in 1868. Following a short course at the Massachusetts Institute of Technology, he obtained an editorial position in 1869 with *The Hub*, a newly founded trade paper published by Valentine & Co., then of Boston, the well-known paint and varnish manufacturers. Despite the lack of a carriage builder's background, he surrounded himself with an able body of correspondents representing every department of the industry, and within a short time elevated *The Hub* to a commanding position among trade journals. As already noted, this paper absorbed Stratton's magazine in 1871, and under a new format became one of the leading magazines of the carriage trade. During his editorship, Houghton compiled *The Hub Dictionary of Carriage Terms*, mentioned in the introduction to this volume, much improving Stratton's earlier work, and earning for himself credit as the *father* of this present dictionary.

Soon after assuming editorship of *The Hub*, he began to agitate for the organization of a trade association. Subsequently, on November 19, 1872, a meeting was held in New York City which resulted in the formation of the Carriage Builders' National Association. An important accomplishment of this association was the establishment of the Technical School for Carriage Draftsmen and Mechanics in New York, an effort to which Houghton devoted long hours. He made several trips to Paris, where his association with the Dupont technical school provided a valuable background for his later work, and in 1880 he was one of seven appointed to the committee to organize the technical school in New York.

In addition to his work with *The Hub* and with the Carriage Builders' National Association, Houghton published six volumes of verse, and was active in the New York Historical Society as well as in several prominent social organizations. He was married in 1886 to Ellen Russel, who bore him two children. On one of his trips to Paris he contracted pneumonia, which left him in a weakened condition and apparently caused damage to his heart. He died at his home in Yonkers, New York, on April 1, 1891, following a two-week illness.

Key to Source Identification Symbols

A — William Bridges Adams, *English Pleasure Carriages*, London, 1837
B — Catalog No. 15 of The Barlow Hardware Co., Corry, Pennsylvania, 1904
C — *Harness and Carriage Journal*, volumes 14 and 15, New York, 1871-1872
F — William Felton, *A Treatise on Carriages*, London, 1796
FR— Fairman Rogers, *A Manual of Coaching*, New York, 1899
G — Catalog No. 11 of the L. J. Kingsley Co., Binghamton, New York, ca. 1911
H — George W. W. Houghton, *The Hub's Vocabulary of Vehicles*, New York, 1892
h — George W. W. Houghton, *The Hub Dictionary of Carriage Terms*, *The Hub*, volumes 21-24, New York, 1879-1882
I — Catalog No. 8 of The Eberhard Mfg. Co., Cleveland, Ohio, 1915
J — J. Geraint Jenkins, *The English Farm Wagon*, Lingfield, Surrey, England, 1961
K — S. D. Kimbark's Illustrated Catalog, Chicago, 1876
M — *The Carriage Monthly*, Vehicle Glossary, Philadelphia, April 1904
O — This symbol covers the many miscellaneous sources that were used, such as catalogs, trade journals, carriage reference works, etc.
R — James Reid, *The Evolution of Horse-Drawn Vehicles*, Institute of British Carriage and Automobile Manufacturers, 1933
S — Catalogs of the Studebaker Bros. Mfg. Co., South Bend, Indiana, (the number following the symbol is the number of the Studebaker catalog)
T — William N. Fitz-Gerald, *The Carriage Trimmers' Manual and Guide Book and Illustrated Technical Dictionary*, New York, 1881
W — Thomas Wilhelm, *A Military Dictionary and Gazetteer*, Philadelphia, 1881
Y — Catalog of Cray Brothers, Cleveland, Ohio, 1910

CARRIAGE TYPES

ACCOMMODATION — The individual name given to the first American carriage specifically built for public service in cities. Built about 1827, the vehicle seated twelve persons inside, the passengers entering at the sides. This forerunner of the omnibus featured a body that was hung on thoroughbraces, somewhat after the manner of the STAGECOACH of the times, for it was in reality a modification of that vehicle. The accommodation was made by Wade & Leverich for Abraham Brower, who ran it on Broadway in New York City, from Bleecker to Wall Street, and at the uniform fare of one shilling a head for all distances. See also its successors, SOCIABLE and OMNIBUS. (H)

ADOPTICON — A six-horse advertising wagon of special design, introduced by a Colonel Haverly, of Chicago. Origin of the name is unknown. (H)

ALBANY CUTTER — An attractive and popular cutter originated by James Goold, of Albany, New York, who began the manufacture of CARRIAGES and SLEIGHS in 1813. His cutter developed over a period of years from a rude prototype into the final swell-side version, which is believed to date from about 1836. The body has a curving sweep of pleasing appearance, which afforded the painter a wide field for the display of his taste, for the panels were often colorfully decorated. The Albany cutter, also known as a SWELL-BODY or SWELL-SIDE CUTTER, and as a GOOLD CUTTER, was widely copied by other builders, and retained its popularity until the end of the horse-drawn era. (H & O)

ALBANY SLEIGH — Both four- and six-passenger sleighs were made, following the lines of the ALBANY CUTTER, in which cases the term "sleigh" was substituted for "cutter." (O)

ALEXANDRA-CAR — A variety of DOS-A-DOS pleasure carriage having a cut-under body to facilitate turning. (H)

ALLIANCE PHAETON — A variety of PARK PHAETON popular in England about 1850. (H)

ALL-NIGHTER — A slang expression applied in New York City to a public CAB that operated after dark. (H)

AMBULANCE — (From the French, *hôpital ambulant*, field hospital.) A two- or four-wheel vehicle used for the conveyance of the sick and wounded. Characterized by easy suspension, it frequently has compartments for carrying medical supplies.

The ambulance was first developed for use in war service, and was simply an adaptation of the ARMY WAGON that was originally used for transporting the wounded to the rear. The original usage of the term *hôpital ambulant* referred to a moving or flying hospital which accompanied an army on campaigns. Often this hospital was a single wagon, outfitted to carry medical and surgical supplies, and to carry surgeons, but it did not at first carry the wounded. Vehicles for this latter purpose are said to have been first designed by Baron D. J. Larrey, Chief Surgeon of the French Army, for use by that army during the 1790s. He designed both two- and four-wheel ambulances, the former used in level country, and the latter in mountainous areas. Both of these ambulances had bodies suspended on braces, and were provided with mattresses for the injured passengers. About a decade after the French, the British Army began to employ a type of light wagon on springs for this same purpose.

In the United States service, ordinary wagons served as ambulances for many years, and it was not until the 1830s that any special vehicles were constructed. Even then the army did not officially adopt any, and throughout the Mexican War (1846-1847) ordinary wagons continued to serve as ambulances. In 1859 a few two-wheel ambulances were built for the service on an experimental basis. These soon proved to be unsatisfactory, and four-wheel types were then developed for use in the Civil War.

The ambulance did not come into common use for civilian service in either Europe or America until the 1870s, primarily because the hospitals of earlier times were not adapted to caring for injured persons. There

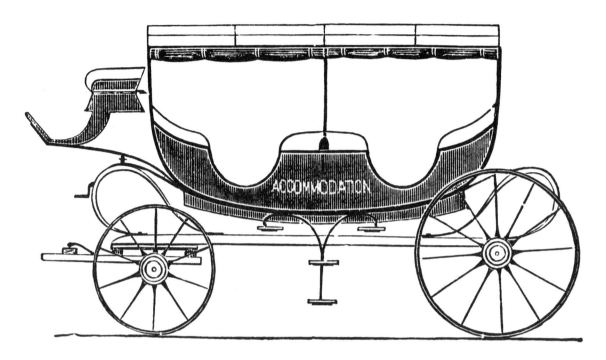

Brower's ACCOMMODATION of about 1827. (Ezra Stratton: *The World on Wheels,* New York, 1878.)

Early version of James Goold's *Albany Cutter,* reportedly built in 1816, and still in use in 1878. This one has an extra, removable seat in front. (Ezra M. Stratton: *The World on Wheels,* New York, 1878.)

ALBANY CUTTER manufactured by The Northwestern Cutter Works, Fort Atkinson, Wisconsin. Overall length, 6 feet; panels, dark green; seat panel may be of light green striped with gold and a fine line of white on each side; runners, dark lake striped with two fine lines of straw color; trimming, green cloth or plush. (*The Carriage Monthly*, August 1880.)

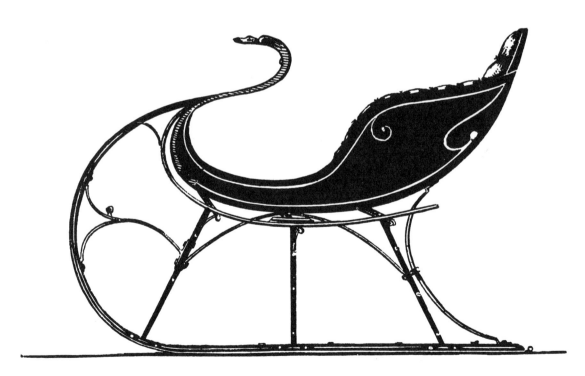

Scroll-side ALBANY CUTTER, 40-inch track, 6 feet overall. Body, dark green striped with gold; runners, vermilion striped with green; trimming, green cloth or morocco. (*The Hub*, July 1881.)

Rucker AMBULANCE used by the army during the Civil War. *The Medical & Surgical History of the War of the Rebellion* stated: "The most serviceable ambulance wagon used during the latter part of the war was that designed by Brig. Gen. D. H. Rucker, and built at the government repair shops in Washington." (Part III, vol. 2, Surgical History, Washington, Government Printing Office, 1883.)

have been many varieties of ambulances in both military and civilian service; in most instances these have been named after the inventors.

In addition to the above ambulances, the term came to be applied to certain types of passenger-carrying wagons. These were rather extensively used in the American West as traveling carriages, and some were equipped with seats which could be arranged so as to form beds. (H & O)

AMBULANCE, HORSE — See VETERINARY AMBULANCE.

AMEMPTON — A variety of LANDAU invented and patented by Edwin Kesterton, of London, about 1855. (H)

AMERICAINE — Houghton states that this term is frequently applied in France to the American BUGGY, and is synonymous with BOGUET. The *New Cassell's French Dictionary* (Funk & Wagnalls Co., Inc., New York, 1962) states that the Americaine is a PHAETON with two interchangeable seats, one of which is hooded. (H & O)

ARABA — A four-wheel springless carriage, sometimes covered, peculiar to Turkey. (H)

ARCERA — A covered carriage of ancient Rome. (H)

ARMY WAGON, FOUR-MULE — See ESCORT WAGON.

ARMY WAGON, SIX-MULE — The all-purpose freighting wagon of the army. It saw extensive use in the Civil War and for some years thereafter. Beginning in 1878 it was gradually replaced by the slightly smaller FOUR-MULE WAGON, better known as the ESCORT WAGON. The SIX-MULE WAGON resembled a heavy farm wagon, with bows supporting a cover, and it was driven by a jerk-line, from the left wheel horse. (O)

AZALINE — A local Pennsylvania term, the derivation of which is unknown, applied to a variety of TUB-BODY BUGGY having paneled sides. (H)

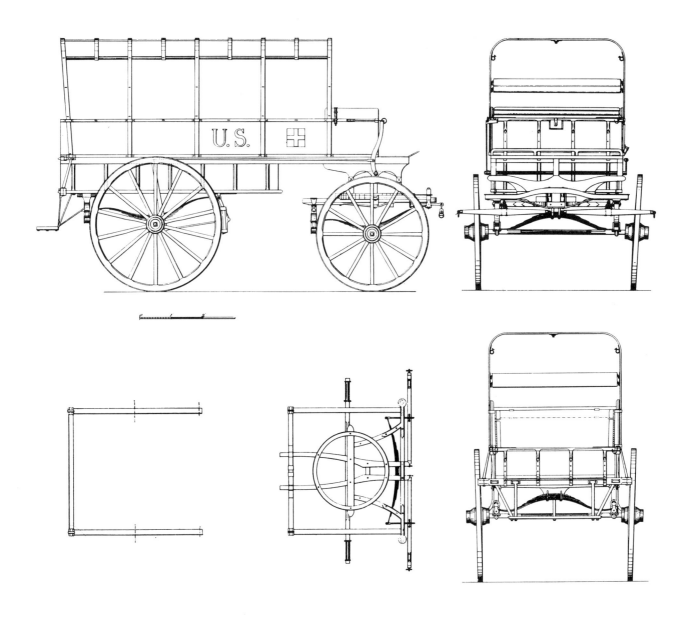

ARMY AMBULANCE. Wheels, 42-inch and 50-inch; body, dark olive green with the exception of the parts supporting the cloth top; all ironwork, black; letters "U. S." bright yellow; red cross on a white ground; gearing, same as the body. (*Specifications for Means of Transportation,* Washington, Government Printing Office, 1882.)

Platform-spring AMBULANCE. Wheels, 33-inch and 45-inch; body, black; gearing, red or green. (Studebaker Bros. Mfg. Co., South Bend, Indiana, 1903.)

VETERINARY AMBULANCE built by Sebastian Mfg. Co., of New York City, for the use of the New York Society for the Prevention of Cruelty to Animals. Wheels, 37-inch and 52-inch; body, carmine striped with gold; gearing, carmine striped with black. (*The Hub,* July 1887.)

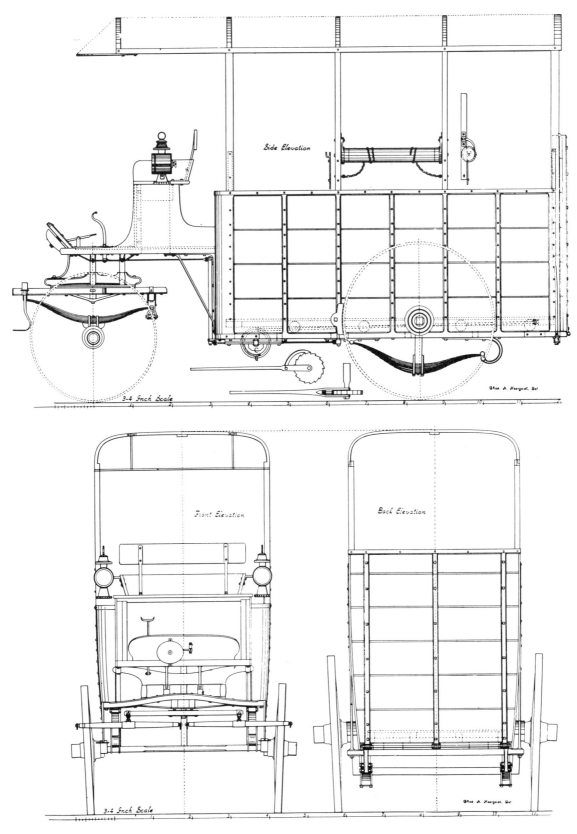

VETERINARY or HORSE AMBULANCE. Wheels, 38-inch and 46-inch; body, deep green; posts, pillars, top rails, and bottomsides, black; chamfers, deep green, and boot, black; gearing, deep green striped black. (*The Carriage Monthly,* August 1898.)

AMERICAINE of French design attempts to modify the lines of the more common *Buggy.* Wheels, 44-inch and 49-inch; body and gearing, black, though the seat could be some other color; trimming, some light color of morocco or cloth. (*The Hub,* April 1888.)

A family ARABA of Turkey. (M. M. Kirkman: *Classical Portfolio of Primitive Carriers,* 1895.)

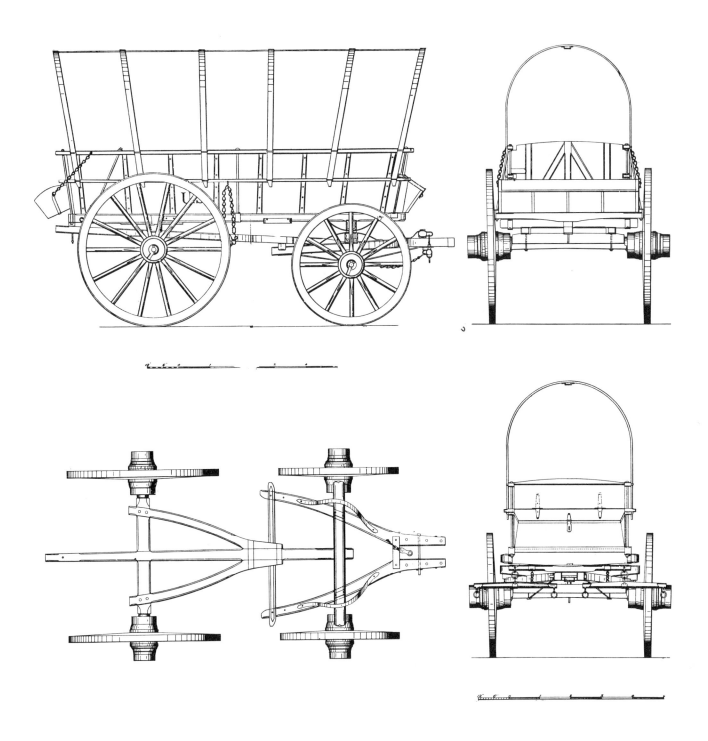

Six-mule ARMY WAGON. Wheels, 46-inch and 58-inch; outside of body and feed box, blue; inside of both, Venetian red; gearing, Venetian red darkened to a chocolate color; letters "U. S." on body, orange; black "U. S." centered on each side of cover, in 4½-inch Doric letters. (*Specifications for Means of Transportation*, Washington, Government Printing Office, 1882.)

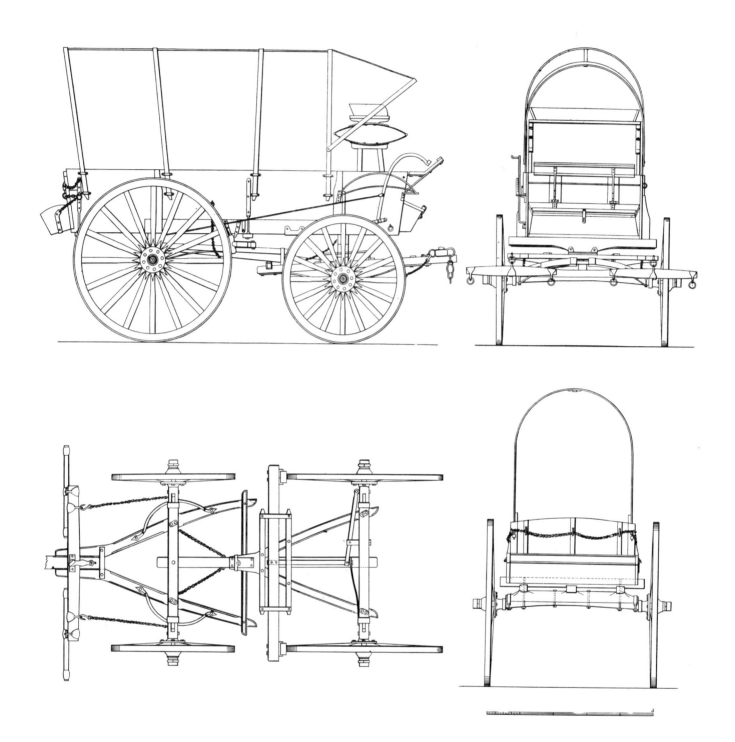

Two-mule or four-mule ARMY WAGON, popularly known as ESCORT WAGON, approved by the Secretary of War November 21, 1878. Wheels, 44-inch and 56-inch; outside of body and feed box, dark leaden blue; inside of both, Venetian red; gearing, Venetian red and vermilion, mixed half and half; all irons and chains, black; letters "U. S." on body, red; black "U. S." centered on each side of cover, in 4½-inch Doric letters. (*Specifications for Means of Transportation,* Washington, Government Printing Office, 1882.)

B

BABY CARRIAGE – Same as CHILD'S CARRIAGE, or PERAMBULATOR. See the latter.

BACK-ACTION – A local term applied in the Pacific states to a small wagon attached behind a larger one, and drawn by the same team, as formerly used when freight was transported by PRAIRIE SCHOONER. (H)

BAGGAGE WAGON – A four-wheel, hand-drawn wagon used around railway depots to move baggage about the area. Also see EXPRESS WAGON. (S226)

BAKERY WAGON – A type of light DELIVERY WAGON, usually on three elliptic springs. Also used for delivery of dairy products, groceries, etc., this type of wagon generally is characterized by a closed, paneled body and sliding doors. It follows much the same construction as a MAIL WAGON, though the latter is generally smaller. (S226)

BALLESDEN CART – See DOG CART.

BAND WAGON – A wagon built for parade purposes, generally, though not exclusively, used by circuses for the purpose of carrying the band. Sometimes the seats are arranged inside the wagon, and sometimes on top of the closed body. In the latter instance, the wagon can be used to carry the instruments and baggage of the band when the circus is moving from town to town. (H & O)

BANDY – A cart drawn by two small bullocks, once the common means of conveyance in India. It is simply a broad platform mounted on high wheels, and covered with mats to keep off the rain and sun. The driver sat astride the cart tongue. (H)

BARCO DE TIERRA – (Spanish, meaning land ship.) A primitive form of OX-CART, with wattled top and sides, and spoked wheels, used on the steppes of South America. (H)

BARGE – A New England term for any large EXCURSION WAGON built after the form of a WAGONETTE, with seats running lengthwise. (H)

BAROUCHE – (Latin, *birotus,* two-wheeled; the application of the name of a two-wheel vehicle is believed to be due to the feature of the hood, which is the characteristic of the Italian BIROCCINO, or CABRIOLET. Equivalents: German, BARUTSCHE; French, CALECHE.) Technically, this four-wheel vehicle is a member of the COACH family, consisting of the under-carriage and lower quarters of a COACH, including the lower halves of the doors. It has a falling top covering the back seat, in place of the full-paneled top, or the landau top. Believed to be of German origin (and sometimes called a GERMAN WAGON), there is evidence that it arrived in England by the 1760s and by the early nineteenth century was becoming popular in the United States. Adapted mainly to town use, this carriage was comfortable and pleasant in warm weather, but was not as agreeable for winter weather. Some types are fitted with a roof and sides of wood, made so as to close against the head, but this arrangement is not convenient. Inside, the seats face one another as in a COACH, the forward seat often being hinged rather than fixed permanently, while the driver's seat is elevated as on a LANDAU. By the 1860s, many American barouches were equipped with extension-tops, which covered both front and rear seats. *The Carriage Monthly* reported in 1875 that the term BAROUCHE had formerly been applied to all extension-top PHAETONS in the area of southern Ohio; possibly the use of the extension-top on some barouches had suggested the application of the name to PHAETONS that were similarly equipped. The BAROUCHE, BRETT, BRITZKA, and CALECHE are all very similar types, and the terms were used almost synonymously by many American carriage builders. The terms BAROUCHE and CALECHE are often applied to the same vehicle in America, the former term having been most commonly used during the earlier period. See also the other types above named (C, H, A, & O)

BAROUCHE LANDAU – A vehicle popular in England in the early nineteenth century, and much used by four-in-hand amateurs. It is a BAROUCHE with a high driving-seat, and a rumble for two servants. (O)

BAKERY WAGON. Wheels, 36-inch and 51-inch; front side curtains, light straw or Naples yellow; reaper painting on rear side curtains; panels, dark lake or carmine; belt, Paris green; gearing, light straw striped with vermilion, fine lined with pea green; nuts and ends of bands, black; moldings on body lined with silver; moldings on picture curtain lined with gold; bevels on belt, silver; scrolls and lines on panels, gold; inside of body, light pink or green; cushion of duck canvas; curtains of oiled duck canvas. (*The Carriage Monthly,* October 1887.)

BAND WAGON. Wheels, 43-inch and 55-inch; body, dark green with gold scrolling on body and seats; gearing, Naples yellow, ornamented with ultramarine blue and gold stripes; trimming, dark green goatskin. (*The Carriage Monthly,* December 1881.)

BAND WAGON constructed by the Schultz Wagon Co., of Dalton, Ohio. Wheels, 40-inch and 52-inch; body and gearing, white, with dragons, scrolls, and lions' heads shaded in darker colors; trimming, leather. (*The Carriage Monthly*, January 1895.)

BAROUCHE, with 44-inch and 49½-inch wheels. (Lawrence, Bradley & Pardee, New Haven, Connecticut, 1862.)

Extension-top BAROUCHE, also known as a "Niagara Hack," as it was commonly seen at the hackstands at Niagara Falls, and at many other summer resorts. Wheels, 36-inch and 46-inch; body, dark green striped with black; gearing, lighter green, black striped; trimming, green cloth; dickey-seat and fall, black leather. (*The Hub,* July 1882.)

BAROUCHE. Wheels, 35½-inch and 45-inch; body, dark blue; moldings, bottom rockers and boot, black, with moldings striped in fine yellow; gearing, dark blue striped with ¼-inch yellow with two fine lines of light blue on either side. (*The Carriage Monthly,* June 1886.)

BAROUCHET — A BAROUCHE with the body shortened by the removal of the front quarter, together with its seat; a DEMI-BAROUCHE. The vehicle commonly called the COUPELET would more properly be called the BAROUCHET. Bridges Adams wrote: "A Barouchet is to a Barouche what a Landaulet is to a Landau." He also reported that it was drawn by one horse, was ungraceful, and not popular. (H & C)

BASTERNA — A closed litter, carried between two mules in tandem, used by the Romans. (O)

BATHING MACHINE — A small dressing chamber, mounted on wheels, used in England at seaside resorts for wheeling bathers into the surf. (H)

BATH WAGON — A wagon for conveying a bathtub. Commonly used in France under the name of VOITURE DE BAINS. (H)

BATTERY WAGON — (1.) A vehicle employed by a battery of artillery to transport the tools and materials needed for repairs. Among the tools were those for carriage makers, saddlers, armorers, and laboratorians' use, spare parts for carriages and harness, and scythes and sickles for cutting forage. The body of the battery wagon is a large rectangular box, having a rounded lid covered with painted canvas, and hinged at the side; to the back part is attached a rack for carrying forage. The bottom of the body is formed of one middle rail and two side rails, resting on a stock and axletree, as in the TRAVELING FORGE. For draught purposes it is connected to a limber in the same manner as other field carriages. It is constructed to have equal mobility with other field carriages, so that it may accompany them wherever they may be required to go. Battery wagons and TRAVELING FORGES are not confined to the service of field batteries, but are also used with siege and seacoast carriages as occasion may require. (2.) This term is also applied to the mobile central office of a United States Army field telegraph system, from which four separate telegraph lines could be worked. It contained tables, stools, a stove, the telegraph instruments, and the batteries required to operate the system. This wagon was mounted on platform springs. Each field telegraph train was composed of one BATTERY WAGON, four WIRE WAGONS, and four LANCE TRUCKS (ca. 1876). (C, H, W, & O)

BATTLESDEN-CART — An English variety of pleasure cart known also as a BATTLESDEN-CAR. (H)

BEACH WAGON — A term peculiar to the states along the Atlantic seaboard, used to describe a simple and primitive form of square-box wagon featuring two or more seats, and generally a standing top and an open front. Suspension varied between elliptic and platform springs. These wagons were used by beach or excursion parties, but sometimes saw service as family wagons. Built at least as early as the 1860s, some varieties had doors between the front and second seats. At a later date, in addition to the square-box types, other varieties were built that closely resembled the CABRIOLET. Beach wagons were especially popular along the New England coast. (C & H)

BEER PATROL WAGON — A variety of beer delivery cart (it is not a wagon, in spite of its name), having a revolving frame to carry the kegs, popularized by the Ohlsen Wagon Co., of Cincinnati, Ohio. (H)

BEER WAGON — A wagon especially adapted with racks, rails, etc., for hauling kegs of beer. Generally, it is an ordinary truck or dray fitted for this purpose, but it is sometimes constructed on the order of an EXPRESS WAGON. (O)

BENNA — A rustic wagon of ancient Rome. It was drawn by two animals and appears to have had a body of twisted grass rope set on a wooden platform. (H)

BERLIN — This term refers specifically to a variety of under-carriage for a four-wheel vehicle. It was the invention of Colonel Philip de Chiese, a Piedmontese who was in the service of Frederic William, elector of Brandenburg, and was so named because it was first built in the city of Berlin. Being sent to France about 1660, de Chiese had built a carriage of his design for making the journey. The mechanical features of the Berlin carriage consist of two perches connecting the hind axletree to the front transom, in place of the single heavy perch, and thoroughbraces running from front to rear for the support of the body. While the body is frequently assumed to be that of a COACH, this is not necessarily the case, for any other type, such as a CHARIOT or PHAETON, may be employed with a Berlin carriage. The perches are placed outside the body, and are either placed sufficiently high, or are curved upward to permit the front wheels to turn underneath. Often the ends of the braces are secured to small windlasses so that the braces can be taken up as the leather stretches with use. In later types, the braces pass over springs mounted at the ends of the carriage. The resilience of the two perches also contributes to the easier movement of the body. The development of the Berlin is one of the more important advances in the earlier history of carriages.

The Berlin carriage did not become popular in Europe until almost a century after its inception. In America the Berlin saw limited use as a private carriage, and suggested the running gear used in the famous CONCORD COACH, and its ancestor, the STAGE WAGON, both of which were thoroughbrace vehicles. Later the term BERLIN came into a vague and incorrect usage, being applied generally to various types of full-paneled COACHES. (H & O)

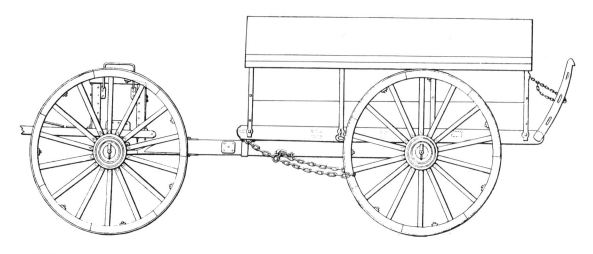

BATTERY WAGON, showing the forage rack attached to the rear and a lock chain secured to the wheel. (John Gibbon: *The Artillerist's Manuel,* New York, D. Van Nostrand, 1863.)

BATTERY WAGON of the U. S. Army field telegraph system. The *Wire Wagon* was nearly identical, except for the furnishings. (*Report of the Board of the U. S. International Exposition, 1876,* Washington, 1884.)

BEACH WAGON of the earlier square-box type, reminiscent of the *Dearborn Wagon* and early *Depot Wagons*. (The George C. Miller Sons Carriage Co., Cincinnati, Ohio, 1883.)

NEWCASTLE BEACH WAGON, showing lines similiar to those of the *Cabriolet*. Without doubt, builders could have been found who would have used the latter term. (George Werner [carriage builder], Buffalo, N.Y., ca. 1900.)

This handsome BEER WAGON, actually a variety of platform-spring *Truck*, was built by Wm. Schukraft & Sons, of Chicago, about 1905. The hooks hanging underneath the body were used to carry several additional kegs of beer, by holding the kegs from the ends. (Smithsonian photo.)

BENNA, shown on the Arch of Trajan, in Rome. (Ezra M. Stratton: *The World on Wheels,* New York, 1878.)

BERLIN COACH — The name given to a vehicle having a COACH body hung on a BERLIN carriage. The term BERLIN may also be compounded with any other type of suspended body, such as LANDAU, CHARIOT, PHAETON, etc. (O)

BERLINET or BERLINETTE — A small or reduced BERLIN, using the term in this instance in the sense of a COACH. Diminutives thus formed by the addition of *et*, masculine, or *ette*, feminine (the former being preferable) are common and entirely proper. (H)

BERLINGOT (or CAROSSE-COUPE) — See COUPE. (H)

BERLIN ROCKAWAY — A ROCKAWAY that is paneled, like a COACH, with glasses in the upper quarters. Here the term BERLIN is not used in its true sense, but according to its later and incorrect application, meaning simply a variety of COACH. (H)

BIAN — A large variety of JAUNTING-CAR, used in Ireland. The invention of Charles Bianconi, of Dublin, its seats ran lengthwise and were arranged back to back. (Straus)

BICYCLE — A two-wheel vehicle, with wheels arranged in tandem, propelled by the efforts of the rider. In 1877, Secretary of the Treasury John Sherman, upon the opinion of the attorney general, decided that a bicycle is a carriage. In 1879, the English Court of the Queen's Bench decided, all of the justices concurring, that a bicycle is a carriage. Because of these rulings, the term is included in this book. (H)

BICYCLE-AMBULANCE — A double BICYCLE intended for conveying sick or wounded persons. (H)

BIER — (Derivation, according to Skeat, is from the Middle English, *beere*, *boere*, Anglo-Saxon, *boer*, meaning to bear or carry.) It was originally a framework, with handles at both ends for the bearers, on which things could be carried. Eventually its use was limited to carrying a dead body, and finally it became the framework on which a coffin is placed. The term is often applied to the entire unit of the coffin and its support. (H & O)

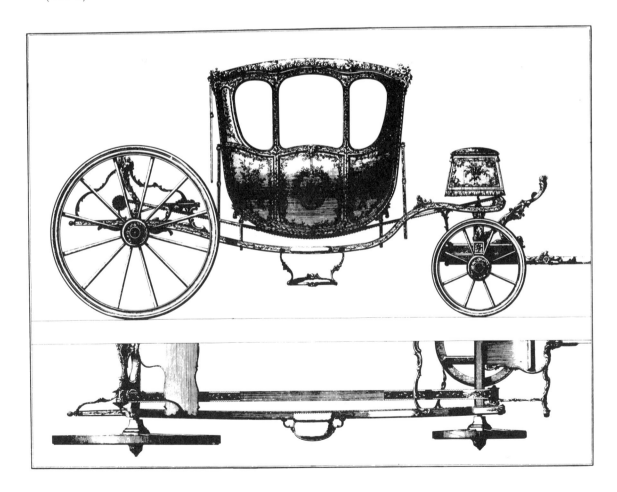

BERLIN COACH, or a COACH with a Berlin running gear. (M. Roubo: *L'Art du Menuisier-Carrossier,* Academie des Sciences, Paris, 1771.)

BIKE WAGON — One of the final developments of the carriage era, following the invention of the pneumatic tire and the safety bicycle in the 1890s, the BIKE-WAGON is a lightweight RUNABOUT or DRIVING WAGON. It is characterized by some or all of the following features: wire wheels, ball-bearing axles, tubular-steel running gear, and rubber tires of either the pneumatic or cushion type. In some instances wooden wheels are used in place of wire wheels, but the tires are almost always of rubber. (Elkhart Catalog, 1907)

BIROTUM — A small, two-wheel Roman vehicle, drawn by one horse. It had a comfortable leather-covered seat, and a rack for carrying luggage behind. Not intended for private use, they were kept at post-stations in the time of Constantine. (Stratton 147)

BISHOP — A term the trade applied to any vehicle that was modified from an old style into a newer or more serviceable style. It might have involved a new, but different body on old gearing, or a new gearing under the old body, or perhaps only a modification of one or both units. This usage came from the earlier usage, meaning to make a horse seem younger by filing or operating on its teeth. (h)

BLACK-AND-TAN — A New York City slang expression, applied to the cheap CAB introduced in that city in 1883, so-named because of its color. (H)

BLACK MARIA — A slang expression for a PRISON VAN. The origin is unknown. (H)

BOAT SLEIGH — A New England term for a large excursion sleigh. The outline of the body often closely resembles that of a boat. (H)

BOBS (or BOB-RUNNERS) — New England terms for short sleds used under vehicle bodies in place of wheels. Also called BOB-SLEDS. (H)

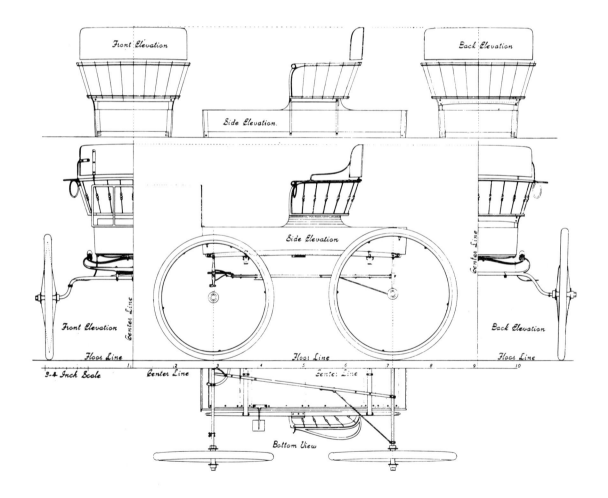

Spindle-seat BIKE WAGON, with ball-bearing axles, wire wheels, and 2-inch pneumatic tires. Wheels, 34-inch and 36-inch; body and seat, black, except seat spindles, which are deep blue; balls on spindles highlighted with light blue; gearing, deep blue striped with two fine lines of yellow; spokes and hubs, nickel plated, with deep blue rims; trimming, blue cloth. (*The Carriage Monthly*, January 1898.)

This BIKE WAGON was available with the regular piano-box body, or with a cut-under body, and was mounted on cushion tires. It had a black body on a red or dark green gear, and was trimmed with dark green or dark blue cloth, whipcord, or leather. The price was $85. (Elkhart Carriage & Harness Mfg. Co., Elkhart, Indiana, 1907.)

Post BIROTUM. (Ezra M. Stratton: *The World on Wheels,* New York, 1878.)

BOB-SLED – (1.) A short sled, used as one of a pair under a vehicle body. (2.) A sled made up of a pair of the former, arranged in tandem and joined by some sort of coupling. Also called a DOUBLE-RIPPER. (3.) The equivalent of BOB-SLEIGH, but especially applied to one used for business purposes, as distinguished from one used for pleasure. (4.) A long, narrow sled used for coasting, the sled being a variation of definition 2, above. (H & O)

BOB-SLED, LUMBERMAN'S – A primitive form of ox-sled, made entirely of wood, and of massive construction. First employed by New England lumbermen for hauling logs over the snow, by 1890 it was in general use for this purpose throughout the northern United States and Canada. It consists of a pair of short runners, connected by a single crossbar near the rear, and surmounted by a heavy bunk or bolster, to which the butt end of a log is chained, and the log is dragged as with LOGGING WHEELS. The runners are shod with hardwood, attached by trenails or wooden pins. (H)

BOB-SLEIGH – A New England term applied to any wheeled vehicle, whether business or pleasure, the body of which is temporarily mounted on BOB-SLEDS or BOB-RUNNERS. It is more commonly applied to a pleasure vehicle, while the term BOB-SLED is more generally applied to business vehicles. (H)

BODY-BREAK – See BREAK.

BOGUET – Houghton states that this term is synonymous with AMERICAINE, and that it is a French term applied to the American BUGGY. Stratton, however, illustrates a rather elaborately suspended two-wheel vehicle which he states is the French boguet. The *New Cassell's French Dictionary* (Funk & Wagnalls Co., Inc., New York, 1962) describes it as a light GIG, or BUGGY.

BOLSTER WAGON – An old form of ROAD WAGON or BUGGY hung on side-bars and bolsters, without end-springs, thus giving a stiff suspension. This construction was frequently used for speeding purposes. Also see BOULSTER WAGON. (H)

BONAVENTIA – A local Pennsylvania term applied to a variety of JENNY LIND, or standing-top BUGGY. (H)

BONESHAKER – A two-wheeled velocipede, driven by pedals on the front wheel, used in America during the late 1860s and early 1870s. (H & O)

BOOBY-HACK – A closed sleigh, kept for hire. (O)

BOOBY-HUT (or HUTCH) – A New England term, used from the mid-eighteenth century to the end of the carriage era, referring to a type of sleigh, having an enclosed body, such as a COACH or CHARIOT body, suspended by leather braces that are attached to the body loops and beams. Later versions have only two knees on each side to support the ends of the beams, the front knees also supporting the driver's seat. (C, H, & O)

BOULNOIS CAB – A two-wheel cab introduced in London about 1832, by William Boulnois, who adapted his patented spring to a vehicle of foreign design. It was a closed carriage having a door at the rear, with seats arranged sideways, and facing one another, for the two passengers, thus causing the slang expression "slice-off-an-omnibus" to be applied to it. The driver's seat was on the front of the roof. This cab was sometimes known as a Minibus. (Moore, page 213, and Adams, page 277)

BOULSTER WAGON – An unidentified type of wagon built by Amos Stiles, a carriage builder of Moorestown, New Jersey, 1812-1821, though there is no reason to believe he was the only builder. The boulster wagon was evidently a specialty of Stiles, for his daybook mentions 140 of them during these years. It appears to have been a light passenger wagon, with either one, two, or three seats, and equipped with shafts or pole. It had a standing top, and was frequently "paneled up behind," sometimes with glasses in the rear, either sliding or fixed. A few had a rear door, but most appear to have had no door. Generally the inside was trimmed to some extent, and some were almost entirely trimmed. Seats were made both with and without cushions, and frequently, though not always, with backs. Bodies were sometimes made with swelled sides, and moldings, and in some instances were varnished. Most often the vehicles were plainly and simply trimmed and finished. They were equipped with curtains, and often an apron was also provided. In no case do the entries in Stiles's daybook indicate that the vehicle was mounted on springs or braces (though the seats were sometimes set on "double springs"), and the name seems to indicate that the body sat on bolsters. Otherwise, the vehicle appears to have been very much like the COACHEE or JERSEY WAGON. Also see BOLSTER WAGON. (O)

BOUNDERS, OXFORD – A term sometimes applied to the extremely high wheels of an OXFORD DOG CART. See same, and also DOG CART and TAX-CART. (H)

BRACKET-FRONT WAGON – A BUGGY or other light carriage made with a bracket-supported footboard which is either continuous, and forms the dash, or forms a semi-dash which is completed by a low additional one of leather. (H)

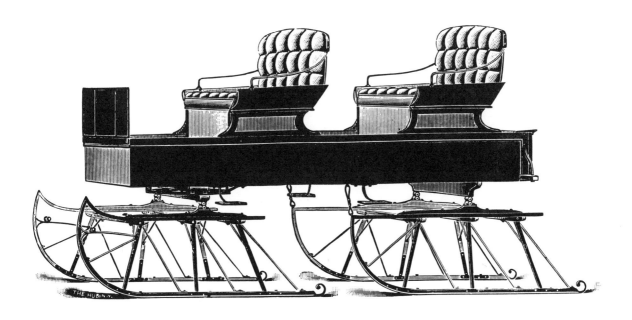

Three-knee BOB SLED with fifth-wheel and oscillating runners. The 7-foot body is identical to the type used on *Spring Wagons*. Trimming, corduroy. (Waterloo Wagon Co., Waterloo, New York, 1901.)

Three-knee BOB SLED for two horses; 2-ton capacity on 1⁷⁄₈-inch by 2½-inch runners; painted vermilion, striped black. (Gray Bros., Cleveland, Ohio, 1910.)

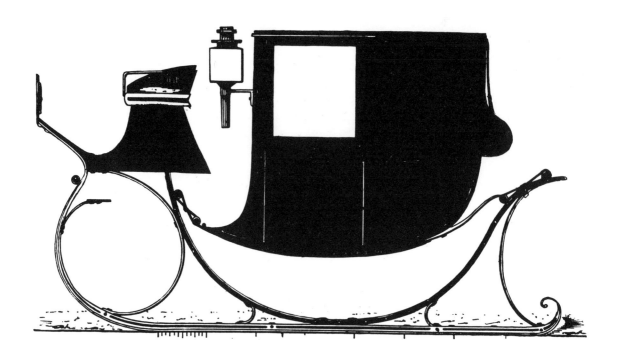

BOSTON BOOBY or BOOBY-HUT. Overall length, 10½ feet. Sleighs were shown in the summer numbers of the trade journals, for it was then that tradesmen were building sleighs for the coming season. The carriage builders' motto was, "Make Sleigh while the sun shines." (*The Hub*, July 1882.)

BOULNOIS CAB. (H. C. Moore: *Omnibuses and Cabs*, 1902.)

38

BRAKE — Same as BREAK.

BREAK — A heavy driving vehicle which obtained its name in England from being used literally to *break in* colts or unruly horses. Its weight, substantial build, and high driving-seat gives the driver the requisite control over the animals. The break has no body, but consists of a heavy running gear with a high driving-seat, behind which is a small platform on which an extra stable-helper can stand. Besides being used for the aforementioned purposes, breaks are also used for exercising four-in-hand teams. The term came to be applied more generally to a heavy variety of PHAETON intended for gentlemen's use, having a high driver's seat, two inside seats, and a raised seat at the rear for grooms; this might more properly be termed a DRAG-PHAETON.

Eventually, the use of the term was extended to cover a great variety of heavy vehicles intended for carrying large parties on excursions and outings. Seating arrangements are varied, and in some cases roof seats are used. One example of this is a so-called TALLY-HO BREAK built by the Briggs Carriage Company of Amesbury, Massachusetts. This had a driver's seat, longitudinal DOS-A-DOS seats, one transverse seat facing to the rear, and four transverse roof seats.

Another type, known as a WAGONETTE BREAK or BODY BREAK, has a high driving-seat, and two longitudinal seats in the rear, like the benches of a WAGONETTE. Some of these have an additional transverse seat behind the driving-seat. (H, C, & O)

BREAKING CART — A two-wheel vehicle, generally two-passenger, used for training colts and for general service around stock farms. Their use was not confined to stock farms, and many were used in much the same manner as ROAD CARTS. These were sometimes called UTILITY CARTS. (S226)

BRETT — A carriage that is somewhat like a BRITZSKA, from which it took its name. It has a calash-top over the back seat, and a lid which, when raised, forms the back to a front seat; the driver's seat is set up on a boot. At one time the term was applied to a BAROUCHE that had a BRITZSKA pillar or some other relic of the BRITZSKA. Also see BAROUCHE. (C & H)

BREWER'S WAGON — See BEER WAGON.

BREWSTER WAGON — A popular variety of American BUGGY, named after its makers, the James B. Brewster Company of 25th Street, New York City. This square-box BUGGY is distinguished from its predecessors mainly by reason of its extreme simplicity of outline and superior finish. The body is suspended on two wooden side-bars by means of a pair of scroll springs (first type, Nomenclature section), while the ends of the bars are supported by a pair of half-springs. (H & O)

BRISKER (or BRISKY) — Corrupted forms of the name BRITZSKA (itself a contraction), formerly much used by coachmen and others. (H & A)

BRITTON WAGON — A lightened version of the GODDARD BUGGY. Designed by John W. Britton of New York, the first of the type was built in 1872 for his own use. (H)

BRITZKA (also BRITSKA, BRITZSKA, BRITZSCHKA, BRITSCHKA) — Commonly pronounced as if spelled *bris-ka*, though most dictionaries give *brits-ka*. The word is said to be derived from the Polish *bryczka*, a light, long traveling wagon; diminutive of *bryka*, a freight wagon. The britzka is a carriage that was in common use in Europe early in the nineteenth century. It features numerous accommodations for traveling purposes, in the form of a variety of contrivances for sleeping, eating, reading, and carrying luggage. Some disagreement exists concerning its migration to England. Bridges Adams stated that it was brought from Germany (where it was a light carriage intended to carry three persons) about 1825, by the Earl of Clanwilliam, while Ralph Straus believed that it was introduced from Austria in about 1818 by T. G. Adams. In England, the britzka rapidly gained popularity following a number of modifications that made the carriage more convenient and adaptable, even though the increased accommodation for both passengers and luggage added considerably to the vehicle's weight.

The coach-length body is mounted on C-springs. Over the rear seat is a calash-top, and a movable knee-flap covers the forward part of the body. A folding glass shutter, fitted into a recess in the head, can be let down to close against the knee-flap, thus shutting the entire front against inclement weather. Inside are two transverse seats, facing one another as in a COACH, the front one folding down under the knee-flap. The interior of the carriage holds four passengers when the knee-flap is open, but only two passengers when the flap is closed. When constructed without a front seat there is sufficient space inside for passengers to recline at full length. Attached to the front part of the head is a hood, which is easily put up and down to facilitate entry into the carriage. The step is outside so that the interior will not be cramped. In front there is a locker having a front board that can be let part way down to form a footboard, and a driving-seat is then placed on top of the locker; or, if the board is let down to a horizontal position it can carry an additional trunk. A boot in the rear can carry one or two servants, or it can be removed entirely and one or two large trunks put in its place. An imperial can be carried on the roof. Being thus equipped for sleeping and carrying large quantities of luggage, the britzka was used as a traveling carriage, and was especially popular in areas where comfortable lodging could not be readily found. A later style of this carriage was made without a perch, suspension being on elliptic springs. Since it was a traveling carriage, the britzka was eventually driven out of existence by the railroads.

Britzkas were not built in America, though the term was applied to American vehicles that differed only slightly from the BRETT. (A, C, & H)

BRITZKA CHARIOT — Similar to a POST CHAISE, this carriage differs in the form of the body, the lower part of which is shaped something like a BRITZKA. In front and rear it is built like the regular BRITZKA, so that either seats or trunks can be mounted. The roof can be made to open. Again like the BRITZKA, this chariot can be made up for sleeping, and it is similarly used as a traveling carriage. According to Stratton, the term BRITZKA CHARIOT is synonymous with EILWAGEN. (A)

BRITZKA PHAETON — The term PHAETON is applied to this carriage because of the mode of hanging, the springs being on the same principle, though not in exact form, as those used on PHAETONS in the eighteenth century. Single elbow-springs are used in front instead of C-springs, in order to leave the space free for ascending the front seat without inconvenience. In the rear, either whip springs or C-springs can be used. The riding motion is not pleasant, as the points of the front springs form a center, causing the rear portion of the body to move up and down in an arclike motion. The body is a modification of the BRITZKA, but of more elegant appearance. Two persons can sit in front, and the shifting seats behind can accommodate two or four persons, depending on whether the head is open or closed. There is no seat behind the body. (A)

BROUETTE — Modern French term for wheelbarrow formerly applied to a French vehicle, the body of which resembled a SEDAN CHAIR mounted on two wheels, and drawn by a man. The brouette was invented in Paris about 1668 by a man named Dupin, who eventually constructed them with two elbow springs beneath the front, one of the first known instances of the application of metal springs to a carriage. The springs are attached to the axletree by long shackles, and the axletree worked up and down in a groove beneath the

A BREAK of the type used for training horses. (F. Rogers: *A Manual of Coaching*, 1899.)

TALLY-HO BREAK, made to carry twenty passengers. (Briggs Carriage Co., Amesbury, Massachusetts, 1893.)

Light BREAK. Wheels, 36-inch and 46-inch; body finished natural; gearing painted in imitation of oak; trimming, drab cloth. (*The Carriage Monthly,* April 1890.)

BREAKING CART with one seat lifted to admit passengers from the rear. Wheels, 52-inch; body, black; gearing, red; trimming, black enamel duck. (Studebaker Bros. Mfg. Co., South Bend, Indiana, 1903.)

CRANE-NECK BRETT with gypsy-top and page-board. Wheels, 44-inch and 50-inch. (Lawrence, Bradley & Pardee, New Haven, Connecticut, 1862.)

This extension-top BRETT differs from the common *Brett* by providing cover for the passengers in the front seat. Wheels, 36-inch and 46-inch; finish not stated. (*The Hub*, April 1888.)

BRITZKA with a boot behind for two persons. The interior had room for four persons when the knee-flap was open, but only held two in inclement weather when the flap was closed. (W. B. Adams: *English Pleasure Carriages*, 1837.)

BRITZKA CHARIOT, in this instance having a seat mounted in the rear and a trunk in front. (Ezra Stratton: *The World on Wheels,* New York, 1878.)

seat. Brouettes are contemporaries of SEDAN CHAIRS, and the proprietors of the latter made efforts to have them prohibited. For a time the attempt was successful, but by 1761 brouettes were in general use in the streets of Paris as public hacks. The ROLLING CHAIRS used at the 1876 Centennial Exposition at Philadelphia and the International Exhibition in Paris were a more recent equivalent of the brouette. Also called a ROULETTE, or a VINAIGRETTE. (H, A, & O)

BROUGHAM — This carriage was originally intended as a gentleman's carriage, and had its beginnings in 1838 when Lord Brougham had a London coach builder named Robinson construct for him a vehicle of his own design, though its lines were based on those of a carriage used earlier on the Continent. It was first called a DROITZSCHKA CHARIOT, but did not entirely satisfy Lord Brougham, who accordingly had Robinson build him an improved version in 1839, this latter carriage quickly becoming known as a BROUGHAM.

The Brougham is a member of the COACH family, with a full-paneled COUPE body, generally with a straight front instead of the swelled or circular front that characterizes the true COUPE, and is a compact, low-hung vehicle with a paneled boot for the driver in front. It differs but slightly from the French COUPE, being usually hung lower, and characterized by English outlines as distinguished from French. The term COUPE was in fact preferred by Americans until late in the century, when the vehicle gradually acquired its English name of BROUGHAM.

Originally they were made for two passengers, though a third could ride with the driver. Similar vehicles were built with an extension-front that contained either a hinged child's seat or a full-framed seat; the former were generally termed COUPES, while the latter were called CLARENCES. In the United States, as these terms became less frequently used toward the end of the carriage era, the term, EXTENSION-FRONT BROUGHAM, was often applied to COUPES and CLARENCES.

Two popular styles of Brougham were the Barker and Peters types, named after the originators, Barker and Company, and Peters and Sons, both of London. The bottom line of the Barker style is a curving line, and the door closes above the bottom line, though in the later period the door sometimes closed on the bottom line in order to simplify construction. The Peters style exhibits the angular bottom line, with a door that closes on the bottom line. The curving lines of the Barker were more popular during the earlier period, while the Peters style enjoyed greater popularity at a later time, yet the Barker was never entirely supplanted by the Peters even in this later period.

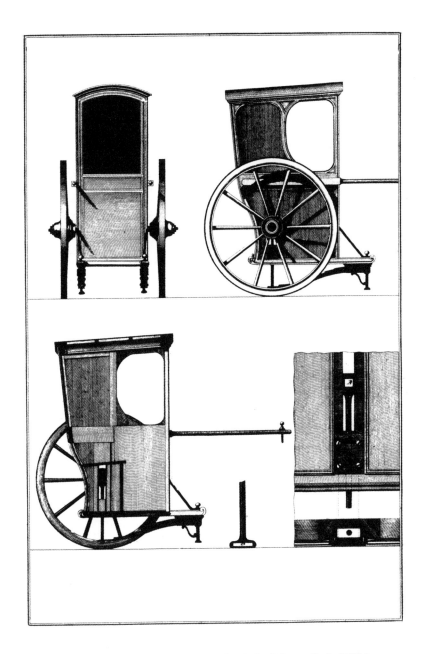

BROUETTE. (M. Roubo: *L'Art du Menuisier-Carrossier*, Academie des Sciences, Paris, 1771.)

    The Brougham is termed, according to size, as *single* or *double* (for one or two horses), and as *miniature* or *ladies*. Although it retained popularity as a gentleman's carriage, it came to be widely used in both England and America as a public CAB. In England it is said to have been the first four-wheel CAB; in America it probably outnumbered all other types of CAB. A greatly admired carriage, the brougham continued in both public and private use until the coming of the motor car. Also see COUPE. (H, M, C, & O)

BROUGHAM-HANSOM — A patented type of public CAB, introduced in 1887, having a driver's seat on the front of the roof, and entrance at the rear, originated by Messrs. W. & F. Thorn, of London. A seat across the door is so constructed that there is no possibility of the door opening until the occupants' weight is off the seat. The interior of this CAB holds three or four persons. (H)

BUCKBOARD — (Sometimes called a BUCK WAGON) — A simply constructed vehicle of American origin, believed to have developed sometime during the first third of the nineteenth century. It became popular

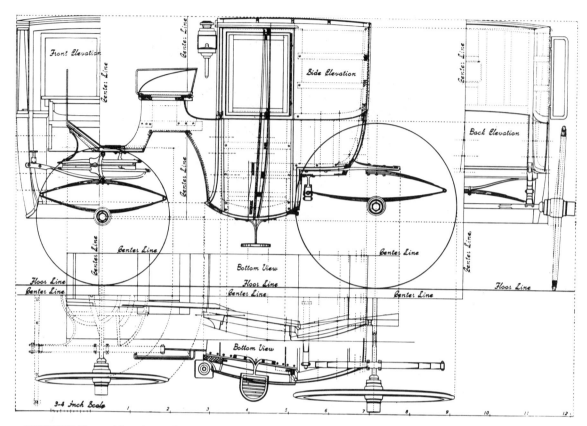

BROUGHAM on rubber tires. Wheels, 40-inch and 48-inch; lower panels, deep blue; moldings, boot and upper part of body, black; moldings on bracket front, blue striped with a fine line of black; upper part of coupe pillar and inside spaces on door pillars, deep blue; gearing, deep blue striped light blue; trimming, blue cloth. (*The Carriage Monthly*, February 1898.)

Barker-style BROUGHAM. Wheels, 34-inch and 43-inch; boot, upper panels and moldings, black; lower panels, deep blue; moldings striped with fine line of yellow; gearing, primrose striped with two fine lines of black, 3/8 of an inch apart; trimmings, green satin; mountings gold. (*The Hub*, October 1890.)

Peters-type BROUGHAM. Wheels, 37-inch and 44-inch; boot, upper panels and moldings, black; lower panels, dark Brewster green; gearing, a lighter shade of Brewster green, striped with ¼-inch line of Naples yellow; trimming, green goatskin. (*The Hub,* October 1890.)

This 1905 BROUGHAM (Peters-style) is in the collections of the Smithsonian Institution.

This BROUGHAM body, believed by some to be from the first vehicle of that type, was photographed on the grounds of Mostyn Hall, Penrith, Cumberland, England, about 1895. A BROUGHAM in London's Science Museum also claims this honor. (*The Carriage Monthly,* September 1895.)

BROUGHAM-HANSOM, built by W. & F. Thorn, and shown at the Sportsman's Exhibition, in London, in 1887. (*The Carriage Monthly,* July 1887.)

throughout the country, particularly in the West and in mountainous areas of the East. A true buckboard is a four-wheel vehicle having a floor composed of one or several planks of ash or some other springy wood, about 1 inch thick, attached to the axles. Near the center of this plank is a seat on risers; in some instances the vehicle was made long enough to accommodate two, or even three seats. The springing action of the board substitutes for steel springs, and on rough roads the comfort furnished is said to have at least equaled that given by ordinary steel springs. Originally there was no dash, but later types frequently had a dash made from a thin board, rather than the conventional leather-covered iron frame. Buckboards were used with either one or two horses.

Later buckboards were frequently modified by the addition of springs to provide greater ease in riding. Sometimes elliptic springs supported the seat, and a variety of ordinary and patented springs frequently were used to mount the board on the axles. One form of buckboard is the slat-bottom variety, sometimes known as a BUCKBOARD WAGON (which see).

The origin of the term is unknown, but as *buck* is an old English word for he-goat, it may possibly mean simply a goat-board, in allusion to its jumping motion (see analogous derivation of the term CABRIOLET). Another possible derivation may be the vehicle's capacity to *buck* against any inequalities in the road. (M, C, H, & O)

BUCKBOARD-BAROUCHE — A popular name applied in the far West to the BUCKBOARDS used as stages. (H)

BUCKBOARD, MAIL — A form of BUCKBOARD WAGON, accommodating from two to six passengers, built by the Studebaker Company of South Bend, Indiana, for mail and passenger service in the West, and especially in Yellowstone Park. (H)

BUCKBOARD WAGON — A modern American form of BUGGY developed from the primitive BUCKBOARD. The feature of a springboard is retained, either in place of springs, or supplementary thereto, but the board is shortened, bringing the wheels closer together, and numerous narrow slats are often substituted for the single springboard, to afford increased elasticity. (H)

BUCKER — Origin unknown. Synonymous with NIGHT-HAWK and OWL, as applied to a street CAB seeking patronage after dark. (H)

BUCKINGHAM WAGON — A name sometimes applied to a variety of four-passenger open PHAETON (1879).

BUGGY (sometimes BUGGIE) — The derivation of this term is unknown, except that it originated in England previous to the close of the eighteenth century. The first known mention of a buggy appears in the *Treatise on Carriages*, published in London in 1796 by William Felton, coachmaker, wherein (volume 2, page 121) he says, "A Buggy is a cant name given to phaetons or chaises which can only contain one person on the seat; they are principally intended for lightness in draught, for the rider to sit snug in, and to preclude the possibility of an associate; mostly used by out-riders. They are built like other phaetons or chaises, and to ascertain their value, is to subtract one-twelfth from the statement of a common-sized carriage, finished to any pattern."

A nicely made BUCKBOARD of about 1890 showing the typical wood dash, and a natural finish that was fairly common on this type of vehicle. (Smithsonian collection.)

The character of the vehicle has since materially changed. The modern buggy is a member of the PHAETON family, and a typical American vehicle of primitive form and simple construction, consisting of a body having accommodation for one or two passengers, and mounted on four wheels. The American buggy developed early in the nineteenth century, evolving through the PLEASURE WAGON from the light work-wagon. It undoubtedly was the most popular carriage ever built, comparable to the famous Model T Ford of a later era. By 1900, methods of mass production had lowered the average price to about $40, while the price of some economy types, complete with top, went as low as $25. Although buggies generally have accommodated no more than two passengers, a great number were built during the 1850s and 1860s that carried three or four passengers by means of various sliding, shifting, turn-out, or jump seats.

There are many varieties, named either from the shape of the body, as COAL-BOX, SQUARE-BOX, PIANO-BOX, etc., or from the modes of suspension, as ELLIPTIC-SPRING, END-SPRING, SIDE-BAR, etc. For descriptions of the various types, see BRITTON BUGGY or BRITTON WAGON, COAL-BOX BUGGY, CONCORD BUGGY, CORNING BUGGY, DEXTER BUGGY, GODDARD BUGGY, MONITOR BUGGY, PIANO-BOX BUGGY, SURREY BUGGY or WAGON, TRAY-BODY BUGGY, WHITECHAPEL BUGGY or WAGON, YACHT BUGGY, etc. (H, C, & F)

BUGGY-BOAT — A novel vehicle intended for amphibious use, patented by Perry Davis of Providence, Rhode Island, in 1859. Essentially, it was a small boat equipped with four wheels. When used in the water the front wheels were detached, and the boat was driven by hand cranks which turned the rear wheels. The spokes of the rear wheels were provided with paddles, thus permitting the wheels to act as paddle wheels.

BUGGY, COAL-BOX — One of a class of BUGGIES, partly cut down at front and rear, so named from a fancied resemblance in shape to the ordinary grocer's coal box. The successor to the YACHT BUGGY, it was the result of an attempt to introduce a radical change, and produce a wagon less sportinglike in its character. It was made with a body about 10 inches high in the rear, and cut down in front as much as possible, leaving only the sill. It had a straight, square dash, and was often made without a top. Designed by James W. Lawrence, of Brewster & Co., of Broome Street, New York, about 1862, it was first introduced as "The Gentleman's Wagon," under which name it was advertised. This induced a rival house to introduce a similar BUGGY, which was advertised as "The Coal-box," intended in derision of the name used by Brewster & Co. By 1874 it had lost much of its popularity, for the *Harness & Carriage Journal* reported that in the city trade, the square-box type was leading ten to one. (H)

BUGGY, COAL-SCUTTLE — An old variety of COAL-BOX BUGGY. (H)

BUGGY, CONCORD — See CONCORD

Slat-bottom BUCKBOARD equipped with torsion end springs, and elliptic springs under the seat. Wheels, 40-inch and 44-inch; body, dark green; gearing, dark green or carmine; trimming, leather. (Studebaker Bros. Mfg. Co., South Bend, Indiana, 1903.)

BUGGY with either a hinged lid on the rear, or a turn-out seat. Wheels, 43-inch and 49-inch. (Lawrence, Bradley & Pardee, New Haven, Connecticut, 1862.)

This half-pillar COAL-BOX BUGGY is an example of a bracket-front wagon. Wheels, 45-inch and 49-inch. A vehicle with a falling top was often used as shown here, with the top lowered but not folded, so that the top covering would not wear so quickly in the folds. (*The New York Coach-Maker's Magazine,* January 1868.)

BUGGY (or WAGON), CORNING — A late form of COAL-BOX BUGGY, with cut-down front, deep sides and molded panels on both body and seat; it is hung on either side bars or elliptic springs. Its form is between the square-box and coal-box patterns, having a modified square-box body, with that part of the side panels forward of the seat cut away. It has no Stanhope pillar, but is always made with a top. When made with a close top, it forms a convenient PHYSICIAN'S PHAETON. It was first built by Brewster and Company, of Broome Street, New York, in 1875. It acquired its name because the first of its type was sold to Erastus Corning, of Albany, New York. (H)

BUGGY, DEXTER — A patented variety of BUGGY, chiefly characterized by its suspension, made by W. W. Grier, of Hulton, Pennsylvania. It was named after the celebrated trotting horse. (H)

BUGGY, DOCTOR'S — The term DOCTOR'S BUGGY was first applied to a variety of physician's PHAETON which had a peculiar top, termed the YANDELL, designed by a physician of that name in Louisville, Kentucky. (M)

BUGGYETTE — A novel design of BUGGY invented and patented under this name by Steuart & Co., coachmakers, of Calcutta, India. (H)

BUGGY, FANTAIL — A term applied to a light variety of BUGGY, wherein the body was cut down at the top, leaving a fantail extension at the rear. (H)

BUGGY, GODDARD — A drop-front BUGGY designed by the noted Boston carriage builder, Thomas Goddard, in the mid-1840s. (C)

BUGGY, HAND — A term sometimes applied to a JINRIKISHA. (H)

BUGGY, IRON-AGE — A patented BUGGY that was introduced in Boston about 1869; its name was suggested by the editors of *The Hub*. This BUGGY was built entirely of metal, with body panels of sheet iron. The defect of inaccurate rattling condemned this carriage. (H)

BUGGY, MONITOR — A variation of the SQUARE-BOX BUGGY, having a deep sunken bottom and concave sides, first introduced by James W. Lawrence, of New York City, about 1859. (H & C)

BUGGY, OGEE — An old form of BUGGY with ogee bottomsides. (H)

BUGGY, PIANO-BOX — Originally this term was applied to a variety of SQUARE-BOX BUGGY, and differed in having round corners. Early versions had a molding at the top and bottom of the body, and sometimes had light moldings affixed to the sides so as to form several small panels. Eventually the sides became plain, and as square corners became the rule on the majority of BUGGIES, the terms *square-box* and *piano-box* came to be applied almost synonymously to square-cornered BUGGIES. The most popular of all vehicles, the piano-box buggy, is believed to have been introduced by R. M. Stivers of New York City shortly before 1855. See SQUARE-BOX BUGGY. (H, M, & C)

BUGGY, SHOW — A no-top BUGGY, highly finished, with wire wheels and pneumatic tires, used for show purposes. (O)

BUGGY, SLIDE-SEAT — A BUGGY, the seats of which are so arranged as to allow conversion to a four-passenger vehicle. The jump-seat is an improvement on this style. (C)

BUGGY, SQUARE-BOX — The most common variety of BUGGY, the name of which is self-descriptive. Having square corners, it was originally distinguished from the round-cornered PIANO-BOX BUGGY. Eventually, as square corners became the rule on the majority of BUGGIES, the terms SQUARE-BOX and PIANO-BOX came to be synonymously applied to square-cornered buggies. See PIANO-BOX BUGGY. (H & M)

BUGGY, STANHOPE — A top BUGGY having a body similar to that of the STANHOPE. Late in the carriage era some manufacturers omitted the word BUGGY from the name, and called the vehicle only by the name of STANHOPE, thus confusing it with the earlier TWO-WHEEL STANHOPE. A few applied the word "Stanhope" only to those four-wheelers with no tops, reserving the compound term, STANHOPE BUGGY, for those with falling tops, but these usages were by no means consistent. The application of the above usages was generally dependent on the resemblance of the body to that of a true Stanhope, or the retention of the outline of the Stanhope pillar; it was infrequently based on the manner of suspension, for most STANHOPE BUGGIES and FOUR-WHEEL STANHOPES employed elliptic springs instead of the original Stanhope suspension. There was also a TWO-SEAT STANHOPE that was somewhat similar to a SURREY, frequently without a top, and again the name was based on the retention of the Stanhope body lines. By the mid-1890s the STANHOPE BUGGY was often being called a STANHOPE PHAETON, though this usage confuses the BUGGY with the true STANHOPE PHAETON. (C & O)

James Lawrence's COAL-BOX BUGGY of about 1862. (Ezra M. Stratton: *The World on Wheels,* New York, 1878.)

CONCORD BUGGY. Wheels, 39-inch and 43-inch; body, black, with green center panels outlined with carmine striping; gearing, Brewster green, maroon, carmine, or yellow; trimming, green or maroon leather. (W. A. Paterson Co., Flint, Michigan, 1910.)

CORNING BUGGY. Wheels, 39-inch and 43-inch; body, black; gearing, Brewster green, maroon, or carmine; trimming, green, blue, or maroon cloth of whipcord. (W. A. Paterson Co., Flint, Michigan, 1910.)

BUGGY, TRAY-BODY — An early form of BUGGY, perfected by George Watson of Philadelphia around 1840. Ten years later Watson exhibited one of this pattern at the World's Fair in London. This BUGGY was so-called because of the resemblance of the body to a serving-tray. The boot was leather covered. The hanging bars and the back bed were elaborately carved, one bed sometimes requiring the labor of a rapid carver for two days. Its successor was the YACHT BUGGY. (H)

BUGGY, TUB-BODY — An old variety of BUGGY, popular in New York about 1840-1850, preceding the piano-box and coal-box patterns, which were introduced about 1855 and 1862, respectively. Its body had a tub-like shape, with round ends, which gave it the name, and the sides and back were covered with embossed leather. (H & G)

BUGGY, YACHT — This was a modification of the old TRAY-BODY BUGGY, in which an attempt was made to follow the lines of a yacht. The leather boot, as used on the TRAY-BODY BUGGY, was supplemented by a paneled boot. It was designed by James W. Lawrence, of Brewster & Company, of Broome Street, New York, and was first built in 1859. It was succeeded by the COAL-BOX BUGGY. (H)

BUILDER'S WAGON — A wagon built much like an EXPRESS WAGON, having low sides and high stakes, for carrying lumber to building sites. It is also called a MILL WAGON. (S226)

BULLOCK WAGON — A traveling vehicle having two solid plank wheels, used with a team of bullocks, still (1891) in common use in Cape Colony, South Africa. (H)

BUS — The popular abbreviation for OMNIBUS. (H)

BUSINESS WAGON — A type of light DELIVERY WAGON, usually being the lightest of the class. Many have bodies resembling, and nearly as short as, the common ROAD WAGON, or PIANO-BOX BUGGY. They are usually on a two-spring gear. (S226)

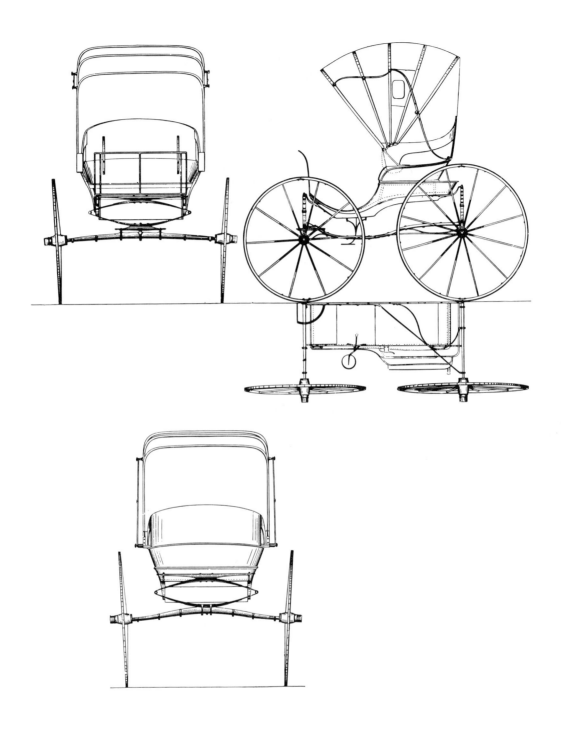

GODDARD BUGGY. Wheels, 42-inch and 46-inch; body and gearing black, with triple lines of cream white on gearing; trimming of cloth, either drab, green, blue, or maroon; back cushions drawn down by black tufts with a red center. (*The Hub,* October 1880.)

New York PIANO-BOX BUGGY. Wheels, 48-inch and 50-inch. (*The New York Coach-Maker's Magazine,* June 1868.)

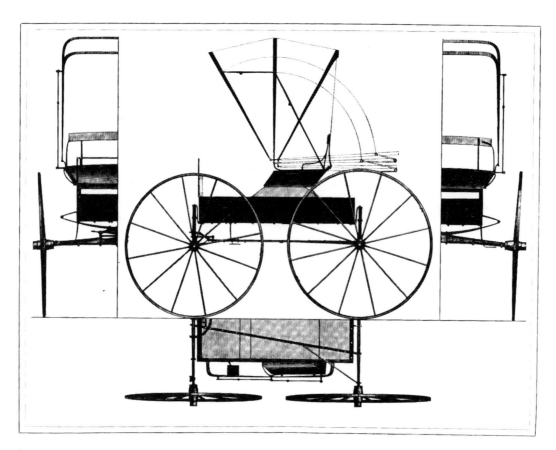

PIANO-BOX BUGGY. Wheels, 46-inch and 48-inch; body color was generally black, with no striping; gearing was commonly red (often carmine and vermilion mixed in equal proportions), yet dark blue or dark green were also frequently used. Gear striping was usually black, and the trimming was commonly dark blue or dark green cloth (green if the gearing was green, but blue if the gearing was red or blue). (*The Carriage Monthly,* June 1884.)

This PIANO-BOX BUGGY was offered with an extra spindle-seat, the substitution of which converted the vehicle into a *Driving Wagon*. The body was black, on a red or dark green gear, and the trimming was dark blue or dark green cloth, though both leather and whipcord were also available. The price was $53.50, plus $15 extra if mounted on rubber tires. (Elkhart Carriage & Harness Mfg. Co., Elkhart, Indiana, 1907.)

Modern SHOW BUGGY built by the Wilform Buggy Works, of Long Beach, California.

New Orleans JUMP-SEAT BUGGY showing jump seat turned out. (Lawrence, Bradley & Pardee, New Haven, Connecticut, 1862.)

New Orleans JUMP-SEAT BUGGY showing jump seat turned in and the main seat moved to the forward position. Wheels, 45-inch and 51-inch. (Lawrence, Bradley & Pardee, New Haven, Connecticut, 1862.)

SLIDE-SEAT BUGGY, showing the additional front seat in position, providing seats for four passengers. (*The American Coach-Makers' Illustrated Monthly Magazine*, April 1855.)

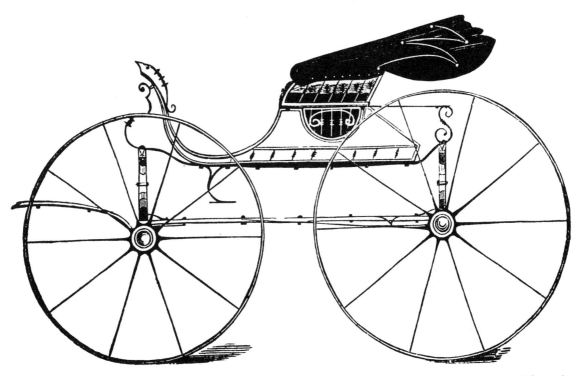

The same SLIDE-SEAT BUGGY, showing the extra seat folded down into the body, with the main seat moved forward over it, converting the *Buggy* into a two-passenger vehicle. Wheels, 49-inch and 53-inch.

This is a STANHOPE BUGGY, though the manufacturer simply called it a *Stanhope*. Having a black body on a dark Brewster green gear, with dark green broadcloth trimming, it sold for $107.50, plus $15 more if equipped with rubber tires. (Elkhart Carriage & Harness Mfg. Co., Elkhart, Indiana, 1907.)

This TRAY-BODY BUGGY, built by G. W. Watson of Philadelphia about 1851, was known as a GAZELLE WAGON. The body was of walnut and hickory, and the gearing of hickory, both having a natural wood finish without varnish, while the mountings were solid silver. It was awarded a medal at the World's Fair of 1851. (B. Silliman and C. Goodrich: *The World of Science, Art & Industry, from Examples in the New York Exhibition, 1853-1854*, New York, 1854.)

Turn-out seat BOX BUGGY with the turn-out seat in the riding position. Wheels, 43-inch and 50-inch. (Lawrence, Bradley & Pardee, New Haven, Connecticut, 1862.)

YACHT BUGGY. Wheels, 46½-inch and 50-inch. (*The New York Coach-Maker's Magazine,* October 1860.)

Platform spring BUILDER'S WAGON. Wheels, 36-inch and 50-inch; body, dark green; gearing, red or yellow; duck cushion. (Studebaker Bros. Mfg. Co., South Bend, Indiana, 1903.)

Side spring BUSINESS WAGON, with a body only 10 inches longer than the average *Piano-Box Buggy*. Wheels, 40-inch and 44-inch; black body on a carmine gearing, striped black. (Studebaker Bros. Mfg. Co., South Bend, Indiana, 1903.)

C

CAB — (Derivation from *cabriolet*, thereby allying the vehicle to the COACH family.) A vehicle available for public hire. The term CAB first came into use in France when the original two-wheel CABRIOLET, which had been introduced from Italy, became the street hack of Paris. Early in the nineteenth century, the CABRIOLET was introduced into London where it began to give some competition to the HACKNEY-COACHES, though it was at first limited both by legal restraint and by the fact that only one passenger could share the single seat with the driver. In 1823 came an improvement by David Davies, whereby a separate seat for the driver was added between the body and the right wheel. Numerous changes followed in the succeeding years, and the term came to be applied to a great variety of public vehicles, whether they had two or four wheels. Some carriage tradesmen, such as Houghton, felt that the term should be reserved for the two-wheel, low-hung, one-horse types, but this usage was not rigidly adhered to. For a thorough study of this type, see *Omnibuses and Cabs*, by Henry Charles Moore. (H, C, & O)

CAB, CRYSTAL — A four-wheel, four-passenger public CAB with glass sides, introduced in New York City about 1870 by A. S. Dodd, of Dodd's Express Company. Many crystal cabs were built by the Abbot, Downing Company, of Concord, New Hampshire, but the popularity of the vehicles was of short duration. (H & O)

CAB, HANSOM — A two-wheel vehicle named after the inventor, Joseph Hansom, an architect of Leicestershire, England, who patented in 1834 an improved form of CAB, the main feature of which was a body hung close to the ground, thereby affording easy access for passengers. Entry was gained through twin doors in the front of the CAB, and the driver's seat was high on the front. The wheels were 7'6" in height, and were attached to the sides of the body by a pair of short axles. The vehicle eventually became the typical public cab of London, but only after Hansom's original design was much improved, first by John Chapman in 1836, and again by F. Forder of Wolverhampton in 1873. Chapman lowered the wheels and mounted them on a cranked axle, introduced the sliding front window, and placed the driver's seat in the rear, thus contributing more to the final design than did Hansom. In its final form, this two-passenger cab has two high wheels, a paneled hood or top, and a skeleton driver's seat in the rear, mounted sufficiently high to enable the driver to see and control his horse. In front is a high dashboard, curved so as to allow the horse to be brought close to the vehicle, and extended to fully protect the passengers from the horse's heels.

The Hansom cab was used primarily as a public vehicle, but was also used to a limited extent as a private carriage. Several unsuccessful attempts were made to introduce the Hansom into the United States before it finally met with approval here late in the nineteenth century. It was most commonly used in New York City. Early in the twentieth century some Hansoms were built with a falling top. These were most common in Berlin, Germany. (H, C, & O)

CAB, KELLNER — A recent (1891) pattern of four-wheel public carriage introduced in Paris by its inventor, Mr. Kellner, a carriage builder of that city. (H)

CAB, MURCH'S CHARIOT — A short, four-wheel, rear-entrance CAB, patented by C. M. Murch, of Cincinnati, Ohio. It carried from four to six persons, and held luggage on the roof and in the front boot. The framework of the body and gear was composed of T and angle iron, and suspension was by Murch's patented springs. (*Scientific American*, September 25, 1880.)

CABOOSE — An American term for the conductor's car on a railway train. The term, however, also was applied to a closed box wagon used by peddlars in some parts of the West. Originally a sea-term, meaning the cook's cabin on a ship. It was sometimes spelled *camboose*, which, according to Skeat, is the more correct form. He adds, "Like most sea-terms, it is Dutch, from *kombuis*, a cook's room, or the chimney in a ship; the etymology is not clear, but it seems to be made up of Dutch *kom*, a porridge dish, and *buis*, a pipe or conduit,

David Davies's CAB of 1823, with curtain drawn. (H. C. Moore's *Omnibuses and Cabs,* 1902.)

so that the literal sense is a dish-chimney — evidently a jocular term." (H)

CAB-PHAETON — (Derivation: a contraction of CABRIOLET-PHAETON, being applied to a modification of the vehicle by that name.) One of a class of vehicles that, during the first third of the nineteenth century, emanated from a fashion of compounding all sorts of vehicles, sometimes with rather odd results. The BRITZKA and the CHARIOT were combined, for example, making the BRITZKA-CHARIOT; and the BRITZKA and the PHAETON were combined, making the BRITZKA-PHAETON. The combination of a two-wheel CABRIOLET as the hind part, with the front portion (including seat) of a PERCH-HIGH PHAETON, resulted in the CABRIOLET-PHAETON. The CAB-PHAETON, developed in England about 1835, was a CABRIOLET-PHAETON having the extreme height of the fore part reduced, along with a general redesigning of the CABRIOLET-PHAETON into a more symmetrical, compact, and harmonious vehicle. Migrating to the Continent, where it was known as the MILORD, the cab-phaeton there degenerated into a public hack by 1850, after which it lost favor among the gentry. In 1869, following a series of modifications (some of which actually had originated in England), the vehicle was reintroduced into England as a VICTORIA. It also contributed to the design of the four-wheel CABRIOLET.

Houghton makes several conflicting statements concerning this type. While once implying that it had a driving-seat, in another statement he says it did not, and compares it to a VICTORIA driven *a la Daumont.* Also, while stating that the CABRIOLET-PHAETON and the CAB-PHAETON are not quite the same, he again implies that they are the same. Possibly the terms are nearly synonymous, with the term CABRIOLET-PHAETON being applied generally to the earlier and high versions. (H & O)

CABRIOLE — An obsolete spelling of *cabriolet.*

CABRIOLET — (Derivation: French, *cabriolet,* through the Italian, *capriolo,* kid, from the Latin *capreolus,* diminutive of *caper,* goat.) A name originally applied to a two-wheel vehicle used in France and Italy during the late seventeenth century, so named because of the capering or goatlike motion given to the vehicle by its long, springing shafts. The body, resembling an exaggerated comma or nautilus shell, was a relic of the cut-down coach body, thus relating the cabriolet to the COACH family. The cabriolet eventually borrowed an additional characteristic from the COACH in the manner of suspension, being hung from whip springs and C-springs. Popular at first among men of fortune, the cabriolet came into common use in Paris as a public hack during the late eighteenth century. Shortly thereafter it was introduced into London where, early in the nineteenth century, it was again put into limited service as a public carriage, and thus was the ancestral form of the two-wheel public CABS.

At about the same time an English coach builder combined the two-wheel cabriolet with the PERCH-HIGH PHAETON to produce a four-wheel vehicle known as the CABRIOLET-PHAETON. Later the vehicle came to the United States where the word *phaeton* was soon dropped from the name, leaving only the word *cabriolet*, though the term now referred to a four-wheel vehicle.

On the Continent, the four-wheel cabriolet was lifted to the dignity of a royal equipage. It was then driven with postilions, of *a la Daumont*, which necessitated the removal of the driver's seat, and at the same time were added an ornate leather dash in front and a rumble seat for the groom behind. The French named this carriage the VICTORIA, in honor of Queen Victoria of England, probably about mid-century. Later, when the vehicle descended to common use, it became necessary to restore the driver's seat. To do this without altering the vehicle, a skeleton boot supported by iron stays was added above and in front of the dash. Thus the social significance in the skeleton boot of the VICTORIA, as compared to the paneled boot of the cabriolet, is obvious. It should be remembered, however, that the VICTORIA properly belongs to the CABRIOLET group, the latter name being applicable to carriages solely because of the resemblance of their bodies to the body of the original two-wheel cabriolet.

MURCH'S CHARIOT CAB weighed 775 pounds, and reportedly could turn in its own length. (*Scientific American*, September 25, 1880.)

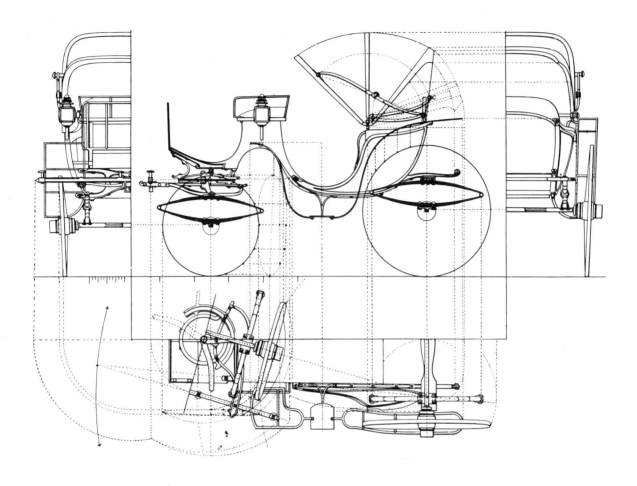

Miniature CABRIOLET. Wheels 33-inch and 44-inch. (*The Hub,* December 1901.)

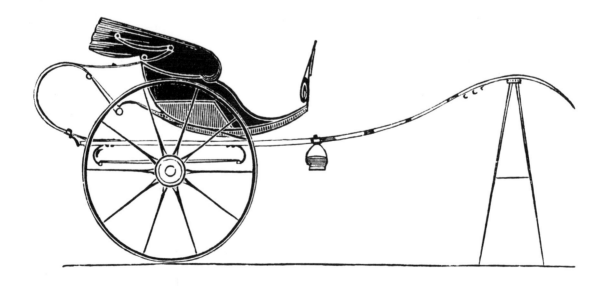

American CABRIOLET of about 1830. (Ezra M. Stratton: *The World on Wheels,* New York, 1878.)

In its final form the cabriolet is a low, four-wheel, four-passenger (including driver) hooded vehicle without doors. Either one or two horses might be used, depending on the size of the carriage. The rear portion is made with a deep quarter, and the driver's seat is on an elevated, paneled boot. Some carriage manufacturers applied the term CABRIOLET if there was only a slight elevation to the driver's seat, while they used the term PANEL-BOOT VICTORIA if there was a greater elevation of the driver's seat. The larger types frequently have a folding child's seat that turns down from the neck behind the driver's seat. The calash-top is generally over the back seat only, but some later types have extension tops that cover the front portion as well. The body usually is suspended on elliptic springs in front and platform springs in the rear, although in some instances elliptics were used in both front and rear.

The terms CABRIOLET, CABRIOLET-PHAETON, CAB-PHAETON, VICTORIA, DUC, and MILORD are often so loosely used as to be nearly synonymous. See these terms for further comparison. Toward the end of the carriage era some builders even applied the term CABRIOLET to their finer SURRIES. (H, C, & O)

CABRIOLET-PHAETON — A carriage resulting from an early nineteenth-century effort to modify the PERCH-HIGH PHAETON, by combining the forepart of this latter vehicle with the two-wheel CABRIOLET as the hind part. The result, however, was inharmonious, and the carriage soon became modified into the CAB-PHAETON. See CAB-PHAETON.

Late in the nineteenth century, the term was applied occasionally to a vehicle nearly identical to the PANEL-BOOT VICTORIA, or CABRIOLET, though the driver's seat generally was not elevated as much above the rear seat. (H & O)

CACOLET — A device for transporting the sick or wounded, similar to a horse-litter. The passengers sit in folding chairs attached to the sides of an animal. Sometimes a litter was hung on one side of the animal, and a cacolet on the other, and occasionally the term was also applied to the litter. The cacolet is reported to have developed during the Crimean War, and was used to a very limited extent during the American Civil War. A few were built by the famed carriage builders, Lawrence, Bradley and Pardee, of New Haven, Connecticut. (O)

CAGE WAGON— A type of CIRCUS WAGON used for carrying wild animals. Essentially a cage, usually brightly ornamented, on a platform running gear.

CAISSON — (From the French, *caisse*, meaning box, or chest.) A carriage used by the artillery for conveying ammunition for a field battery. It is a four-wheel carriage, consisting of two parts, the foremost of which is a LIMBER similar to that of a gun carriage, and connected in a similar way by a wooden stock and lunette. On the axle-body of the rear part, and parallel to the stock, are placed three rails upon which are fastened two ammunition-boxes, one behind the other, and similar to the one on the LIMBER. Thus, the caisson has three ammunition-boxes which will seat a total of nine artillerymen. The interior compartments of the ammunition-boxes vary according to the nature of the ammunition with which they are loaded. On the rear of the last box is placed a spare-wheel-axle of iron, with a chain and toggle for securing the wheel at the end of it. On the rear of the middle rail is placed a carriage hook similar to the pintlehook, to which the lunette of a gun carriage whose LIMBER has become disabled may be attached, and the gun carried off the field. The caisson has the same turning capacity and mobility as the gun carriage, so that it can follow the gun in all its maneuvers, if necessary. Besides the spare wheel the caisson also carries a spare pole, pick, shovel, etc. (W, H, O, & C)

CALASH — (From the Czech *kolesa*, meaning wheels, carriage.) This term causes much confusion because of the spellings, *calash* and *caleche*, and the various types of vehicles to which the names are applied. While some have attempted to assign each spelling to a specific type, it does not seem feasible to do so at this time, since those in the trade applied the terms somewhat synonymously in their own day. Generally, the term CALECHE (preferred spelling) refers to a four-wheel carriage that is also known as a BAROUCHE, yet the CALASH spelling is also used for the same vehicle. In the United States, CALECHE was the favored term by about 1850, while BAROUCHE was more frequently used in the earlier part of the century. During the 1850s, *The American Coach-Makers' Illustrated Monthly Magazine* showed illustrations of carriages that were almost identical to the VICTORIA, but were called CALASHES. *Calash* has also been used to designate a CHAISE having a small seat in front of the dash for the driver, while the word CALECHE was used in Canada to designate a similar chaiselike two-wheel carriage that has a driver's seat on top of the dash. To confuse the matter further, Ephraim Chambers' *Cyclopedic Dictionary* (London, 1741) states that CHARIOTS that "are very gay, richly garnished and have five glasses . . . are called Calashes." The term CALASH also came to be applied to any single folding carriage top, because of the similarity to the falling top of Calash carriage. (C, H, & O) See BAROUCHE.

CABRIOLET-PHAETON (a late nineteenth-century usage of the term), showing the slight elevation of the front seat above the rear. (*The Carriage Monthly,* February 1898.)

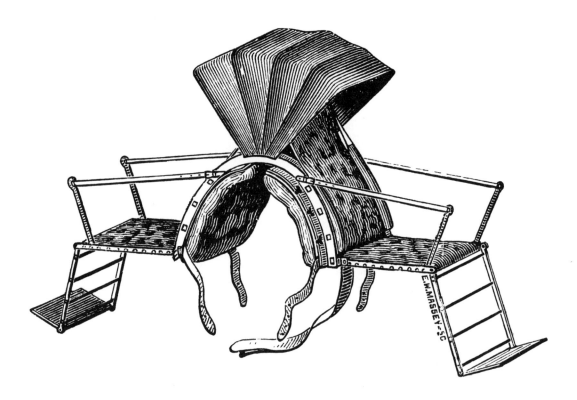

CACOLET made by the carriage builders, Lawrence, Bradley & Pardee, of New Haven, Connecticut, in 1861. *(Medical & Surgical History of the War of the Rebellion,* Part III, vol. 2, Surgical History, Washington, Government Printing Office, 1883.)

CALECHE – (1.) A BAROUCHE.   (2.) A two-wheel, chaiselike vehicle used in Canada. It has a small seat on top of the dash for the driver, and the inside seat holds two passengers. Also see CALASH. (C & O)

CALECHE COACH – A variety of COACH with a fixed roof, but having sides or front quarters that can be shifted or removed at will. (O)

CALECHE, NANTUCKET – A local term applied to the primitive, one-horse, two-wheel, springless carts peculiar to the Island of Nantucket, Massachusetts. They had no seat boards, but passengers were accommodated by three or more high-backed, rush-bottomed chairs lashed by cords to the sides or gunwales of the cart, and covered by figured counterpanes. They were used prior to the introduction of CHAISES, which made their first appearance in Nantucket about 1800 (H)

CALESA – A kind of one-horse, two-wheel vehicle used in the Philippines. (H)

CALESIN – Diminutive of CALESA. A small, one-horse, hooded GIG, with a seat behind for the driver, used in the Philippines. (O)

CALESSO – (An Italian word, from the old Italian, *calesca, caresca*; Latin, *carrus*.) The name of a modern (1891) Italian two-wheel, hooded GIG. (H)

CALIFORNIA WAGON – A wagon built in varying styles for either farm or delivery use. It features a spring seat which is elevated above the sides of the box on risers, a toolbox across the front, and either ordinary side boards or stake-racks which are generally removable only up to the seat risers. (S226)

CALIFORNIA WOOD-SPRING WAGON – A variety of COAL-BOX BUGGY, hung on wooden springs and thoroughbraces, invented and introduced by the Kimball Mfg. Co., of San Francisco, and widely distributed by them throughout the Pacific states about 1867-1870. (H)

CALLIOPE – A steam-operated musical instrument used in circus parades. It is mounted on wheels and consists of a steam boiler, a set of steam whistles, and a keyboard with connections running to the whistles. There are usually either 21 or 32 whistles, which operate on a pressure of about 120 pounds. The music produced is rather harsh. Pronounced ka/lié/o/pea by Webster, but kál/e/ope by circus employees. (O)

CAMEL CARS – Railway cars drawn by camels, used in central Asia late in the nineteenth century. (H)

CAMP WAGON – A term sometimes applied to a CHUCK WAGON, or ROUND-UP WAGON. (O)

CANOE-LANDAU – A LANDAU in which the lower body line forms a continuous sweep, resembling the outline of a canoe, as distinguished from a LANDAU having an angular quarter and center. (H)

CAPE CART – A local variety of pleasure cart developed in South Africa, and also used in Australia. Heavy in proportions, usually it was made with a pole and a falling top, and carried four persons. (H & O)

CAR – (From the Latin, *carrus*, according to Skeat, a kind of four-wheel carriage which Caesar first saw in Gaul; a Celtic word. Old French, *car, char*; modern French, *char*, a car or vehicle. Allied to Latin, *carrus*, a chariot, and *currere*.) In late nineteenth century usage the term in America was applied mainly to a railway vehicle; in Ireland it was confined to the JAUNTING-CAR, and in England it was applied to the various forms of VILLAGE-CARTS, such as ALEXANDRA-CAR, GOVERNESS CAR, PRINCE VICTOR-CAR, RALLI-CAR, etc. It is now applied to passenger automobiles. (H & C)

CARAVAN – (1.) An English vehicle of the sixteenth century. Adams, in his *English Pleasure Carriages*, page 44, says, "Spenser, in his *Faery Queen*, uses the words, *Wagon, Coche* and *Charat*. Long wagons for passengers and goods, afterwards called *caravans*, were introduced, though of somewhat clumsy construction to accommodate them to the badness of the roads and the number of persons they carried."   (2.) A covered vehicle, such as one equipped as traveling living quarters, or as a traveling photographer's studio.   (3.) A type of family wagon used in the American colonies late in the eighteenth century. Drawn by either two or four horses, there is the possibility that this little-known carriage may have been similar, if not identical, to the COACHEE.   (4.) A column of vehicles traveling together. (H & O)

CAR, BAGGAGE – An American railway car, equivalent to the English term GOODS-VAN or LUGGAGE-VAN. Modern French equivalent, *fourgon*. (H)

CAR, BOBTAIL – A small streetcar, wherein the services of a conductor are dispensed with. (H)

CARETTA – An ancient Italian vehicle of unknown description. Thrupp, in *The History of Coaches*, page 27, says, "In 1267, Charles of Anjou entered Naples, and his queen, Beatrice, rode in a Caretta, the outside and inside covered with sky-blue velvet, powdered with golden lilies; and, in 1273, Pope Gregory X entered Milan in a Caretta." (H)

CARETTE – A type of vehicle used in Australia for conveying passengers from the city to the suburbs, as an OMNIBUS was used. It was built by Duncan & Fraser in Adelaide, Australia, beginning about 1895. The driver sat upon a slightly elevated seat, with the roof from the body extended over him. The rear portion is divided into two sections, the forward one being open and having two transverse seats, for smokers, and the

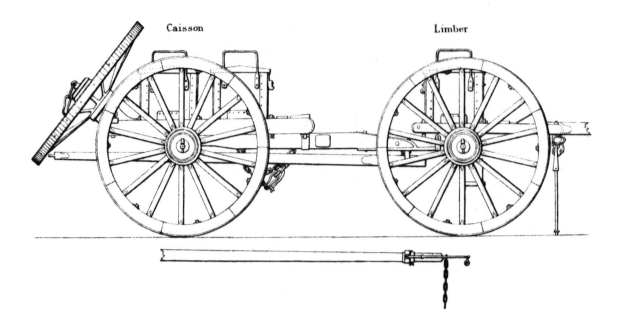

CAISSON with its LIMBER attached, and a spare wheel on the rear. (John Gibbon: *The Artillerist's Manual,* D. Van Nostrand, New York, 1863.)

CALECHE, fitted with a half top. At a slightly earlier time in America this would have been called a *Barouche.* Wheels, 42½-inch and 51-inch. (Lawrence, Bradley & Pardee, New Haven, Connecticut, 1862.)

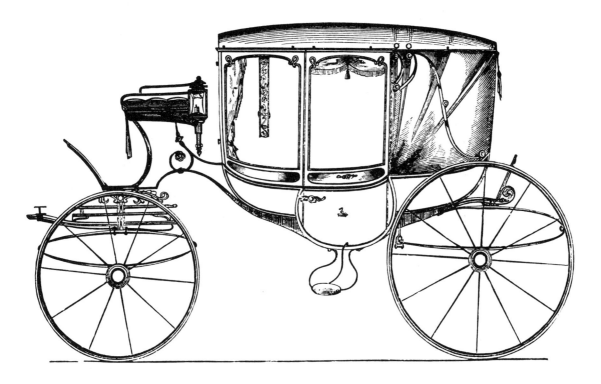

Full CALECHE, on which the shifting front quarters, front portion of the roof, and tops of the doors could be removed at will, and the falling top made to extend forward as on an ordinary *Barouche*. Wheels, 41-inch and 48-inch. (*The American Coach-Makers' Illustrated Monthly Magazine*, October 1856.)

Canadian CALECHE, shown in *Scribner's Monthly,* September 1871.

CALECHE COACH. Many, such as this example, omitted the mock joints, but they were retained on other vehicles as ornaments. The upper sides of this coach were removable. Wheels, 43-inch and 49-inch; body and gearing, plum color, striped black; trimmed with coteline. (*The New York Coach-Maker's Magazine,* July 1858.)

CALIFORNIA WAGON with stake rack, suitable for farming or freighting. Wheels, 44-inch and 54-inch. (Mitchell & Lewis Co., Racine, Wisconsin, ca. 1891.)

The Two Jesters CALLIOPE is in the permanent collection of the Ringling Museum of the Circus, Sarasota, Florida.

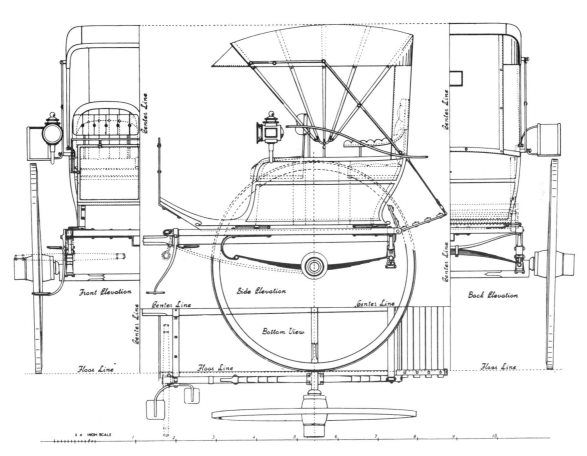

CAPE CART. Wheels, 54-inch; panels, deep green; moldings, black; gearing, deep green striped with black; trimming, green cloth. (*The Carriage Monthly,* September 1898.)

CAPE CART with body built of stink wood and finished natural, with fine white striping on the moldings; gearing, dark blue with white striping; trimming, maroon leather, tufted with silver buttons. (*The Hub,* December 1894.)

closed rear portion has longitudinal seats with a door in the rear. Doors to the front portion are in the sides. The body sits very low upon the gear. (*The Hub,* May 1898.)

    This term was also applied to a public vehicle introduced in Chicago about 1890, this being essentially the body of a street-railway car set on crank-axles and wooden wheels. Like the OMNIBUS, it was used on streets where there were no railway tracks, but had the advantage of being hung low for easier access. Because of the lack of this feature the OMNIBUS lost some of its former popularity by the 1890s. (*The Hub,* November 1890.)

CAR, FUNERAL — A HEARSE with open sides, specially designed for pageantry. Also see CATAFALQUE. (H)

CAR, HORSE — A street-railway car drawn by a horse, equivalent to the English TRAM-CAR. According to Forney's *Car-Builder's Dictionary*, the term HORSE-CAR is also applicable to a railway car that is specially fitted for carrying horses. The first street railway line in the world was the New York & Harlem, incorporated in 1831. The first cars (built by John Stephenson) were run over this line in November 1832, from Prince Street to Harlem Bridge, each car being made in the form of three stagecoach bodies, with side doors only, carried on a single set of wheels, with a platform at each end for the driver. (H)

CARIOLE — (1.) A two-wheel vehicle somewhat resembling a CABRIOLET, formerly used in Norway. It originally had no springs, the long shafts providing some resilience, but springs were later introduced. The bodies were often narrow so that only one person could be seated. Behind the body was a rack or platform for the accommodation of luggage. A two-wheel cariole was also used in eighteenth-century America, but it is not known to what extent it may have resembled the Norwegian cariole. (2.) A small, low sleigh popular

CARETTE. (*The Hub,* May 1898.)

HORSE-CAR in Hartford, Connecticut, in 1891. (Smithsonian photo.)

among the French Canadians, frequently painted red, and decorated with unusual floral designs. Drawn by one horse, they were often driven over the ice on frozen rivers. Some were drawn by dogs. (C, H, & O)

CAROCHE — A luxurious and stately carriage used by the French from the fourteenth to sixteenth centuries. (O)

CAROSSE-COUPE — See COUPE.

CARPENTUM — (1.) An ancient Roman vehicle, known to have been in use during the first century A.D. It was named in honor of Carmenta, the mother of Evander, and was sometimes called the covered litter, which it appears to have replaced. CARPENTUM seems to have been the generic term for different types of covered two-wheel vehicles. They were frequently lined with costly fabrics, and profusely ornamented. Entry was gained through a door in the rear. In rare instances these carriages were on four wheels, but these apparently were used exclusively by royalty. (2.) This term also was applied to a modern vehicle that was invented by the English coach builder David Davies. It had but a short existence in England about 1840. (H & O)

CARRETELA — A kind of two-horse, covered HACK used in Manila in the Philippines. (H)

CARRIAGE — (From the Latin, *carrus*, wagon, through the Old French, *carier*, *charier*, to cart. Equivalents: French, *voiture*; German, *wagen*.) The term is generic, and is applied not only to all wheeled vehicles, but to anything that carries or supports, including litters, pack-saddles, certain machine parts, etc. In a more restricted sense it is applied to four-wheel pleasure vehicles, and more specifically to those in the medium and heavy classes, as distinguished from CARTS, BUGGIES, and light PHAETONS. Technically, the term is applied to the running part of a vehicle, on which the body is hung or supported. (C, H & O)

CARRIAGE-BUILDER'S DELIVERY WAGON (or TRUCK) — A specially constructed vehicle used by carriage builders to deliver carriages to customers, or to the railroad for shipment. They usually have either low bodies, or a body sloping downward at the rear, similar to a truck. Some are equipped with a winch for drawing the carriage onto the wagon, and the rear end gates are high, and hinged at the bottom, so that they may be lowered to the ground to form a ramp for loading and unloading the carriages. (O)

CARRIAGEM — A Portuguese term applied generally to any kind of wheeled vehicle, for either pleasure or business purposes. (H)

CARRI-COCHE — Literal meaning, cart-coach. A primitive form of two-wheel passenger vehicle used in Buenos Aires. Adams, in his *English Pleasure Carriages*, page 16, says, "The Carri-Coche consists of a close framed body, painted and lined, with sliding glasses, and a door to open behind, the whole suspended on long braces or twisted cords of untanned hide. When used in towns, it is intended to be drawn by one or two horses, with a postillion, and to carry six persons, three on a side, like an omnibus." (H)

CARRINHO — A Portuguese term (diminutive of *carro*), generally applied to a CART. (H)

CARRIOLE — A provincial variety of passenger-carrying cart, peculiar to northern France. The term seems related to the Norwegian CARIOLE, though the vehicles themselves are quite different. (H)

CARRO — A Portuguese term applied generally to any kind of wheeled vehicle, whether for pleasure or business purposes. (H)

CARROCA — A Portuguese term applied to a primitive form of open wagon. (H)

CARROCCIO — A large four-wheel vehicle used in certain Italian cities in medieval times, corresponding to the Ark of the Israelites. It bore aloft on a mast the city's standard and a crucifix. Painted vermilion, it was drawn by two, four, or six bullocks covered with scarlet cloths. During a battle it served as a rallying point for an army. (H & O)

CARROCIM — A Portuguese term (diminutive of *carro*), generally applied to small pleasure vehicles. (H)

CARRO DE BOES — The primitive ox-cart peculiar to the Azores, described in Walker's *The Azores*, page 279, "A relic of the past may occasionally be seen, and more often heard, in the rough country parts, probably introduced and used in Portugal by the Romans. They consist of a solid flat framework of strong wood, with ponderous pole, all in one. The linchpin, also of wood, being not too firmly wedged, and fixed onto the center of the two equally solid wheels (also as solid and heavy as the stone discs still used in the central provinces of India), and revolving with them, makes an indescribably creaking noise heard far and wide. These Luso-Romano carts are always drawn by two or more oxen, yoked by means of a heavy wooden frame, and they are said to like the singing noise of the vehicle. Round the floor of the cart are placed uprights which support a wattled structure holding the contents, and in these receptacles heavy loads of grain and produce of all kinds are carried." (H)

CARROMATA — A light, two-wheel vehicle used in the Philippines. Passengers sit in a boxlike body having a standing top, and there is a seat under the same roof for the driver. It is usually drawn by a single pony. (O)

CARRUCA — (From the Latin *carrus*.) An ancient Roman vehicle of very high quality, said to have been ornamented with gold and precious stones. (H)

A Norwegian CARIOLE, driven by a boy riding on the luggage rack. (Smithsonian photo.)

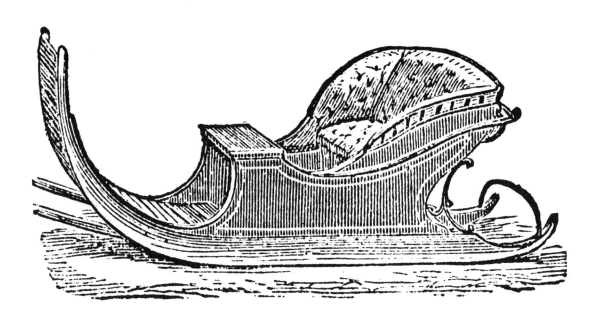

Canadian CARIOLE of about 1870.

CARPENTUM. (Ezra M. Stratton: *The World on Wheels,* New York, 1878.)

CARRIAGE DELIVERY TRUCK. Wheels, 33-inch and 38-inch; front track, 4 feet 8 inches, rear track, 7 feet 2 inches. (*The Hub,* July 1881.)

CARRI-COCHE. (W. B. Adams: *English Pleasure Carriages,* 1837.)

CARRUS — A cart of wain built in varying forms, much used by the Romans and Gauls for carrying heavy baggage. According to some authorities neither the term nor the vehicle are of Roman origin, but are said to be Gallic in origin. (H & O)

CARRYALL — A New England provincialism applied to a light ROCKAWAY, undoubtedly so named because of its capacity as a family carriage. Webster's derivation from CARIOLE seems remote and improbable. Also see DEARBORN WAGON. (H)

CART — The origin of this term is uncertain, but Littré suggests that it may have come from an early Celtic word. The cart is a two-wheel vehicle that is made in numerous varieties, primarily as a freight carrier, though eventually many types of passenger-carrying carts evolved. Most frequently used with one animal, many were also constructed for use with two or more animals. The wheel, believed to have been developed in the Middle East about 3500-3000 B.C., was first applied to the rude carts used in that area. There is evidence that the Greeks and Assyrians first used carts about 1800-1600 B.C.

Freight-carrying carts are the most primitive form of wheeled vehicles. Through the last several centuries there have been several principal types, one having a box body, while the other had no sides other than a few stakes, and often no floor other than the framework. The shafts were often continuations of the side framings of the body. The former type was in more recent years often made with a dumping body, the body being hinged to the shafts or axle, and held down in front by some fastening device. Such carts were most versatile for farm and construction work. The variety using stakes in place of the paneled sides could more properly be called a two-wheel DRAY or TRUCK, and carried boxes, bales, barrels, and other large or heavy objects. There were also carts with spindle sides, and side rails, these being used to carry such things as hay. Freight-carrying carts rarely employed spring-suspension.

Passenger carts could be divided into two categories: sporting vehicles, and those intended for ordinary personal transportation. Examples of sporting carts are DOG-CARTS, T-CARTS, and TANDEM-CARTS. Occasionally some of these terms may be applied to four-wheel vehicles, but the four-wheel types are more properly assigned to the PHAETON class. Of passenger carts there are the common ROAD CARTS, much

A CARRUS loaded with spears and shields. (Ezra M. Stratton: *The World on Wheels,* New York, 1878.)

like the SULKY, and PHAETON-BODY CARTS having trimmed seats and falling tops. Most of these carts are provided with springs.

There are numerous compounds formed with the word "cart," many of which are more or less self-descriptive. Examples of this are: BUTCHER-CART, CAPE-CART, COAL-CART, CURATE-CART, DAIRY-CART, DUMP-CART, GO CART, GROCER'S-CART, HAND-CART, HAY-CART, HOSE-CART, ICE-CART, MARKET-CART, MILK-CART, MULE-CART, NIGHTSOIL-CART, OX-CART, PUSH-CART, SCRAP-CART, SURREY-CART, TAX-CART, TIP-CART, TO-CART, WATER-CART, WHITECHAPEL-CART, etc. (C, H, & O)

CART-CAR — A term sometimes applied to the TROTTLE-CAR.

CART, NEWPORT PAGNELL — The local name for a variety of pleasure cart built at Canterbury, England. (H)

CART, PONY — A small-size pleasure cart that is most suited for use with a small horse. (H)

CASKET WAGON — An undertaker's wagon, sometimes open like an EXPRESS WAGON, but more often closed, like a panel DELIVERY WAGON, though in both instances they are usually more finely constructed and finished. Used to carry a casket at times other than the actual funeral procession.

CATAFALQUE — (From the Italian *catafalco*, meaning a scaffold or canopy.) A funeral-car with open sides, usually decorated with paintings and sculptures, used in funeral solemnities. The term is also applied to an elaborate and ornamented structure on which a body lies in state. (H)

CATALONIAN CART — A local term, applied to a primitive cart that was peculiar to Catalonia, Spain. (H)

CATTERICK — A name sometimes given to a WAGONETTE having no top.

CELERITY WAGON — (The name suggests speed) A variety of stage wagon reportedly developed for the Butterfield Overland Mail line by the well-known Albany, New York, firm of James Goold, originator of the ALBANY CUTTER. Somewhat similar in appearance to the MUD WAGON, the celerity wagon developed in the late 1850s, was suspended from jacks on leather thoroughbraces, and was furnished with three inside seats having backs that could be let down, so as to form a bed. This feature was also seen in the DOUGHERTY WAGON, causing one to speculate as to whether the celerity wagon's design influenced the design of the latter. The celerity wagon apparently did not enjoy a long existence. (O)

CHAIR — (From the French *chaire*, Old French *chaiere*, Latin *cathedra*, Greek *kathedra*, meaning seat or chair.) A light, one-horse, two-wheel vehicle used extensively in England and America in the eighteenth and early nineteenth centuries. It is frequently confused with, and sometimes synonymous with, the CHAISE. Some authorities make the distinction that the two were often identical, except that the CHAISE had a top while the chair did not. This usage was certainly adhered to by many American tradesmen, and it seems desirable to do so now. It must be remembered, however, that others made no use of the term CHAISE, but used the term CHAIR to describe a vehicle with a top. Evidence indicates that these latter tradesmen applied either the term SULKY or SOLO to the vehicle that is frequently called the WINDSOR CHAIR, or RIDING CHAIR. Thus the chair can have, according to the various eighteenth-century American usages, a CHAISE body made with or without a top, or it can have the smaller one-passenger body of the SULKY or SOLO. One popular form of the SULKY or SULKY-CHAIR was the WINDSOR CHAIR, the body of which was an ordinary Windsor chair set on a platform. The SULKY or SULKY-CHAIR could also have a top, and on both this type and the CHAISE-type the top could be either the standing or falling variety.

Most chairs had some sort of lining or trimming, except the more simple types such as the WINDSOR, and even here a separate cushion could be laid upon the seat. The bodies of some chairs had no suspension other than being set directly on the shafts, which in themselves offered a springing action. More often thoroughbraces were used, or a combination of braces with steel springs, or wooden cantilever springs. One method of suspension, thoroughbraces with wooden cantilever springs, was common to the New England area, and is believed to be an American invention. Also see CHAISE.

The term CHAIR can also apply to the SEDAN-CHAIR, which see. (C, H, & O)

CHAIR, ICE — A chair on runners, with handles at the rear, pushed across the ice by a skater. (H)

CHAIR, ROLLING — A chair, mounted on four wheels, and pushed from behind by a man. Rolling chairs saw extensive use in the late nineteenth century at the various exhibitions and fairs. They are the modern equivalents of the old French BROUETTES, and might properly be so called. (H)

CHAISE — (Corrupted from the French *chaire*.) A two-wheel vehicle used in England and America during the eighteenth and early nineteenth centuries. Felton assigns all two-wheelers drawn by one horse to the CHAISE family, the GIG, WHISKEY, and CHAIR being included in this grouping, while all two-horse types he designates as CURRICLES. The two-passenger body was equipped with either a standing or falling top, which feature is sometimes used to distinguish the chaise from certain varieties of the CHAIR. Some chaises

CARRYALL built by Charles Caffrey, of Camden, New Jersey. Wheels, 42-inch and 46-inch; body, black; gearing, carmine, striped with ¼-inch black stripe, but no striping on spokes; trimming, blue cloth. (*The Carriage Monthly*, May 1879.)

CASKET WAGON. Body, 7 feet 2 inches long from door to front of seat. (Geneva Wagon Co., Geneva, New York, 1906.)

had bodies set directly on the shafts, while others were mounted on leather thoroughbraces or a combination of steel or wooden springs with braces. The wooden cantilever springs and thoroughbraces used on the New England chaise are believed to have been an American development. The chaise was one of the most popular vehicles used in colonial America, and did not pass from the scene until mid-nineteenth century. It was widely used by all social classes. In New England the chaise was popularly called a SHAY.

Occasionally, early American signs or broadsides are seen (showing the rates of toll for various vehicles on turnpikes or bridges) that list both two- and four-wheel chaises. Since Felton advises that "the bodies of two-wheeled carriages are exactly the same as bodies of phaetons," it is believed that these four-wheel chaises were, in reality, PHAETONS

The chaises used in eighteenth-century France had CHARIOT-like bodies with a single door in front that was hinged at the bottom and opened downward. Originally mounted on two wheels, these vehicles were later provided with four wheels, and evolved into the POST-CHAISE.

See also, CHAIR and POST-CHAISE. (C, H, & O)

CHAISE-A-PORTEURS — Contracted to PORTE-CHAISE. A French term, meaning literally *chair by porters*, synonymous with the English term SEDAN CHAIR. (H)

CHAISE DE JARDIN — A light chair, with either two or four wheels, similar to those now frequently used by invalids, intended to be drawn or pushed by hand. (H)

CHAISE DE POSTE — The two-wheel French carriage, or CHAISE, that contributed both its name and, to a degree, its design, to the English POST-CHAISE. See POST-CHAISE.

CHAISE, QUAKER — See GRASSHOPPER-CHAISE.

CHAISE, ROULANTE — This French term, meaning literally *a rolling chair*, was the name of a vehicle used during the eighteenth century in the courts of Europe for short trips across courtyards, or from palace to palace, by the courtiers and maids of honor. (H)

CHAMULCUS — An ancient Roman vehicle of unknown description. (H)

CHAR-A-BANC — (Of French derivation, meaning literally a vehicle with benches or long seats.) Varied descriptions of this vehicle can be found, which can only be explained by the statement by the Duke of Beaufort, in his book, *Driving*, wherein he states: "Char-a-bancs are more various in form than most other carriages; they are generally high and strongly made, to carry a good many persons. Some have four seats, each carrying three or four persons, on the top of a high and long boot; the seats are reached by convenient folding and sliding steps concealed in the boot and shut in by a small door. Others have the central seats kept low; the four persons sit as in a coach, facing one another; doors and folding steps provide easy access. The front driving-seat is made high in this class of carriage, and frequently the hind seat for the grooms is also high, being carried, as in the case of drags, on strong ornamental irons; at other times this seat is kept low, and the grooms sit with their backs to the horses. Most of the large carriages of this type are used with four horses and are suspended in various ways, some on the perch under-carriage with mail springs, others have in addition under-springs, while others again have four ordinary elliptic springs. Some are now made on a smaller scale and go well with a pair of horses. A char-a-bancs is essentially a carriage for a 'grande maison,' and for country use, and it is rarely found where a coach-house has not room for more than four carriages."

The term CHAR-A-BANCS is sometimes used in a singular sense, and sometimes plural; as shown above, the Duke used it both ways. In other references, the singular is often given as CHAR-A-BANC.

While most sources describe the seats as facing forward, some writers, including Houghton, state that the seats are installed lengthwise, and compare the arrangement to a WAGONETTE or OMNIBUS. Houghton (1892) says that it is "nearly equivalent to the English term Wagonet." Descriptions also vary concerning the top of these carriages, some being open, while others are covered. Stratton illustrates a French char-a-banc with a permanent top, and Straus states, "Awnings, permanent or temporary, are generally provided."

The char-a-banc was originally a French vehicle, patterned after the national carriage of Switzerland; it was eventually used in both England and the United States. The term has been applied more recently in England to the sight-seeing motor bus. (H & O)

CHARE (or CHAR) — Old English forms of the words, *chair, chaire,* or *chaise.* In Old French, it is somewhat synonymous with *charette*. (H & O)

CHARET (or CHARETTE) — A general term applied to many forms of wheeled vehicles. In Biblical use, a WAR-CHARIOT. In Old French it was applied to wheeled vehicles used to transport either persons or goods, such as CARTS, WAGONS, CHARIOTS, or carriages generally. In nineteenth-century French, CHARETTE was a two-wheel vehicle, while CHARIOT was a four-wheeler. In England the terms CHARET and CHARIOT

American RIDING CHAIR of about 1795. Gear, chocolate with red striping; body, green with yellow striping. (Model from Smithsonian collection, from original *Chair* of Lord Fairfax, now exhibited at Mount Vernon, Virginia.)

New England CHAISE of about the 1770s. Because of the driver's seat, it could also be called a *Calash*. (Smithsonian collection.)

New England CHAISE of about 1830. (Smithsonian collection.)

were often confounded and even used synonymously. The former term was obsolete by mid-seventeenth century, but survived as a pronunciation of CHARIOT until the nineteenth century.

In the United States the term CHARETTE was applied in the last quarter of the nineteenth century to a variety of VILLAGE-CART. This usage was restricted to certain areas, particularly to Pennsylvania. (O)

CHARIETTE — A type of litter on four wheels, having no springs, that was used in England in the thirteenth century. (R)

CHARIOT — The derivation of this word is somewhat obscure because of the early and indiscriminate use of such terms as *char, chare, charet, charette*, etc., which appear to have been applied not only to passenger vehicles, but also to wagons.   (1.) In the classic sense, the term describes an ancient two-wheel vehicle that developed about 2800-2700 B.C. in Mesopotamia. Evidence suggests that chariots may have been first used for warfare or hunting, but eventually came to be used for racing, funerals, religious ceremonies, and other public occasions. Its use spread to Assyria, Greece, Egypt, and Rome at an early date. Ancient chariots were known as Biga (two-horse), Triga, Quadriga, etc., up to Decemjuga (ten-horse), according to the number of horses used with them.   (2.) A member of the COACH family. The term CHARIOT is believed to have been applied about the middle of the seventeenth century to the French COUPE, which had just come into common use in England. It was probably so named because of the fashion of the time of assigning classic titles to many things. The chariot was a sort of half-coach, with a similar body that was cut off just in front of the doors, the front being closed by a panel below and glass above. The carriage was like that of the COACH, but shorter. The single seat inside held two or three passengers, and some had a small folding seat in front to accommodate children. A detached driver's seat was ornamented with a hammercloth, and the vehicle was attended by footmen.

Chariots were somewhat more popular than COACHES, particularly when greater seating capacity was not essential, as they were slightly less expensive, lighter in weight, more maneuverable, and because of the glass in front, more light and airy. It was also considered to be the most attractive of carriages. In America, chariots began to appear about 1700, and were symbols of rank or wealth, as were COACHES. General

Braddock, coming to America in 1755, purchased the chariot of Governor Sharpe, of Maryland, for use on his ill-fated expedition, but later found that he could not use it on the hastily constructed roads that the army cut through the forests. George Washington is not known to have owned a COACH (though Martha had one), but preferred chariots, four of which were owned by him at different times.

The chariot was gradually superseded by the BROUGHAM, and the modern COUPE, yet even in the 1890s it was frequently seen in London at such functions as drawing-room receptions and court levees. It was, in fact, still built at that time, with the modifications of later inventions and mechanical improvements, by the leading builders of London, Paris, and New York, and used as a court or dress carriage. Its use, however, in more democratic times, robbed it of much of its former glory. The footmen were dispensed with, a pair of ornamental straps being substituted, crossed at the rear of the vehicle in mute token of their departed grandeur, but sometimes the hammercloth was retained in a last desperate effort to cling to the aristocratic past. Also see COUPE, POST-CHAISE, and CHARET. (C, H, & O)

CHARIOT, GALA — A state carriage, characterized by a high degree of ornamentation, common in Europe from the seventeenth to the nineteenth centuries. (H)

CHARIOT, SCYTHE — A war chariot of the ancient Persians, and believed to have been previously used by the Assyrians. The scythe chariot had long blades, similar to scythe blades, attached to the wheels, axle, or body in order to inflict injury on the enemy. (O)

CHARIOTEE — A provincial term, formerly common in the southern United States, that was applied during the early part of the nineteenth century to a vehicle resembling the extension-top PHAETON of the latter part of the century. (H & C)

CHARVOLANT — A curious four-wheel vehicle, with kite attachment, intended to be drawn along the roadway by the force of the wind. It was patented and demonstrated in England in 1826, by a Colonel Viney and

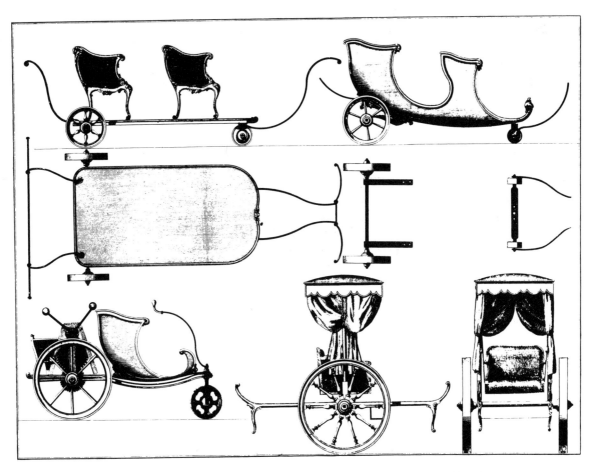

A variety of CHAISES DE JARDIN, showing one (lower left) that was mechanically propelled by the rear passenger, and steered by the front passenger. (M. Roubo: *L'Art du Menuisier-Carrossier,* Academie des Sciences, Paris, 1771.)

CHAR-A-BANCS built on the *Wagonette* plan. Wheels, 34½-inch and 44½-inch; body, natural wood finish; gearing, red with black striping. (*Le Guide du Carrossier,* February 1902.)

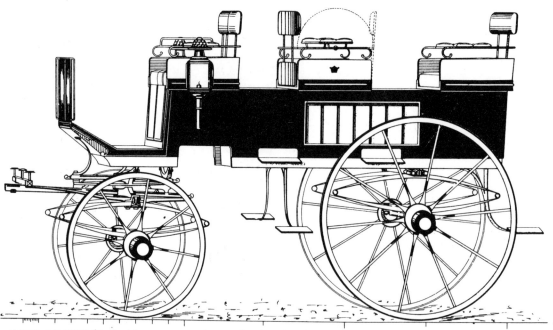

CHAR-A-BANCS fitted with a center seat that can face in either direction. Wheels, 34½-inch and 49½-inch; body, black, except for those lighter areas in the illustration, which are light green; gearing, light green with black striping. (*Le Guide du Carrossier,* February 1902.)

A popular type of CHAR-A-BANC. (F. C. Underhill: *Driving for Pleasure*, 1897.)

George Pocock. Attempts were made in both England and the United States to market this vehicle, undoubtedly with little or no success. (O)

CHATELAINE — A term, of unknown derivation, applied by A. T. Demarest & Co., of New York, to a miniature form of STANHOPE GIG. (H)

CHESS WAGON — A military term applied to part of the equippage of a canvas pontoon train. The specially constructed chess wagon carried the chesses, or planks used to form the flooring of a pontoon bridge. (O)

CHIHUAHUA WAGON — A local term applied to a peculiar style of Mexican traveling wagon of heavy build. (H)

CHILD'S CARRIAGE — Same as BABY CARRIAGE, or PERAMBULATOR, which see. (H)

CHILESE CAR — A primitive form of CART used in Chile. (H)

CHUCK WAGON — See ROUND-UP WAGON.

CIRCUS WAGON — A term applied to any of the several types of highly decorated WAGONS used by a circus, such as TABLEAU WAGON, CAGE WAGON, BAND WAGON, and CALLIOPE. They are used for the dual purpose of carrying circus baggage, and making up the circus parade in each town where the circus plays. They are very heavy wagons, built for hard service, and are generally mounted on platform springs. Decoration is by painting, carving, mounted figures, etc., and both bright colors and gold leaf are freely used. Some distinctive parts of circus wagons are sky boards, which are the top panels of the wagons, usually cut in fanciful shapes and very highly decorated, and sunburst wheels. The latter are ordinary wheels, with spokes grooved on both sides so that wedge-shaped pieces of wood can be inserted between the spokes, thus presenting a solid wheel to the brush of the decorator. The ends of these wedges do not touch the felloe, but end several inches short of same, and the ends are cut in some decorative manner. See the various types of WAGONS mentioned above. (H & O)

CISIUM — A two-wheel Roman vehicle. Some writers claim the word is derived from the Latin word, *cito*, meaning quick; others maintain it comes from *scissum*, cut, a hole being cut through the sides of the body. These vehicles were both light and fast, though the wheels are said to have been larger than those of other Roman vehicles. The body was generally fixed to the frame, or shafts, though the better types are believed

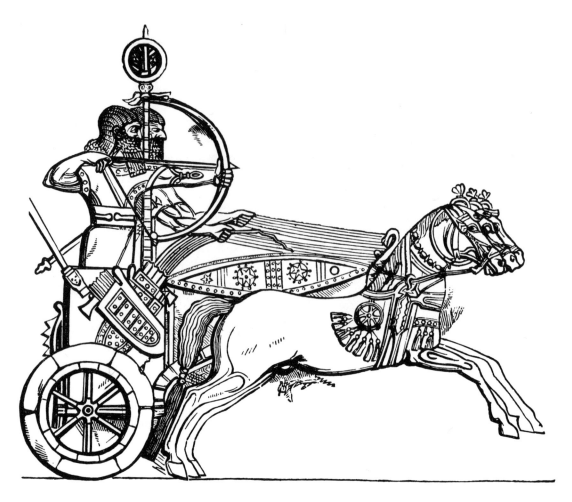

Assyrian WAR CHARIOT. (Ezra M. Stratton: *The World on Wheels,* New York, 1878.)

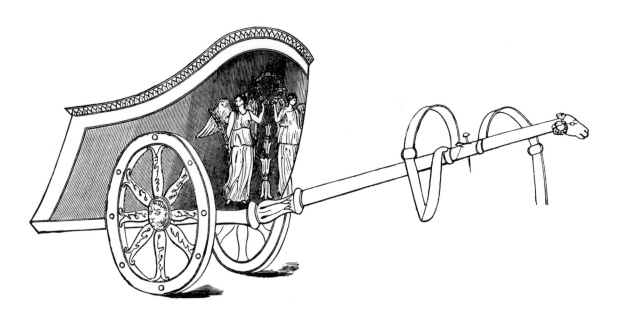

Greek RACING CHARIOT. (Ezra M. Stratton: *The World on Wheels,* New York, 1878.)

Roman CIRCUS CHARIOT. (Ezra M. Stratton: *The World on Wheels,* New York, 1878.)

Persian SCYTHE-CHARIOT. (Ezra M. Stratton: *The World on Wheels,* New York, 1878.)

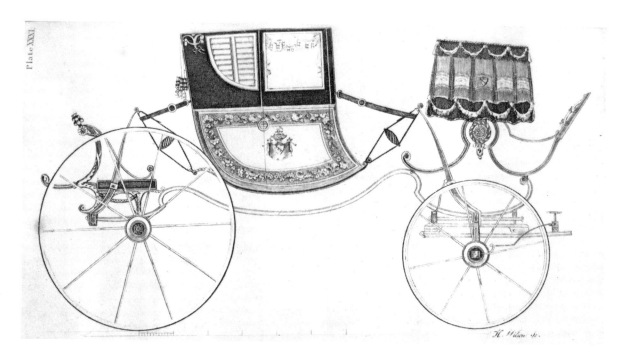

An elegant CHARIOT, extensively used in grand processions. Lower body panels, maroon; upper panels, black; silver-plated moldings; Venetian blinds and glass-frames, pink; floral pattern, pink, green, and gilt; gearing, maroon striped with gilt; carvings and hub bands, gilt; hammercloth, pink; trimming, morocco. (W. Felton: *Treatise on Carriages,* 1796.)

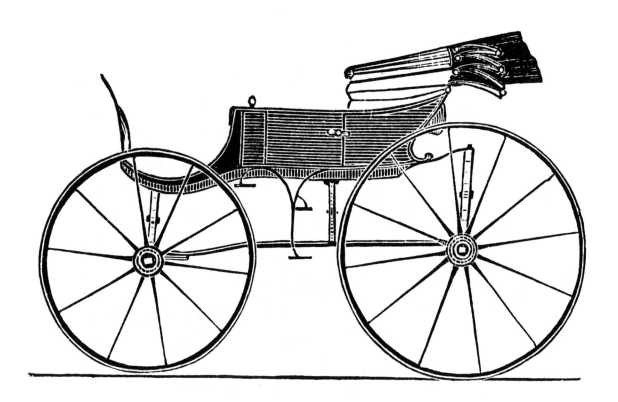

American CHARIOTEE of about 1827. (Ezra M. Stratton: *The World on Wheels,* New York, 1878.)

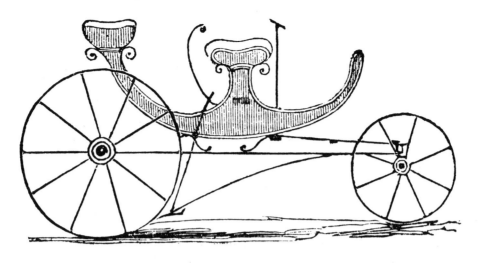

CHARVOLANT of 1826. (From original brochure in Smithsonian collections.)

This magnificent "Twin Lion" TABLEAU WAGON, built in England around 1860, was last owned by the Sir Robert Fossett Circus, of Northampton, England. The upper section of this 17½-foot-high wagon telescopes down into the lower part to reduce the height when not in parade. Obviously, this wagon was designed in the days before utility wires. (In the collection of the Circus World Museum, Baraboo, Wisconsin.)

This "Cinderella" FLOAT was one of seven nursery-rhyme floats built during the 1880s for the Barnum & Bailey Circus. With the only other survivor of the seven floats, it is shown at Circus World Museum, Baraboo, Wisconsin.

This late type TABLEAU WAGON, built during the 1930s for the Hagenbeck-Wallace Circus, is equipped with sunburst wheels. (In the collection of the Circus World Museum, Baraboo, Wisconsin.)

This elegant BAND WAGON, now owned by Circus World Museum, Baraboo, Wisconsin, was ordered in 1903 by Pawnee Bill's Wild West Show. Note the sunburst wheels.

Circus CAGE WAGON owned by Circus World Museum, Baraboo, Wisconsin.

to have had bodies suspended on straps or braces. Cicero claimed that it was possible to write in them, so smooth was the ride. The back of the body was closed, and entry was gained through the front. One or two horses were usually employed, and sometimes three or four, the additional animals being more for show than necessity. These vehicles were commonly used to carry mails from one town to another, and eventually were used throughout Italy and Gaul. They carried passengers from one post to another in much the same manner as did STAGECOACHES in nineteenth-century America. Those that were privately owned were often provided with cushions to offer greater comfort to passengers. (O)

CLARENCE — Named in honor of the Duke of Clarence, this vehicle is a member of the COACH family, and is sometimes described as being midway between a BROUGHAM and a COACH. After the popular demand for democratic simplicity, compactness, and economy was appeased by the BROUGHAM in 1839, several efforts were made to restore the dignity of its predecessor, the CHARIOT. One of these was an attempt by Bulwer, the novelist, to initiate a style of BROUGHAM that retained the hammercloth and footman's board, but this effort proved a failure. Another effort was more successful, and resulted in the carriage known as a Clarence. Introduced in London about 1840 by Messrs. Laurie and Marner, the Clarence remained in fashion as a private carriage until the 1870s, when it passed into the hands of hackmen. Its decadence is clearly shown by the fact that only one Clarence was displayed at the Paris Exposition in 1878. The Clarence is an extra large COUPE, having a semicircular or extended front glass, with either paneled or glass quarters, and two seats inside for the accommodation of four passengers. The body is hung on four elliptic springs, or on elliptic springs in front, with various combinations behind employing elliptics, C-springs, and platform springs. It is always made without a perch. See COUPE. (H & O)

CLOSE COACH — A COACH having all the quarter panels made without lights. (C)

CLOTHES BASKET — A general term applied in England to any PHAETON having a wickerwork or basket body. The name was also applied by the London *Daily Telegraph* to the RALLI-CAR. (H)

COACH — A four-wheel, enclosed CARRIAGE used for the conveyance of passengers. It is characterized by a roof which forms part of the framing of the body, as distinguished from vehicles with either folding tops, or canopy tops supported by rods or pillars. An additional feature always associated with coaches is a suspended body, though there is some reason to believe that not all of the earliest coaches were so equipped. Passengers, four to six in number, face one another from two transverse seats.

The term is believed to have been derived from the Hungarian town of Kocs, where the earliest and most primitive form of coach is said to have originated from the German agricultural WAGON. As early as 1457 one of these CARRIAGES was sent by King Ladislaus of Hungary to the Queen of France, and the fact that it was described as *branlant* (trembling) causes many historians to believe that the body was in some manner suspended. During the next hundred years the use of the coach spread throughout Hungary, and into Germany, Italy, and The Netherlands, yet France and England were slow to follow this lead. In 1560 but three coaches were known in Paris, and in England the first coaches had been introduced only a few years before. There was a general prejudice against the use of coaches during this period, and they were seldom used by any but royalty and the aristocracy.

One of the earliest coaches of which we have reliable information was that in which Louis XIV entered Paris about 1650, and it was suspended by short braces extending from upright pillars or posts to the corners of the body. Several mechanical improvements took place during the second half of the seventeenth century. About 1660 another method of suspension was developed in Berlin, that city lending its name to the type of carriage devised. The front and rear portions of the running gear were connected by two perches, in place of the extremely heavy single perch usually employed, and leather thoroughbraces extended from front to rear, with the body resting upon them. Nearly a century passed, however, before this construction came into popular use. About 1670, elbow-springs began to come into use, these being fastened underneath the four corners of the body, and the leather braces suspending the body were in turn attached to the ends of these springs. Bodies also underwent change, for after mid-century they began to lose the traditional squareness of WAGON bodies, and were carved, paneled, and more tastefully shaped. Glass windows also came into use during this period.

Around 1750 the four corner pillars — from which hung the braces — were constructed with a slight curve and a springing action, thus evolving into whip springs or S-springs, and by 1790 these further evolved into the popular C-springs which continued in use until the end of the nineteenth century. The use of these springs, decreasing the effect of road shocks on both the body and the running gear, made possible the lightening of the previously excessively heavy bodies. The perch, by successive developments, gradually as-

CISIUM shown on a Roman column at Ingel, near Treves, France. (Ezra M. Stratton: *The World on Wheels,* New York, 1878.)

CLARENCE, with 38½-inch and 47-inch wheels. (Lawrence, Bradley & Pardee, New Haven, Connecticut, 1862.)

A French and English COACH used by Queen Elizabeth in 1582, from a print by Hoefnagel. (Ezra M. Stratton: *The World on Wheels,* New York, 1878.)

An English COACH of 1688, from a print by Romaine de Hooge. (Ezra M. Stratton: *The World on Wheels,* New York, 1878.)

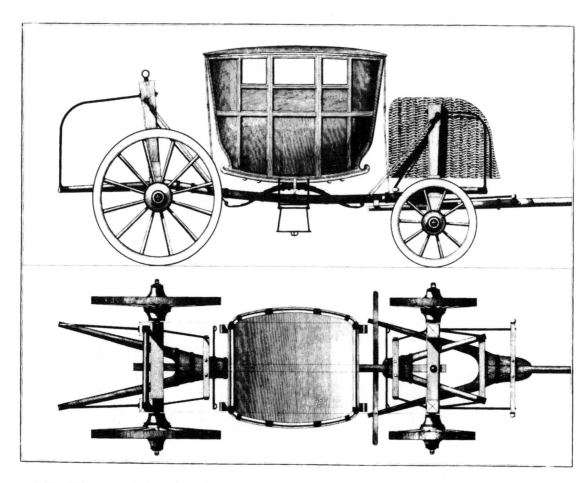

A COACH hung on a single-perch running gear. (M. Roubo: *L'Art du Menuisier-Carrossier,* Academie des Sciences, Paris, 1771.)

sumed the form known as *crane-neck*, making possible a greater maneuverability in crowded areas, by permitting the front wheels to pass under the perches.

Improvements of the nineteenth century were the elliptic spring, invented in England by Obadiah Elliot in 1804, this feature permitting the elimination of the perch; and the framing of the boot into the body, a change which occurred in the early part of the century.

The coach came to America late in the seventeenth century, and was at first used almost exclusively by the various colonial governors, apparently as a mark of their high office. During the eighteenth century other persons of wealth or high rank began to keep a coach, but the vehicle was rarely put to common use, even in the nineteenth century. Records show that Massachusetts, in 1753, had only 6 COACHES and 18 CHARIOTS, yet numbered over 1300 CHAISES and CHAIRS. Coaches were apparently more frequently used in Virginia than in the other colonies.

See entries under other coach types. (H & O)

COACH-AND-FOUR (or COACH-AND-SIX) — A COACH driven with a four-in-hand or six-in-hand team. (H)

COACH, CARDINAL'S — A state COACH, intended for use by a Cardinal, the distinguishing feature being that it was painted red. (H)

COACHEE — An American vehicle, generally used as a family carriage, that appears to have developed late in the eighteenth century. It is described by Isaac Weld (*Travels Through the States of North America*, published in London in 1800), who in 1795 saw a number of them in use in Philadelphia. He wrote, "The coachee is a carriage peculiar, I believe, to America; the body of it is rather longer than that of a coach, but of the same shape. In the front it is left quite open down to the bottom, and the driver sits on a bench under the roof of the carriage. There are two seats in it, for the passengers, who sit with their faces towards the horses. The roof is supported by small props, which are placed at the corners. On each side of the doors, above the pannels, it is quite open, and to guard against bad weather there are curtains, which are made to let down from the roof, and fasten to buttons placed for the purpose on the outside. There is also a leathern curtain to hang occasionally between the driver and passengers." He also stated that it was a well-finished vehicle, with varnished panels and doors in the sides (some apparently had a door in one side only), differing from the STAGE WAGON only because of the doors and superior finish. Because the driver and passengers were under the same roof, the coachee may be considered the ancestor of the ROCKAWAY.

Several surviving carriages of the early nineteenth century correspond with Weld's description, and some of these differ by having a single door in the rear. Suspension is on thoroughbraces that are secured to jacks in several instances, while the example in the Smithsonian Institution employs thoroughbraces and wooden C-springs. Other references have been found which seem to indicate a rather broad usage of the term coachee. Several early nineteenth-century documents in the Smithsonian reference collection indicate such features as a hammercloth and a front boot, neither of which apply to the type of vehicle Weld described, but seem to suggest a vehicle that was more nearly like the COACH, but probably lighter in weight. At a later date this usage certainly does apply. Lieutenant John Harriott, writing of the 1790s, indicates that the term was applied to a MAIL COACH, which is the same vehicle that Weld likened to the coachee, but slightly differentiated.

There is also evidence indicating that a number of terms were used rather synonymously to describe the same type of vehicle. The descriptions of Amos Stiles's BOULSTER WAGONS in New Jersey seem to agree with that of the coachee. Robert Sutcliff, traveling in America in 1804-1806, both describes and illustrates vehicles that must certainly be identical to the coachee, yet he calls them JERSEY WAGONS. *A Dictionary of American English* (University of Chicago Press, 1960) quotes early references which indicate the terms JERSEY WAGON, DEARBORN, and CARRYALL were used synonymously, assigning the various words to different sections of the country, yet no descriptions have been found that can justify the synonymous use of the last two terms with COACHEE. Thus, while similarity undoubtedly existed between the coachee and JERSEY WAGON, and also between the latter and the DEARBORN and CARRYALL, it seems likely that the coachee differed too greatly from the DEARBORN and CARRYALL to warrant the use of those terms synonymously.

By 1840 (and probably much earlier) the term was being applied to a vehicle that might be compared to a light-weight, curtain-quarter COACH. Some of these had a detached driver's seat, while others had a seat framed to the body. The upper quarters were closed by leather curtains. Suspension was variously on C-springs and braces, or on elliptics. These carriages were used throughout the East, but were especially popular in the Southern states until about 1860, when their use began to decline.

The term COACHEE was also used as a slang expression for a coachman. (C, H, & O)

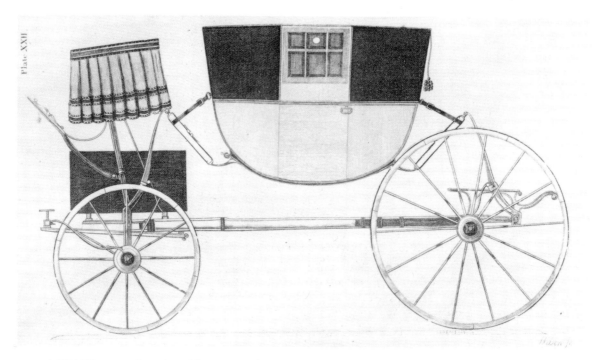

A COACH on a perch carriage, "the painting of any color," with the moldings striped with but one color. (W. Felton: *Treatise on Carriages,* 1796.)

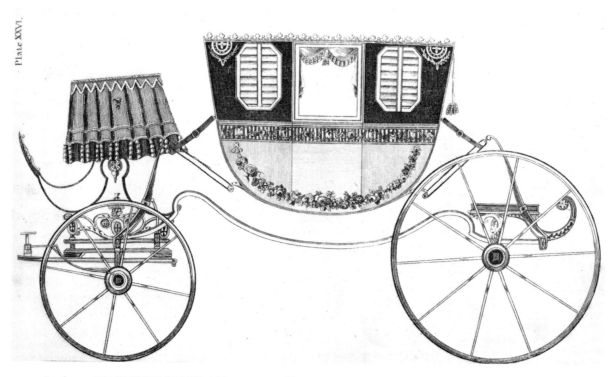

An elegant CRANE-NECK COACH, richly ornamented by carvings, gilt, paintings, and silk and velvet trimmings, formed the principal part of any grand procession. Lower body panels, rich, medium blue; middle panels, lavender; upper panels, black; Venetian blinds and glass-frames, pale yellow; floral pattern is pale yellow, lavender, pale blue, and green; silver plated moldings. Gearing, lavender striped with the blue of the body panels; carvings, gilt; hub bands and pole-hook, silver plated; hammercloth, pale yellow. (W. Felton: *Treatise on Carriages,* 1796.)

COACHEE of about 1810. The term, JERSEY WAGON, could likewise be applied. (Smithsonian collection.)

Parker's COACHEE built in New York City about 1840. (Ezra M. Stratton: *The World on Wheels,* New York, 1878.)

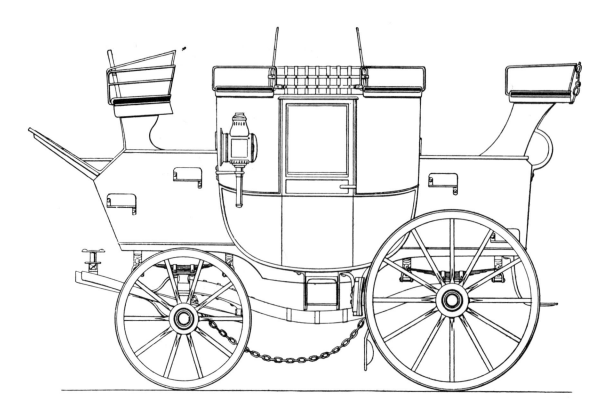

PUBLIC, or ROAD-COACH, by Brewster & Co., New York. (F. Rogers: *A Manual of Coaching,* 1899.)

STATE COACH of England, built for George III in 1762. It supposedly cost the equivalent of nearly $37,000. (Ezra M. Stratton: *The World on Wheels,* New York, 1878.)

COACHEE-ROCKAWAY — A variety of ROCKAWAY modeled after the COACHEE. (H)

COACH, FLYING — A public COACH or WAGON of the eighteenth century, the name implying that it was fast traveling. Ezra Stratton quotes an eighteenth-century author whose writings suggest the use of the term around London as early as 1669. Houghton quoted as follows from Chamber's *Book of Days*, volume 2, page 228: "In 1754, a company of merchants in Manchester, England, started a new vehicle called the 'Flying Coach,' which seems to have earned its designation by the fact that it proposed to travel at the rate of four or five miles an hour. The proprietors, at the commencement, issued the following remarkable prospectus: 'However incredible it may appear, this coach will actually (barring accidents) arrive in London four and a half days after leaving Manchester.' Three years afterwards, Liverpool established another of these 'flying machines on steel springs,' as the newspapers of the period called them, which was intended to eclipse the Manchester one in the matter of speed. . . . Sheffield and Leeds followed with their respective 'Flying Coaches;' and before the last century closed, the whole of them had acquired the respectable velocity of eight miles an hour."

These FLYING COACHES were the predecessors of the first MAIL COACH that was tested in 1784 in accordance with the plan of John Palmer. See MAIL COACH.

In America the term was applied to the STAGE WAGONS of John Mercereau, running between New York and Philadelphia at least as early as 1771. (H & O)

COACH, FOUR-IN-HAND — A heavy, four-horse COACH, the use of which alternated, during the nineteenth century, between business and pleasure. Four-in-hand driving became popular about 1800 when George, the Prince of Wales, made four-horse COACHES one of the attractions of his residence at Brighton. Since that time, it has been in and out of fashion a number of times. The first four-in-hand association, known as the Benson Driving Club, was organized in London in 1807. The Four-horse Club, sometimes erroneously called the Four-in-hand Club, was founded a year later, in 1808, and broke up about 1826. In 1838, the Richmond Driving Club was founded, lasting but a short time. In 1856, the Four-in-hand Club of London was formed; and in 1870, a second English driving association, the Coaching Club, was organized. In 1875 the movement was introduced into the United States by the formation of the Coaching Club, which had its headquarters in New York. See MAIL COACH, REGULATION COACH, and ROAD COACH. (H)

COACH, GALA — A state CARRIAGE, characterized by a high degree of ornamentation, common in Europe from the seventeenth to the nineteenth centuries.

COACH, HAMMERCLOTH — Any COACH having its driver's seat provided with a hammercloth. (H)

COACH, MAIL — A public COACH having a contract to carry mails. The English mails were first sent by coach in 1784, in accordance with the plan of John Palmer, whose coach ran from London to Bristol on August 8, 1784, starting at 8:00 A.M., and reaching its destination at 11:00 P.M. The benefits to the public quickly became obvious, and the government entered into a contract with Palmer, engaging to give him two and one one-half percent on the savings effected in transmission of the mails. This saving soon amounted to 20,000 pounds a year, but Parliament declined to fulfill the bargain and made Palmer a grant of 50,000 pounds instead.

A procession of all royal mail coaches took place in London each year on the King's birthday, and in 1824 there were twenty-seven of these in line. The speed attained at that time was 10 miles an hour.

The mail coach had a full-paneled body, seating six or nine passengers inside, with boots before and behind, and accommodations on the roof for passengers and luggage. The running gear consisted of stout, low wheels, heavy axles known as "mail axles," and thoroughbrace suspension that was later changed to telegraph springs. The FOUR-IN-HAND-COACH, or DRAG, is the modern development of the old MAIL COACH, adapted for private use.

In the United States, the term was often applied to any public COACH or STAGE that carried the mails, and was, in fact, the name given by the Abbot, Downing Company to their own famed product, popularly known as the CONCORD COACH. (H & O)

COACH-OMNIBUS — A public vehicle possessing the combined characteristics of a COACH and an OMNIBUS, popular in the Southern states during the mid-nineteenth century. (H & M)

COACH, PUBLIC — Synonymous with ROAD COACH and, according to Fairman Rogers, more strictly correct. (FR)

COACH, REGULATION — The Coaching Club, of New York, assumed the right to define the requirements of a true DRAG, or GENTLEMAN'S ROAD COACH, adopting as a standard the most approved English design. No person was admitted to membership unless his coach came up to this standard, known as the REGU-

LATION PATTERN, with perch, roof-seats, and outlines not too far removed from the old English MAIL COACH. (H)

COACH, ROAD — Fairman Rogers wrote in his *A Manual of Coaching*, as follows:

"Road-coach has come to be an accepted name for a public-coach or a stagecoach, but public-coach is more strictly correct.

"As at present built, either for public or private use, the coach is essentially the same as that which existed at the time when coaches in England were superseded by railroads.

"A distinction is made, however, between a drag, built for private use, and a road-coach, or public-coach, intended to carry always a full load, and to be driven at a high speed over long routes. The drag is made lighter than the coach, but between the two extremes of weight and of finish there are many grades, depending upon the taste of the owner.

"Some men, living in the country and liking to drive long distances, use their coaches like road-coaches, at high speeds and with changes of horses; certainly the most 'sporting' way of doing the thing. A coach for this purpose should be built almost exactly like a public-coach. Other men use their coaches only about home, or, if they live in a city, principally in park driving, with small loads, at a moderate pace, on good roads, and such a coach should be two or three hundred pounds lighter than a public-coach, and may be slightly ornamented with plain mouldings worked on the edges of the undercarriage timbers, and with a little carving on the ends of the splinter-bar and the futchells, which in a public-coach are always perfectly plain.

"There are, however, between the two kinds of coaches, some essential differences which should be observed. In a public-coach, the rumble is supported on the hind boot by a solid wooden bench, and seats three or four persons, including the guard; in a drag, the rumble holds only two persons, usually the grooms, and is supported by open irons. It is quite easy to have both kinds of rumble fitted to a coach, so that it can be used either as a drag or as a public-coach. On a road-coach there is an iron rod running between the side-irons of the roof-seats along both sides of the coach, and this usually has a net of leather straps connecting it with the roof, so that wraps thrown on the roof cannot fall off. This net should be omitted in a drag, although for long trips one may be made with buckles in such a way that it can be taken off. The door of the hind boot of a public-coach is hinged on the off side; that of a drag is hinged at the bottom. The public-coach is not trimmed inside, but is usually finished in hard wood. This hard-wood finish is, however, a modern fashion, as in old coaching days the inside passengers paid higher fares than those outside and were made as comfortable as possible. The interior of a drag is trimmed plainly in morocco, cloth, or cord. The general finish of a drag may be higher than that of a public-coach without being elaborate; it should be about the finish of a plain, first-class brougham.

"The reader should be reminded that a drag is a *sporting* vehicle; it is not at all a *voiture de luxe*, and in all its appointments it should retain the sporting character. Elaborate harness or unnecessary ornament of any kind about a drag is in bad taste; a drag is nothing more than a well 'turned out,' neat, public-coach, and the showy features of a lady's carriage should be avoided.

"Down to about 1870, drags were made to take only three persons on each roof-seat, and these seats, like those of a mail-coach, did not extend beyond the edge of the roof; now they are always made long enough to accommodate four persons." (FR)

COACH, SALISBURY-BOOT — Any COACH characterized by the feature named. (H)

COACH, STATE — A COACH used in Europe by royalty and persons of high official position, distinguished by being drawn by six or more horses, with postilions, outriders, and numerous footmen and attendants. The bodies of state coaches, most of which were built no later than 1840 if not much earlier, are often elaborately carved, gilded, and painted by eminent artists. They have a hammercloth driving-seat, footmen's board, open quarters decorated by costly curtains, and suspension that is generally primitive, with long braces, giving their ponderous bodies a swaying motion that is most uncomfortable to the occupants. State coaches are almost unknown in America. (H)

COAL CARRIAGE, MOORE'S — See MOORE'S HIGH-WHEEL CARRIAGES. (H)

COB CART, or COB PHAETON — A small-size vehicle, adapted for use with a cob, or small horse. Usually it was distinguished from a PONY CART, or PONY PHAETON by being less aristocratic in its appointments. (H)

COCKING CART — An English variety of SPORTING CART, originally designed to accommodate fighting cocks, in the same manner that the DOG-CART, in its first form, was intended to carry hunting dogs to cover. Both of these vehicles eventually developed into aristocratic pleasure carts, with little or no trace of their

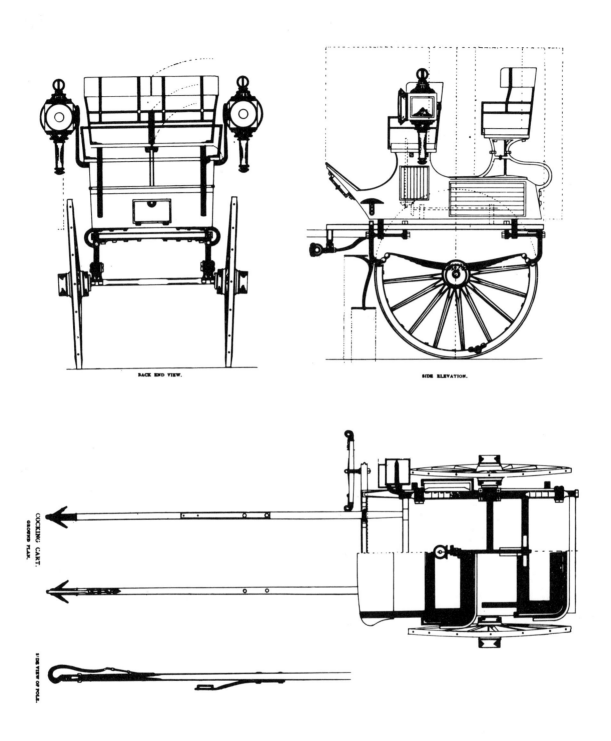

A COCKING CART with two poles, designed for use with three horses abreast. Wheels, 67-inch; panels, black; slats, yellow; gearing, yellow striped black; trimming, Bedford cord. (*The Hub*, February 1892.)

extraction, other than the names. C. P. Kimball & Co., of Chicago, built a high, six-passenger cocking cart that was drawn by three horses abreast, about 1892. (H)

COMBAT WAGON — This general-purpose, freight-carrying vehicle was developed by the army during World War I, and an unsuccessful attempt was made to replace the ESCORT WAGON with it. It was built in two parts, like a CAISSON, and the rear part had a stock and lunette by which it was connected to the front section. The front portion carried a cushion for the driver, but the seat had no springs. This wagon was more expensive and heavier than the ESCORT WAGON, weighing 3000 pounds as compared to 2145 pounds for the latter. By November 11, 1918, 7099 had been built, while 38,613 ESCORT WAGONS had been built during the war years. (O)

COMFORTABLE — A term applied in Austria to a one-horse open HACK that accommodated from one to three passengers. (H)

CONCORD COACH — A popular type of public COACH used as a HOTEL COACH and a ROAD COACH, and occasionally for carrying excursion and sightseeing parties. Concord coaches were built by the Abbot, Downing Company, of Concord, New Hampshire, who referred to them as MAIL COACHES. The first of the type was built by J. Stephens Abbot — then an employee of Lewis Downing, but later a partner — in July 1827, and production continued until early in the twentieth century. Abbot has been credited with the design of this COACH, yet there is evidence that the style actually developed slightly earlier in Troy, New York (See TROY COACH). Though this COACH was built and made famous by Abbot, Downing, it was copied by several other carriage builders.

The Concord coach was generally built in six-, nine-, and twelve-passenger sizes (this does not include roof passengers), though company records reveal that a few four-passenger and sixteen-passenger sizes were built between 1858 and 1864. Passengers were seated on two transverse, facing seats, in the usual coach fashion, and on one or two additional benches between the fixed seats, and on one or more roof-seats. Suspended on a three-perch, thoroughbrace carriage, the body could accommodate a large amount of baggage, for there was a rack on top, another at the rear, and space in the front boot for the purpose. These COACHES were painted in bright colors, and then highly decorated with painted scrollwork, oil paintings, ornate

COMBAT WAGON of 1918. (National Archives photo.)

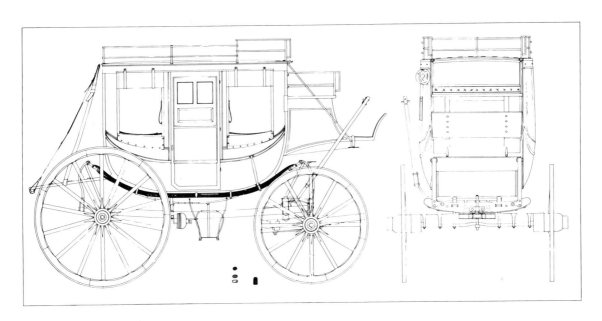

This small hotel-size CONCORD COACH was built to accommodate six inside passengers, two on each of the three seats. Another style was wider, and held three each on only two seats. Either type was priced at $1000 in the late 1880s, though $20 could be deducted if no ornamental painting was wanted. Considering the ornate art work that went into many of these coaches, the $20 figure is somewhat surprising, but is an indication that most of the cost was due to the quality of the construction. The *Coach* from which this drawing was made is in the Smithsonian collections, and is the earliest known *Concord,* having been built in the shops of Lewis Downing, Concord, New Hampshire, in 1848. The ornately decorated body and gearing are both a pale yellow.

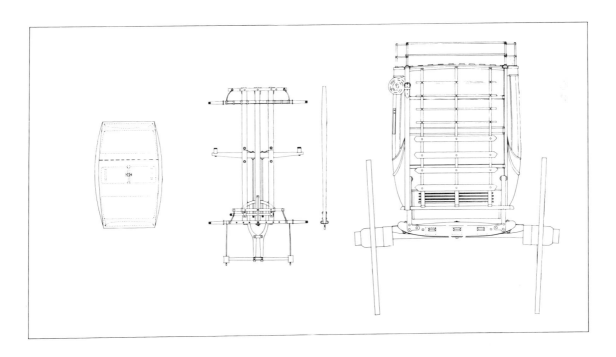

CONCORD COACH.

lettering, and gold leaf. The most common color was a red body on a pale yellow carriage, but varied combinations of green, red, orange, white, blue, yellow, olive, black, maroon, etc., were also used. These last-named colors were more commonly used on HOTEL COACHES, while the red-yellow combination was fairly standard for ROAD COACHES.

Concord coaches were widely used, not only in all parts of the United States, but in South America, South Africa, and Australia. During the 1870s they cost from $1000 to $1200. Less expensive types were also made by the firm, but these were not properly called Concord coaches. The terms PASSENGER WAGON, OVERLAND WAGON, and AMBULANCE WAGON were generally applied to these cheaper vehicles. Harness for most of the Abbot, Downing vehicles was made by James J. Hill & Company, another Concord firm. (H & O)

CONCORD WAGON (or BUGGY) — A very light modification of the New England PLEASURE WAGON, developed by Lewis Downing of Concord, New Hampshire, following his settlement there in 1813. It has a shallow body, either straight or slightly curved, and frequently the cantilevered seat of the PLEASURE WAGON. The body was mounted on the famous Concord gear, composed of three reaches and steel side springs. Built by numerous carriage manufacturers, its use continued until the end of the carriage era. (O)

CONESTOGA WAGON — A covered, freight-carrying wagon that was used largely in the area of Pennsylvania, Maryland, Virginia, and Ohio circa 1750 to 1855. While the wagon's origins have never been authenticated, it is believed to have acquired some of the characteristics of both English and German wagons, through the immigrants from those countries who had settled in the areas of Pennsylvania north and west of Philadelphia. Its name apparently came from the Conestoga River valley, in Lancaster County, Pennsylvania, which was one of the first areas that came to depend on the wagon for the movement of farm produce to Philadelphia, and other commodities back to the farming communities. The earliest known use of the term CONESTOGA WAGON is in 1717, though it appears more likely that this did not yet identify a specific style of wagon, but was only applied to a single wagon that was serving the storehouse of a merchant in the Conestoga area.

Sometime prior to 1750, however, a definite style of wagon developed to which the Conestoga name was applied, for in that year advertisements began to appear in the *Pennsylvania Gazette* that mentioned a Philadelphia tavern known variously as the Conestoga Wagon or the Dutch Wagon. The use of these wagons was

CONCORD COACH, made for excursion purposes. Designed to carry twelve passengers inside, it could carry fourteen more and a driver outside. Wheels, 45-inch and 60-inch.

This heavy CONCORD MAIL COACH carried nine inside passengers, and was the type widely used for road service throughout the United States. During the 1880s, these sold for $1100. Wheels, 45-inch and 60-inch.

CONCORD COACH with French windows, built for six inside passengers. Wheels, 41-inch and 58-inch.

confined for a while to the Eastern settlements because of the lack of adequate wagon roads to the West, but by the close of the Revolutionary War, a substantial trade to Pittsburgh had begun.

While most areas of Pennsylvania, Maryland, Virginia, and Ohio were serviced by Conestogas, the bulk of the trade moved between Philadelphia and Pittsburgh, over the Pennsylvania Road, and between Baltimore and Pittsburgh, over the National Road. From Pittsburgh, the trade eventually spread to Wheeling, and finally into Ohio. The most active wagon freighting occurred between about 1820 and the early 1850s, at which time the extension of rail service to Ohio rapidly brought the wagoning era to a close. The rates charged for Conestoga haulage varied, of course, according to time, location, conditions, etc., but a rate of one dollar per hundred pounds per hundred miles was fairly common, while at other times it might range as high as two dollars.

The wagoners most frequently drove their own wagons, though there were occasional small companies, these often being operated by the more successful wagoners, who used additional wagons with hired drivers. The teams, too, were customarily owned by the individual wagoners, and consequently made the through-trip, regardless of distance, with the wagon, and without change. On the other hand there were companies that operated line teams, wherein wagoners, and other interested persons, put their teams into a pool, the teams being stabled at intervals along the road so that a fresh team could be put in at times, just as stages changed teams. The distances covered in a day varied according to the roads and weather conditions, averaging perhaps 15 miles a day. However, a distance of twenty miles was not uncommon under good conditions, while on a bad winter day a team often covered but a few miles.

The wagoners were of two classes, the "regulars" who were constantly on the road, as a profession, and the "militia" or "sharpshooters," who were farmers by profession, but took their wagons and teams onto the road a few times a year, as their seasonal farm work permitted, to earn a supplemental income. The teams they drove numbered from four to seven horses. A militia wagoner often used no more than four horses with his lighter wagon, while a regular's larger wagon would preferably have six. Since a single lead-horse was a relatively common practice, many five-horse teams were in use, and a few seven-horse teams could likewise be seen. The wagoner had three choices for driving positions. He might ride the saddle horse, which was the left wheel-horse, he could walk alongside, or he could ride the "lazy-board" which projected

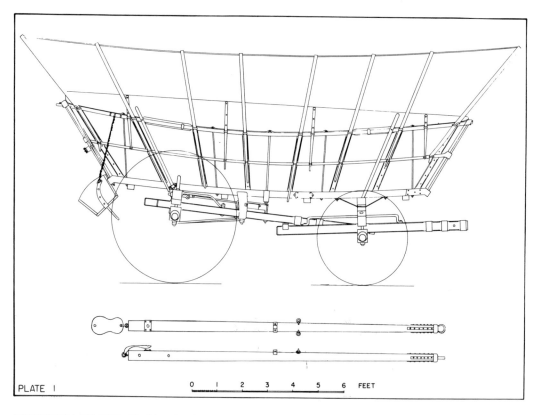

CONESTOGA WAGON. (Smithsonian drawing by Donald W. Holst.)

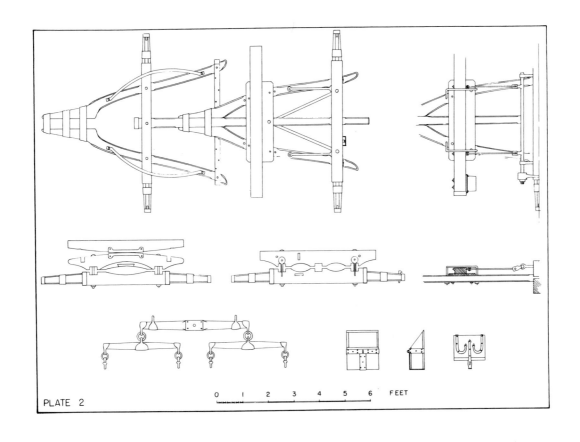

PLATE 2

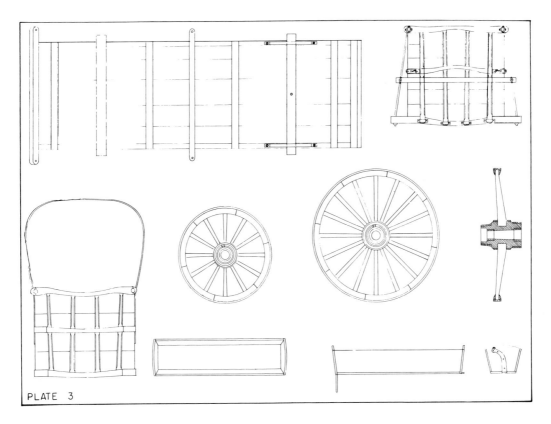

PLATE 3

CONESTOGA WAGON.

from the left side of the wagon. He did not ride in the wagon. He drove by means of a single jerk-line, which attached to the lead rein of the left lead-horse; this line was readily accessible from any of his three driving positions. Many teams were equipped with bronze bells which hung from iron arches attached to the tops of the hames. These bells were both for show and for warning of approach, for an ascending wagon had the right-of-way on a mountain road, and the descending vehicle was required to pull aside when he heard bells or a coach horn announcing the approach of an ascending vehicle.

A type of draft horse, known as the Conestoga horse, was developed for use with the wagon, though this animal was not usually accorded the distinction of a separate breed. This horse's ancestry is uncertain; it averaged between 16 and 17 hands, and weighed around 1600 to 1700 pounds. When the wagoner stopped for the night, the horses were fed from a feed-box which hung from the rear of the wagon when not in use, but was attached to the tongue of the wagon for feeding. The horses were then turned into pasture, stabled, or tethered to the iron rings on the wagon, according to the facilities of the tavern at which the wagoner had stopped for the night.

The characteristics which distinguish the Conestoga wagon from other covered wagons are mainly its attractively curved lines, whereas most other wagons have comparatively straight lines. The body has a pleasing curve, with a low center and ends curving upward, with end-gates that flare outward. Further, the body is not only of planking, as are some wagons, but consists of three horizontal side rails together with a number of upright standards, and thin planking. The wooden bows (eight to twelve in number) supporting the linen cover follow the angle of the end-gates, to give the cover a considerable overhang beyond the end-gates. Removable sideboards are fixed to the tops of the sides, to support bulky loads that rise above the top rails. An attractive toolbox is mounted on the left side of the body, near the center, and near the same point the lazy-board pulls out from underneath the body. Just behind the toolbox is a long brake lever, a feature that was not added to the wagon until about 1830. The body is frequently twelve to thirteen feet long on the bottom, and sixteen to seventeen feet at its top (exclusive of bows), and an average load is perhaps three to four tons, though some carry as much as five tons. Front wheels average forty-five inches in diameter, while those in the rear are often sixty inches or more.

The running gears of the wagons were painted red, though the shade varied to orange if red lead was used, while the bodies ranged from dark blue to gray-blue. The unbleached linen cover was supposedly white, but in reality was tinted by road dust, and became a tattletale gray after a hard rain. The color schemes, then, were standardized, yet there was no uniformity of the shades.

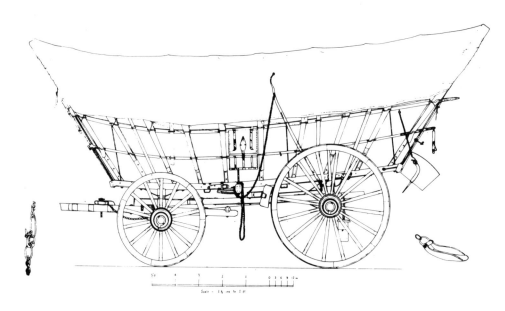

Drawings above and on the facing page were made by John Thompson from the original wagon in the American Museum in Britain, at Claverton Manor, Bath, Somerset, England. The A's in the lower plan on the facing page indicate top views in drawings of the underside.

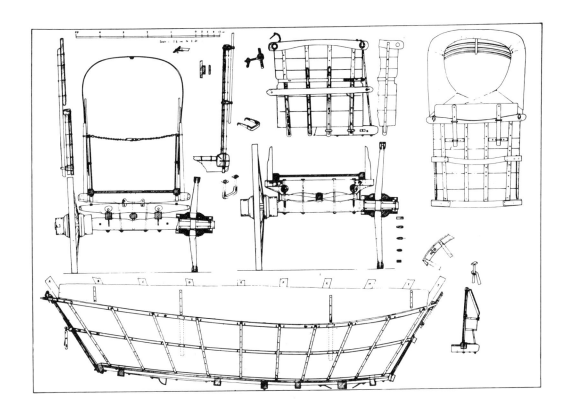
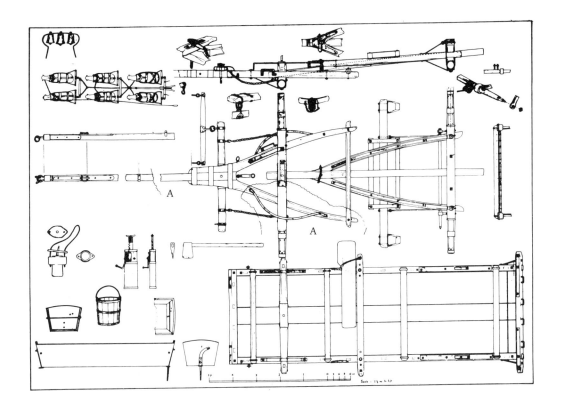

This fine example of a heavy, 12-bow CONESTOGA FREIGHT WAGON is equipped with two tool boxes. Normally there was only one, on the left side of the wagon. The feed box can be seen projecting at the rear. (Smithsonian photo.)

An extremely light 9-bow CONESTOGA WAGON, in the collection of the author.

A number of misconceptions have existed concerning the Conestoga wagon. It was not an immigrant wagon, but was primarily a freighter. It did, however, carry the goods of anyone who wished to move toward the West, and obviously, a wagoner who was relocating his home, perhaps in Ohio, would move his family in his own wagon. Also, the Conestoga was not a Western wagon, though in the earliest days of Western settlement, a few did go West. The Western wagon lacked the curves of the Conestoga. Finally, the wagon was not made boat-shaped so that the seams could be calked, and it could be floated across rivers. Because of the manner of construction — the removable end-gate, and the flexibility of the body — making a Conestoga wagon water-tight would have been almost an impossibility. While the curve of the body did tend to shift the load toward the center, it is believed that the true reason for this curvation was just that it was an attractive design, which is one of the reasons for the wagon's antiquarian popularity today. (O)

CONTRACTOR'S CART — A heavily constructed DUMP CART, especially adapted to hauling earth and stone. (S226)

CONTRACTOR'S WAGON — A variety of DUMP WAGON of simplified construction, built by Studebaker, wherein the sides and floor boards are removable for the purpose of dumping the load. (S226)

CONVENIENCY — An old slang expression for COACH. The first use of it known occurs in Scott's *Waverly Novels*. Stratton, on page 408 of *The World on Wheels* wrote: "This latter gentleman (Robert Murray), who belonged to the Society of Friends, and resided on Murray Hill (New York City), . . . in a measure to avoid the scandal of being thought proud and vainglorious in an age when coaches were treated with scorn, called his merely a 'leathern conveniency,' to the amusement of 'the world's people.' " (ca. 1775.) (H)

CONVEYANCE — A general term, even more comprehensive than "vehicle," applicable to anything used to convey either passengers or freight, including CARRIAGES, WAGONS, RAILWAY CARS, SEDAN CHAIRS, BARROWS, DRAGS, BALLOONS, etc. (H)

COPAICUT — Variant spelling of COPICUTT.

COPICUTT — A small compact PHAETON with two seats. The back seat closes down as a lid, and the back end of the body slants forward from the bottom bar. The origin of the term is unknown. (H & C)

CORACLE — A Welsh term for a light, round, wicker boat, covered with hide. The term was suggested by *The Hub* for application to a round-body basket PHAETON. (H)

CORBILLARD — Modern French for HEARSE. When Roubo wrote, in 1771, the term was applied to a primitive form of curtain-quarter Coach, of which he says: "These coaches are the most ancient of French vehicles of which the form is accurately known to us." See illustration in Roubo's *L'Art du Menuisier-Carrossier*, plate 172 (also shown in Stratton's *The World on Wheels,* page 224). (h)

COSTERMONGER'S CART — (From the English *costard*, a kind of apple, and *monger*, seller.) Originally this was an apple seller's CART, but eventually came to be applied in England to the CART of any itinerant peddlar. (H)

COUCOU — A primitive public carriage used in Paris during the eighteenth century. It is described by Ralph Straus, in his *Carriages and Coaches*, who quotes M. Ramee: "Figure a box, yellow, green, brown, red or sky blue, open in front, having two foul benches which had formerly been stuffed, on which were placed six unfortunate voyagers. In the sides it had, right and left, one or two square openings, to give air during the day or in summer. While the interior was sufficiently open to the world, there was built an apron in front, framed in woodwork and covered with sheet iron. Upon this apron was thrown a third bench, on which were seated the driver of the *coucou* and two passengers who were termed *lapins* (rabbits)."

COUPE — (From the French *coupe*, cut, signifying a COACH cut in two.) The term *coupe* technically describes all the forms of cutting down COACH bodies, but it eventually became identified with one specific type of the vehicles thus cut down, which will be described later.

An important era in the history of carriages is preserved by this word. While one form of cutting down the COACH body originated with the Germans, and has come down to us in the LANDAU and LANDAULET, another system of cutting down originated with the French, and is described by the term *coupe*. When the BERLIN was naturalized in France, it was there cut down in many ways, with a view to greater elegance of shape, and superior compactness and convenience. It was cut longitudinally and latitudinally; it was halved and quartered; the front was cut off, the top cut off, and in fact it was *coupe'd* in every manner which fashion could devise. Thus Roubo informs us (1771) that, to render the BERLIN lighter, they were *coupe'd* close to the door in front, in such a manner that the front door-pillar became the front corner-pillar. This carriage was called the CAROSSE-COUPE or BERLINGOT, and its body corresponds with that of the later coupe. When the BERLIN was cut in half longitudinally, the resulting vehicle was called the VIS-A-VIS, in which form it accommodated only two persons, who sat facing each other. When the process of

CONTRACTOR'S CART on 58-inch wheels. Blue body on red gearing. (Studebaker Bros. Mfg. Co., South Bend, Indiana, 1903.)

This CONTRACTOR'S WAGON had bottom boards 3 inches thick, and side boards 1¾ inches thick. Wheels, 37-inch and 51-inch; body, blue; gearing, yellow striped black. (Studebaker Bros. Mfg. Co., South Bend, Indiana, 1903.)

coupeing was carried still further, and the VIS-A-VIS was also cut down latitudinally (with the same result as cutting the CAROSSE-COUPE or BERLINGOT longitudinally), the BERLIN was so dismembered that only one quarter of it remained. This vehicle, accommodating only one passenger, was appropriately called a DESOBLIGEANT, or DISOBLIGER. The CAROSSE-COUPE, or BERLINGOT, when first used for travel, was called a DILIGENCE, on account of the speed at which it performed the journey from Paris to Calais; its equivalent was afterwards known in England as the CHARIOT. Roubo also tells us that the DILIGENCE was *coupe'd,* and became the DILIGENCE-COUPE, or DILIGENCE COUPE EN BIROUCHE, the *coupeing* in this instance being the removing of a portion of the front panel, allowing the roof to remain supported by the door pillars.

In France the term *coupe* eventually widened in significance, and came to be applied, thusly: (1.) To the front boot of the modern DILIGENCE. (2.) To the front compartment of a railway coach. (3.) To the vehicles known in America as COUPES and BROUGHAMS. In England, the term *coupe* has never been used to any extent, although the English borrowed from the French one of the early forms of the COUPE, which they retained throughout the nineteenth century under the name of CHARIOT, and a later and more compact form under the name of BROUGHAM.

In America, the terms COUPE and BROUGHAM were formerly used synonymously. Later there was an effort to distinguish between the two, by applying the term BROUGHAM to that class of peculiarly English build that were small, low-hung, and having a straight front without child's seat, while the term COUPE was applied to the larger class, having an extended or curved front, and a hinged child's seat. When the front was sufficiently large to admit a full-framed seat in place of the hinged seat, thus giving four permanent sittings inside, it was called a CLARENCE. (H)

COUPE-DORSAY — Contracted to DORSAY, which see. (H)

COUPE, DOUBLE-SUSPENSION — Same as COUPE-DORSAY. (H)

COUPELET — Diminutive of COUPE. A carriage made like a COUPE, except that in place of a paneled top, a large calash-top covers the back seat. Though the term is sometimes applied to the LANDAULET, the COUPELET differs in having no standing front. This vehicle would be more precisely called a BAROUCHET. (H & C)

COUPE-ROCKAWAY — See ROCKAWAY.

COVINA — A variety of heavy war CHARIOT, provided with scythes, used by the ancient Britons. The word is said to be of Celtic origin. Also see ESSEDA. (H)

CRAB — A slang expression applied in America to a type of bicycle having the smaller wheel placed in front. The Star bicycle is an example of this. (H)

CRANE-NECK CARRIAGE — Any carriage in which the underside of the front, or boot, is gracefully curved upward like the neck of a crane. This feature permits the front wheels to turn sharply under the body, allowing greater maneuverability in crowded areas. (C)

CRIOS — A Greek war engine, consisting of a battering ram mounted on wheels. (H)

CROCK-WAGON — See JOLT-WAGON. (H)

CURATE CART — An English variety of pleasure CART. (H)

CURRICLE — (From the Latin *curriculum,* an early Roman racing CHARIOT.) Believed to be of Italian origin, the principal features of the curricle — paired draft and two wheels — were copied from the early CHARIOTS. This vehicle was in reality a CHAISE, or CHAIR, differing by having in place of shafts, a pole for use with two horses. The curricle was also of slightly heavier construction, and was generally more elegant in its finish and appointments. It was most frequently used by persons of eminence. The back of the horse supported the shafts of an ordinary two-wheel vehicle, but with the curricle, a single heavy pole replaced the shafts. The pole was supported from the backs of the two horses by a stout strap hanging from the center of a transverse metal bar (known as a curricle-bar), the ends of which rested on the horses' saddles. Several arrangements were used with the curricle in addition to the single pole. In the late eighteenth century there were curricles having both a pole and shafts, while others featured a convertible arrangement so that either shafts or pole could be employed for use with one horse, or a pair. Originally these fashionable vehicles were frequently preceded or followed by mounted grooms, but later this practice was discontinued, and some curricles were equipped with a sort of rumble that was attached to the rear portion of the frame. This feature acted as a counterbalance, taking some of the weight of the vehicle from the backs of the horses. To ease passengers from the movements of the horses, the curricle-bar was eventually provided with arms that could slide freely in and out of the main body of the bar; also, a taut rope was run underneath the pole, from the axle to the pole-crab, and the strap hanging from the curricle-bar was attached to this rope, to minimize the effects of the trotting movements of the horses.

CORBILLARD. Wheels, 32-inch and 46½-inch; body and gearing black. (*Le Guide du Carrossier,* 1902.)

An octagon front and an inside folding seat in front make this a COUPE instead of a *Brougham.* Wheels, 37-inch and 41-inch; body, black, except lower panels, green, with straw-color striping on moldings; carriage, green, striped with three fine lines of carmine; trimming, green morocco. (*The Hub,* June 1887.)

COUPE-DORSAY. (F.C. Underhill: *Driving for Pleasure*, 1897.)

A British COVINA, or variety of SCYTHE CHARIOT. (Ezra M. Stratton: *The World on Wheels*, New York, 1878.)

CURRICLE equipped with a rope which provides additional comfort and safety. The rope is attached to the front of the pole by a double brace, and by a single brace to a jack on the axletree; when drawn tight, the pole is released of the vehicle's weight, and the elasticity of the rope provides more riding ease. The leather band buckled around the pole and rope maintains a proper distance between the two. The curricle bar runs through the loop of the strap above the pole. (W. Felton: *Treatise on Carriages*, 1796.)

A CURRICLE on C-springs, exhibited by Studebaker Bros. at the World's Columbian Exposition in Chicago, Illinois, in 1893. Wheels, 54-inch; body, deep blue, with black moldings; the latter striped with two lines of carmine; carvings, carmine touched up with black; gearing, deep blue striped with a ¼-inch line of carmine having two fine lines on each side; trimming, drab cloth. (*The Carriage Monthly*, September 1893.)

The curricle was in use in England and America at least as early as the mid-eighteenth century. In America the terms POLE-CHAIR and DOUBLE-CHAIR were frequently used, and in fact were probably used even earlier than the terms CURRICLE. By 1850 the curricle had lost most of its popularity, but even as late as 1893 the Studebaker Bros. Mfg. Co. exhibited one at the World's Columbian Exposition in Chicago. (H & C)

CURRICULUM — An early Roman racing CHARIOT.

CURRICULUS — Diminutive of CURRUS. A small four-wheel carriage drawn by a single animal, and used as a plaything by Roman children. (O)

CURRUS — Latin word for CHARIOT.

CUTTER — An American term, first applied around 1800 to a light SLEIGH having but a single seat-board, for the accommodation of two or three passengers, and usually drawn by one horse. The name should not be used for double SLEIGHS, yet sometimes a SLEIGH having a full and a three-quarter seat-board (commonly called a C-seat) is referred to as a FIVE-SEAT CUTTER, which usage is objectionable. The term JUMP-SEAT CUTTER is more legitimate, this type having a small seat before the principal seat. The former seat can be folded down, and the latter jumped or slid forward into its place, thus transforming a four-passenger SLEIGH into a cutter.

There are many varieties of cutter, but the principal types are the PORTLAND and the ALBANY, the former being the most popular. The cutter retained its popularity into the twentieth century. See ALBANY and PORTLAND CUTTERS. (H & O)

CYCLE — (From Greek, through the Latin *cyclus*, meaning circle or ring, and hence a wheel.) Later applied to a peculiar form of vehicle, and more recently used in compounds describing various forms of VELOCIPEDES such as UNICYCLE, BICYCLE, and TRICYCLE. See BICYCLE, and VELOCIPEDE. (H)

A CURRICULUS with a pair of rather elderly looking Roman children. (Ezra M. Stratton: *The World on Wheels,* New York, 1878.)

QUEEN'S BODY CUTTER, with folding child's seat as indicated by dotted lines. Length, 6 feet overall. Panels and deck, green; gear, and front of dash, red; ironwork, black; trimming, green cloth; plumes, black with red tops. (*The Hub,* June 1903.)

SHIFTING-SEAT CUTTER, with a front seat that could be either jumped out of the way, or entirely removed, making a two- or four-passenger *Sleigh* as desired. Overall length, 8 feet. (*The New York Coach-Maker's Magazine,* September 1868.)

D

DAK — A system of transporting passengers and mail in India, by relays of men and horses. The term was also applied to the boats and carriages used in this service. The carriage was a rather primitive four-wheel vehicle, somewhat similar to certain PASSENGER WAGONS used in the United States. (O)

DANDI — A kind of litter used in India, consisting of a cloth hammock hung from a bamboo pole. (O)

DAUMONT, A LA — Driven by outriders or jockeys, the carriage having no coach-box or driving seat, a small box called a *coffre* being substituted upon the front gearing. This term is appended to the class names of the vehicles driven in this fashion, such as COACH a la Daumont, COUPE a la Daumont, etc. Four horses are employed, with two jockeys, and generally two footmen behind. This method of driving was introduced in France under Napolean I, by an eccentric nobleman, the Duke of Daumont. (H)

DAYTON WAGON — A variety of SPRING WAGON having raved sides, two or more removable seats, and generally a standing top. They are used for both business and pleasure. (O)

DEARBORN WAGON — A light SQUARE-BOX WAGON having two seat-boards and a standing top, usually drawn by one horse. This carriage was developed early in the nineteenth century, and is said to have acquired its name because General Henry Dearborn used one in the field. During the 1820s, a number of Dearborns were used to carry freight over the Santa Fe Trail. Their adaptability to this purpose seems questionable, and they were, in fact, very shortly replaced with Conestoga-type wagons. No accurate description of the earlier Dearborns has been found, but it is believed that they were of slightly heavier build than the later types, and most likely drawn by two horses. It is also known that springs were employed in their construction, but it is not clear whether these springs supported the entire body, or merely the seats. The Dearborn wagon eventually enjoyed a modest popularity in many sections of the country, continuing in use beyond 1900.

The later Dearborn wagon has a body of about six feet long, suspended in most instances on two elliptic springs. The standing top is supported by eight slender pillars, except on that variety in which the two front pillars are omitted to give the carriage a lighter appearance. Each side is closed by three curtains, while a single curtain closes the rear. The rear end-gate is sometimes hung on hinges, and the seats are often made to slide in either direction, making the Dearborn useful for carrying both passengers and baggage. Some of these carriages have a door in each side — merely a frame with a roll-up curtain — while others have none, and the passengers stepped over the side, as in mounting a SPRING WAGON. Some have a dash in front, while others are open, having a footboard projecting slightly beyond the body.

Early in the nineteenth century the terms DEARBORN WAGON, JERSEY WAGON, and CARRYALL were used somewhat synonymously (*A Dictionary of American English*, University of Chicago Press, 1960). Late in the century the Dearborn was often called a DEPOT WAGON, while the term CARRYALL was also still used on occasion. (O)

DECOMEO — An early variety of cut-under COAL-BOX BUGGY. The origin of the name is unknown. (H)

DELIVERY WAGON — A term applied to a great variety of light vehicles intended for delivery purposes. Some are made with open bodies, very much like SPRING WAGONS, while others have closed, or paneled bodies. They were generally built with a single perch and three elliptic springs, although some employ platform gears, or other types of suspension. In some instances bodies were constructed to resemble such things as milk bottles, cigars, etc., for advertising purposes. Intended for local delivery of light parcels to the homes or business places of customers, they are the lightest class of delivery vehicles. (S226)

DELIVERY WAGON, PANEL TOP — A delivery wagon with a closed, paneled body.

DEMI-BAROUCHE — See BAROUCHET.

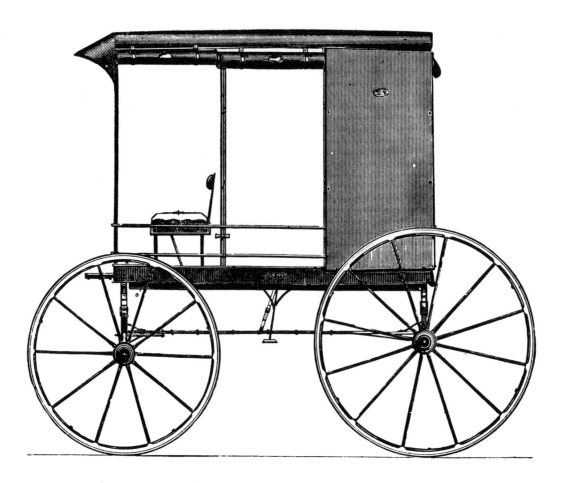

DEARBORN WAGON. Wheels, 43-inch and 49-inch; body, black; gearing, olive green striped black; trimming, blue cloth. (*The Carriage Monthly*, 1879.)

DEMI-COACH — A term occasionally applied late in the nineteenth century to a COUPE or CLARENCE, having either a hinged or full-framed second inside seat. The term carried over to the automobile era, with a few of the earliest autos being called DEMI-COACHES. (O)

DEMI-LANDAU — Synonymous with LANDAULET.

DEMI-MAIL PHAETON — See PHAETON, DEMI-MAIL.

DEMOCRAT WAGON — A SQUARE-BOX WAGON similar to the SQUARE-BOX BUGGY, but larger, with from two to four movable seat-boards on the same level, accommodating two persons on each seat. It is frequently an open WAGON, but can have either a canopy or extension top. The body is set on either platform, elliptic, or some form of patented springs, and the term is, in fact, synonymous with SPRING WAGON. It was called a Democrat wagon because of its unassuming and democratic character. (H & C)

DENNETT (or DENNETT GIG) — A variety of GIG which came into use in England during the early part of the nineteenth century, about the same time as the STANHOPE GIG and the TILBURY. Its primary feature, and that from which it takes its name, is the Dennett spring, which is a variation of the STANHOPE suspension. The STANHOPE is mounted on four springs arranged in pairs, shackled together at the ends, and precisely like the mail spring; whereas the Dennett has three springs (two at the ends and one cross-spring behind), resembling the modern platform spring. The body of the Dennett is similar to that of the TILBURY and STANHOPE, of which types it is a variety. Bridges Adams said of it in 1837, "This vehicle is easier for the horse as it is lighter, and the shafts rest on the side-springs at their front points . . . . But they are uneasy to the passengers, on account of the unequal motion; and, if the horse falls, the danger of being thrown out is greater than with a Stanhope."

Adams also quotes a story that was circulated at the time concerning the origin of the term, "the three springs were thus named after the three Miss Dennetts, whose elegant stage dancing was so much in vogue about the time the vehicle was first used." Another story, equally probable or improbable, attributes the name to a carriagemaker of Finsbury, named Bennett, in whose name the letter B was, at the west end of London, changed to a D. Also see DENNETT spring. (H & C)

DEPOT WAGON – A popular type of family carriage that appears to have evolved from the DEARBORN WAGON, but eventually became a ROCKAWAY. The most primitive form of this WAGON was in use at least as early as 1860, and was similar to the DEARBORN. The term DEPOT WAGON was not yet applied, however, and it was more likely to be called a DEARBORN, or simply, a TOP WAGON, as shown in the 1862 catalog of Lawrence, Bradley & Pardee. This catalog also shows a carriage that at a later date could be called a depot wagon, but was then termed a ROCKAWAY. The *Dictionary of Carriage Terms* in the *Harness and Carriage Journal*, of 1871, describes the depot wagon as a light SQUARE-BOX WAGON similar to the DEMOCRAT, or SPRING WAGON, having a bow-top that could be lifted off bodily, but could not be folded down. The earlier carriages to which the term *Depot* was applied generally fit this 1871 description.

Possibly because of the similarity between the ROCKAWAY and the variety of DEARBORN in which the front pillars were omitted, the depot wagon soon became a vehicle nearly identical to the ROCKAWAY. Houghton, in 1891, still gives the definition of 1871, even though the Rockaway type of DEPOT WAGON had been developed a number of years earlier. Because of the careless manner in which carriage builders applied the two terms to their products, almost synonymously, it is often difficult to assign specific features to either type. The most notable difference that generally prevails is that the ROCKAWAY body is cut away underneath the rear seat, similar to a COUPE or BROUGHAM, while the depot wagon has bottomsides that

DECOMEO. Wheels, 46-inch and 49-inch. (*The New York Coach-Maker's Magazine,* December 1863.)

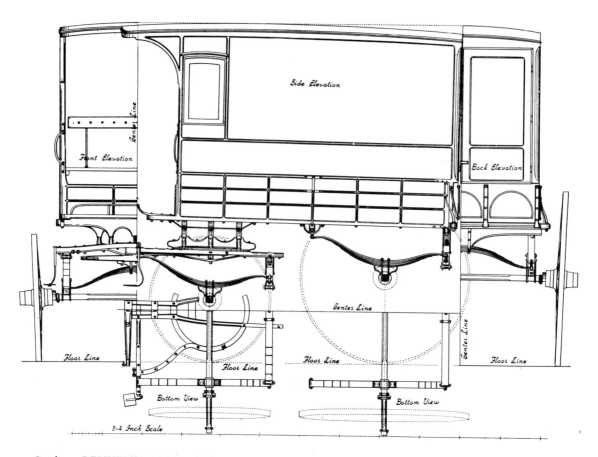

One-horse DELIVERY WAGON. Wheels 37¼-inch and 51½-inch; body, dark green striped gold; upper panel, red; gearing, carmine, striped black. Or, body, red striped gold; center panel, yellow; gearing, yellow striped red. Or, body and gearing all orange, striped black. (*The Carriage Monthly,* April 1896.)

are straight, or nearly so, reminiscent of the square-box lines of the DEARBORN. The straight bottomsides enabled quarter lights to drop, if the carriage builder desired this construction, while those of the ROCKAWAY were almost always stationary. The depot wagon, like the ROCKAWAY, could be entirely glass-enclosed, curtain-enclosed, or a combination of both.

As the name implies, this vehicle was frequently taken to a depot to carry off passengers and their baggage. Some were built with a removable back seat, and a rear end-gate that could be lowered, so that a large quantity of baggage could be accommodated. Occasionally, the compound term DEPOT ROCKAWAY was used for that type that most nearly resembled the ROCKAWAY; however, the earlier DEARBORN type was not entirely replaced by the ROCKAWAY type, but the two shared their popularity until the end of the carriage era. The term STATION WAGON was also applied to this carriage, this usage becoming common after 1890. Houghton, in 1891, had not yet entered STATION WAGON in his dictionary, but still used the earlier expression, DEPOT WAGON. (C, H, M, & O)

DESOBLIGEANT — (A French word, meaning *disobliging*.) One of the methods of cutting down a COACH, described under the COUPE entry, resulted in this carriage. If the COACH was cut longitudinally, the VIS-A-VIS resulted, and if cut latitudinally, the DILIGENCE or BERLINGOT was the result. Either of these vehicles, if halved again, formed a quarter-COACH, accommodating one person only, and this vehicle was given the name of desobligeant. This body characterized a number of vehicles which, differently mounted (on two or four wheels), and in various countries, went under different names. In England, on four wheels, they were called SULKIES (William Felton's *A Treatise on Carriages*, volume 2, page 66); on four wheels, in France, they were called desobligeants, as above stated. If on two wheels, in France, they were a variety of POST-CHAISE. Laurence Sterne, in his *Sentimental Journey*, notices these desobligeants, and tells us that when he was in Calais and a lady wished to travel in the same vehicle with him, he had to *disoblidge* her,

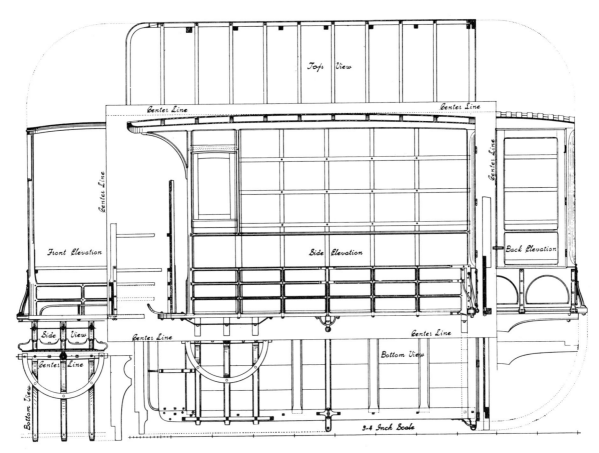

One-horse DELIVERY WAGON.

This is an interesting variety of a special-purpose DELIVERY WAGON, built for delivering and advertising cigars. Wheels, 36-inch and 46-inch; gearing, carmine striped black and white; seat panels, carmine; cigar, tobacco brown; label, red, or any desired shade. (*The Hub*, January 1892.)

Half-platform DEMOCRAT WAGON, so-called because only the rear is on platform springs, while the front is supported by an elliptic spring. (Eberhard Mfg. Co., Cleveland, Ohio, 1915.)

DEPOT WAGON built by George W. Moore, of Shoemakertown, Pennsylvania. Wheels, 45-inch and 48-inch; body, black; gearing, black striped with two fine lines of dark brown, and one fine line between of light brown, glazed with carmine; trimming, black enameled leather, with head lining in blue cloth. (*The Carriage Monthly,* October 1879.)

owing to the want of capacity of the DISOBLIGER, as he calls it. (H)

DEXTER CUTTER — A light pattern of PORTLAND CUTTER popularized by R. M. Stivers, of New York City. It was named after the celebrated trotting horse. (H)

DIABLE — (A French word, meaning devil.) The name was applied in France to a vehicle described by Roubo in 1771 (*L'Art du Menuisier-Carrossier*, fig. 1, plate 212): "This represents a vehicle called the *Diable*, which is to the Caleche what the Diligence is to the Berlin; that is to say, it has the front door pillars cut away above the waist-rail." As represented in the cut referred to, it resembles the modern COUPELET, though the top of the diable does not fold down. (H)

DICKEY COACH — A full round COACH having the driver's seat made separately from the body, that seat being hung upon iron loops. (C)

DILIGENCE — A French word meaning speed or promptness. The term appears to have been first applied, during the early part of the eighteenth century, to a STAGECOACH running from Paris to Lyons, which journey was customarily accomplished in five days in summer, or six days in winter, and which, said Roubo, "is the promptest and most commodious of our public vehicles." This particular COACH seems to have been more elaborately constructed than the ordinary COACHES of that day, being mounted on half-elliptic side-springs attached to braces, whereas Roubo says, "all the Coaches do not have springs like these." The same term apparently was applied, either at the same time or later, to another vehicle, for the same writer advises that the BERLIN, when cut down, was called a "*Carosse-Coupe* or *Berlingot*, or more ordinarily *Diligence*"; the DILIGENCE shown in his plate 210 is not a traveling vehicle, but an elegant COUPE, holding two persons only. Eventually the term seems to have returned to nearly its original significance, as it was used to describe large COACHES which performed stated journeys in France and Switzerland. This diligence was a peculiar variation of the common STAGECOACH and consisted of the ordinary COACH body, with a boot in front that was enlarged to the form and dimensions of a COUPE body, and known as the *coupe*, with a driver's seat immediately on top of it. The boot in the rear was surmounted by a hooded seat, known as the *banquette*. The term was rather generally applied in France to numerous types of heavy public vehicles resembling the COACH, or sometimes similar to an OMNIBUS, and having both inside and outside seats. Some had a driver's seat, while others were driven by a postilion. See COUPE. (h, C & O)

DILIGENCE COUPE (or DILIGENCE COUPE EN BIROUCHE) — See COUPE. (H)

DIOPHRAMAXA — A late English variety of WAGONETTE-BREAK. The derivation of the term is unknown. (H)

DIOROPHA — A variety of LANDAU that was invented and patented by James Rock of London, and introduced as a novelty at the London Exhibition of 1851. (H)

DIPHRON — A two-wheel, Greek, war CHARIOT. (O)

DISINFECTING CART (or WAGON) — See SPRAY CART.

DISOBLIGER — The English equivalent for the French term *desobligeant*. (H)

DOG CART — Originally, a cart for dogs; this variety of CART was first used for shooting purposes, and carried hunting dogs. Felton (vol. 2, page 90) describes an English SHOOTING PHAETON of his day (1796) which was a variation of the common SHOOTING GIG, and which he says "is much to be preferred," as "being more steady and carrying more conveniences." This SHOOTING PHAETON was intended for shooting from, and had conveniences for carrying the guns, the game, and the dogs. In the SHOOTING GIGS referred to, the dogs were carried in a well, which was provided at the sides with slats, giving the dogs the necessary ventilation. Since that day the term DOG CART has been applied to a large and miscellaneous class of two-wheel vehicles. This tendency was in part due to a natural desire on the part of the people to escape the oppressive tax on pleasure carriages. "In 1843 the Chancellor of the English Exchequer," says Mr. Sidney, "desirous of throwing a sop to the ever-distressed and discontented agricultural interest, exempted all two-wheeled carriages not costing more than 21 pounds, from the assessed tax, provided the name of the owner was painted in letters, not less than four inches in length (nothing was said about the breadth) on a conspicuous part of the vehicle. This exemption created a new and large class of two-wheeled vehicles, which, although the exemption has been repealed, flourish to this day, under the name of Dog-Carts, Malvern Carts, Leamington Carts, Whitechapels, Norfolk Shooting Carts. The first crop were actual Dog-Carts, constructed to carry four persons (each pair being, instead of vis-a-vis, dos-a-dos), mounted on very high wheels, sometimes called 'Oxford Bounders,' with ample room for the conveyance of dogs or luggage. Long letters of attenuated shape made the names of the owners almost illegible. A bill which, duly receipted for 20 pounds 19 shillings, satisfied the tax-gatherer, was followed by another bill for extras, in the shape of seat-cushions, rugs and lamps, which brought up the total cost to from 25 to 30 pounds, ... "

In due course of time the original dog-cart design was altered to conform to the requirements of ordinary

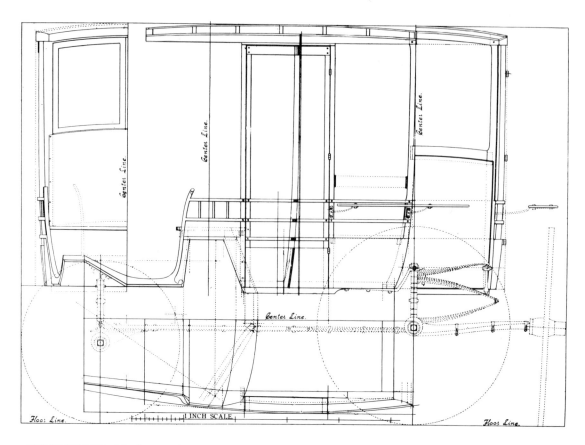

DEPOT WAGON. Wheels, 36-inch and 43-inch; body, black; moldings striped with a heavy line of Indian red; gearing, dark green, striped with three lines of Indian red; trimming, green cloth. (*The Carriage Monthly,* April 1893.)

family use, and carriagemakers consulted the wants and wishes of family customers of both sexes, who, requiring a cheap carriage, would submit to the title of dog cart, although they kept neither greyhounds nor pointers, but decidedly objected to perching on lofty wheels at the mercy of a stumbling horse. For them was devised an endless variety of two-wheel CARTS — BALLESDEN, LEAMINGTON, NOTTINGHAM, WORTHING, WORCESTER, etc. — suited to every size of animal, from the curate's pony to the reckless COACH horse, running so low to the ground as to make the worst calamity of a tripping horse the breaking of his knees. Thus dog carts became so universal that the expression, "I will send the Dog Cart to meet you," became a common postscript to letters. The dog cart eventually became an aristocratic vehicle for gentlemen's driving, retaining much of its original form, but often having a painted imitation of slats in place of the ventilators formerly used. For tandem-driving the dog cart came to be considered as the vehicle *par excellence.*

The dog cart is a two-wheel vehicle for one horse (unless driving tandem), and is especially adapted for gentlemen's driving. It has a square-box body, accommodating four passengers, the gentleman driver and his companion facing the horse, with one or two grooms seated dos-a-dos to them, and resting their feet on the lowered end-gate which is held in place by straps or chains. It is suspended on mail springs, or half-elliptics, and supplied with a mechanical device for shifting the body backward or forward to balance the load carried. The so-called FOUR-WHEEL DOG CART is properly a DOG CART-PHAETON, yet for many years a number of manufacturers did apply the term DOG CART to a four-wheel vehicle. (H)

DOG CART PHAETON — A DOG CART transformed into a PHAETON by mounting the body, usually a cut-under, upon four wheels. This change was a frequent one, the most familiar example being that of the WHITECHAPEL CART, afterwards developed into the WHITECHAPEL WAGON, or SURREY. When thus changed, the name should be changed to correspond, by forming a compound word incorporating the word PHAETON, or WAGON. The term FOUR-WHEEL DOG CART is an undesirable usage, but the fact cannot be ignored that this usage was common, even among the carriage builders.

In the 1862 catalog of Lawrence, Bradley & Pardee, FOUR-WHEEL DOG CARTS are shown, bearing some resemblance to the TRAPS of a later period. The back seats could be folded down out of sight, and some had reversible backs so that the passengers in the rear could face either forward or backward, but these carriages did not feature front access to the rear seats as did the TRAPS. Some models had a large closed box under the front seat. (H & O)

DOG CART-SLEIGH — A four-passenger SLEIGH having a body like a DOG CART. (O)

DOLGUSHA — A Russian hunting wagon. (H)

DORMEUSE (or DORMEUSE-CHARIOT) — (From the French, meaning sleeper.) An old form of French traveling carriage, much used in Europe during the pre-railway period. The body had extensions, either front or rear, or both, enabling the occupants to have a bed made up and recline at full length. Thus, the travelers could sleep in their carriage, eliminating the necessity of staying at inns, which were often very poor. (H)

DORSAY — Contracted from COUPE-DORSAY, and named after Count D'Orsay. A COUPE or BROUGHAM

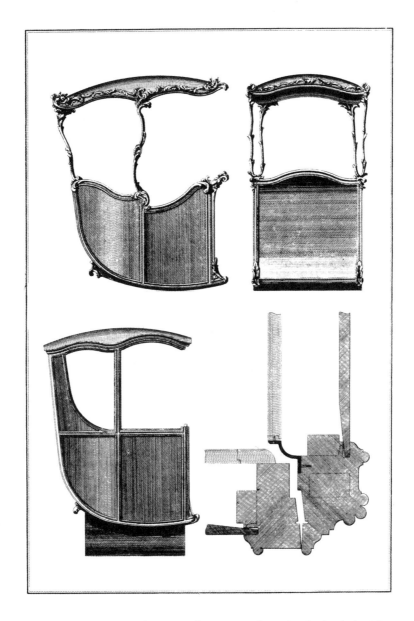

The DIABLE described by Roubo. (M. Roubo: *L'Art du Menuisier-Carrossier,* Academie des Sciences, Paris, 1771.)

Todd & Wright showed their DIOPHRAMAXA at the London Sportsman's Exhibition in 1887. The rear seats revolve, so that the riders may readily face away from the wind. (*The Carriage Monthly,* July 1887.)

English DOG-CART on 54-inch wheels. This *Cart* had an extra outrigger attachment that could be bolted to the drawbar in place of the step, thus allowing the vehicle to be drawn by a pair of horses. (Lawrence, Bradley & Pardee, New Haven, Connecticut, 1862.)

DOG CART-PHAETON. *The Carriage Monthly* actually titled this plate "Dog Cart", and in another two years the term, "Trap," would be coming into use. Wheels, 36-inch and 45-inch; body and gearing painted in imitation of oak; trimming, drab cloth. (*The Carriage Monthly*, April 1890.)

DOG-CART SLEIGH. Overall length, 8 feet; body and seats, vermilion; gearing, yellow striped carmine; trimming, blue cloth. (*The Hub,* May 1892.)

hung on four elliptic and four C-springs; otherwise known as a double-suspension COUPE or BROUGHAM. (H)

DOS-A-DOS — (A French term, meaning back to back.) It is applied to a vehicle, generally a DOG CART, wherein the rear seat is reversed, so that the passengers on that seat face toward the rear. In America, the vehicle that came to be known as a TRAP was occasionally called a dos-a-dos before it acquired the former name. (H)

DOUBLE-CHAIR — A term sometimes applied in America to the CURRICLE, since that vehicle was equipped with a pole for use with two horses. (O)

DOUBLE-RIPPER — See BOB-SLED.

DOUBLE-RUNNER — A local New England term. This vehicle consists of two SLEDS, surmounted by a plank placed lengthwise upon them, with the foremost SLED pivoted to the plank for steering purposes. Used for coasting, this SLED came to be known as a BOB-SLED in the twentieth century. (H)

DOUBLE-SUSPENSION COUPE (or BROUGHAM) — See DORSAY.

DOUGHERTY WAGON — A variety of PASSENGER WAGON of simple design, built much like the OVERLAND WAGON of the Abbot, Downing Company, but hung on springs instead of braces. The Dougherty wagon, supposedly originating in St. Louis and named after the builder, was widely used in the West, and many were used by the army. They had side doors as an OVERLAND WAGON, a luggage rack on the rear, but none on top, three seats including the driver's, and four elliptic springs. The driver's seat was generally on a level with the passenger's seats, and under the same roof. The seat-backs could be lowered to a horizontal position, so that together with the seats a bed might be formed. This type is shown in the army specifications for 1882, yet an 1885 Signal Corps photograph in the collections of the National Archives shows a Dougherty wagon with a driver's seat like that of an OVERLAND WAGON. The 1930 *Handbook for Quartermasters* shows one of these WAGONS with a wheel-house, and platform springs. (H & O)

DRAG — A sporting vehicle, very similar to a ROAD-COACH, which see.

DRAG-PHAETON — See BREAK.

DRAISINE — A two-wheel velocipede invented about 1816 by a German, Baron von Drais, and introduced into England a short time later. The rider simply straddled the machine and pushed it along with his feet. Also known as a HOBBY HORSE, or PEDESTRIAN CURRICLE. (H & O)

DRAY — The heaviest class of commercial freight carrier, with the exception of certain special-purpose vehicles. Generally restricted to use in and around cities, it was used for any heavy hauling, but especially by industry for movement of heavy machinery, castings, etc. The dray was built very much like the TRUCK, except that the floor was straight and generally level, and the rear wheels were often smaller than a TRUCK's so that they were below the floor, enabling either the floor, or large and unusual objects to project over the wheels if necessary. The bodies often had no sides, but stake, rack, and box bodies were also common. Suspension was sometimes on truck or platform springs, but frequently construction was of the dead-axle type, without any springs. These vehicles were sometimes called FLOATS, or TRANSFER WAGONS.

This term was also applied to two-wheel vehicles used for the same purpose. Throughout the eighteenth century, and much of the nineteenth, this centuries-old, two-wheel type appears to have been the mainstay of commerce and industry in America, and can be seen in countless early prints and lithographs. They were often built as a simple framework, sometimes having no floor, other than the several cross-members. Short stakes generally formed the sides. (H, O, & S226)

DRIVING WAGON — A BUGGY, or ROAD WAGON, made without a top. Synonymous with RUNABOUT.

DROITZSCHKA — A vehicle contemporary with the BRITZSCHKA in England and on the Continent, during the first half of the nineteenth century. Its name was probably derived through the German from the Russian DROSHKY. In his *English Pleasure Carriages* (page 232), Adams wrote, "A Droitzschka, or, as it is commonly called, a Drosky, was, in the outset, of Russian origin, being, in fact, in its simplest form, an improvement on the Sledge, by adding springs and wheels, the single passenger sitting with his legs on each side of the perch, as he would sit on a horse. But the Droitzschka, as made in England, is a modification of the elliptic-spring Britzschka, by placing the passengers' seat nearly at the level of the hind axle, and sinking the central part of the body below the level of the axle for the legs. This carriage can only carry two persons inside, and two on the driving seat. . . . The principal utility of the Droitzschka is for languid, aged, or nervous persons, and children, as it is low on the ground, and consequently easy of access and difficult to turn over. It is made for one horse, or two ponies, but is very heavy to draw." (H)

DROITZSCHKA CHARIOT — See BROUGHAM.

DROSHKY — (Also spelled drosky; droshka, -ke, -ki; droska; droskcha.) A four-wheel vehicle peculiar to Russia, of simple construction, consisting of a bench extending between the axles, on which the passengers (Adams

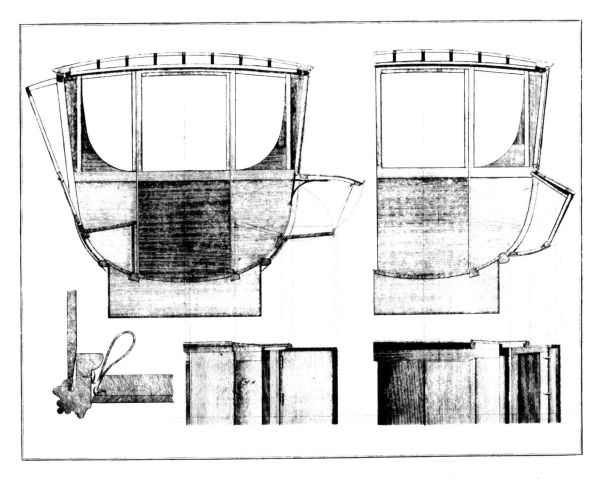

The body of a DORMEUSE; figure 1 shows how sufficient length for sleeping was gained. (M. Roubo: *L'Art du Menuisier-Carrossier,* Academie des Sciences, Paris, 1771.)

says but a single passenger) sit astride, as on a saddle, with their feet resting on bars near the ground. The term came to be generally applied to various types of two- and four-wheel public vehicles used in Russia. (H & C)

DUC — (From the French word for duke.) A Duc might be defined as a PONY-VICTORIA; in other words, it is a cross between the PONY-PHAETON and the VICTORIA, and may be described either as a VICTORIA reduced to the dimensions of a PONY-PHAETON, or a PONY-PHAETON having the characteristics of the VICTORIA — the skeleton or movable boot. A Duc proper, like the VICTORIA, is a royal equipage, driven by outriders, and has properly no boot; as commonly used, however, the skeleton-boot is substituted. (H)

DUC-PHAETON — A DUC, driven neither by outriders nor by a coachman from the skeleton-boot, but by its occupant (commonly a lady), with attendant groom in the rumble behind, thus being self-driven, and hence properly termed a PHAETON. (H)

DUMP-CART — A CART with a body so constructed that it can be tipped over backwards to assist in emptying the contents. The front end of the body is held down against the shafts by some simple mechanical device, which, when released, allows the hinged body to raise from the shafts and tip over backwards. It is also called a TIP CART. (H)

DUMP WAGON — A Wagon constructed so as to automatically dump its load without the use of hand labor. A wide variety of these WAGONS existed, some being bottom-dump and others being end-dump WAGONS. The bottom-dump variety had a door, or doors swinging downward when the driver released the mechanism from his seat. Many of them could also be closed from the seat by the use of chains which were drawn up by such devices as ratchets, or gears. The end-dumping types were often constructed so that the brakes could be set on the rear wheels, the body released from the rear axle, and the team backed against the front wheels, shortening the wheelbase, thus causing the unbalanced body to tip over backward. A special type

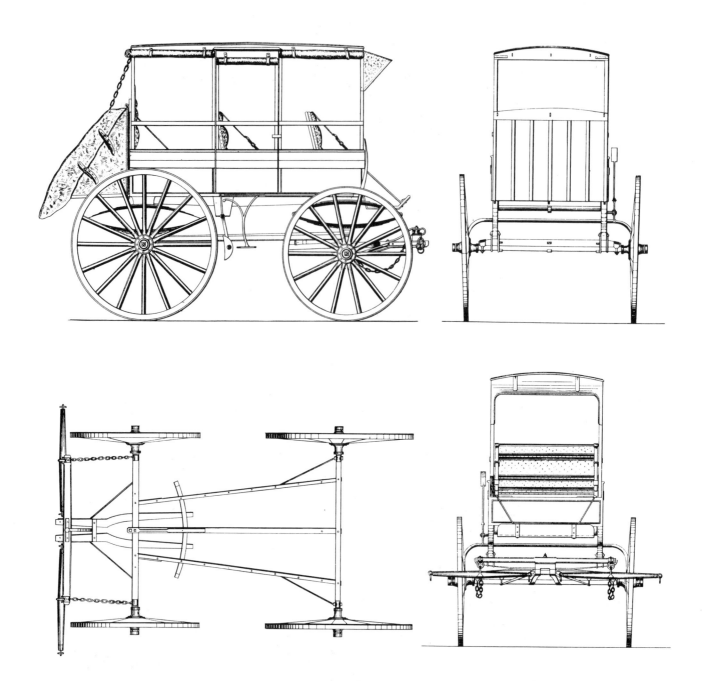

DOUGHERTY SPRING WAGON. Wheels, 44-inch and 50-inch; body and gearing black; gearing striped with 3/8-inch stripes of dark green; curtains and boot cover, yellow ochre; 3-inch letters "U. S." in Doric style centered in lower panel of each door; inside, drab, except back of seats, black; trimming russet leather. (Specifications for Means of Transportation, Washington, Government Printing Office, 1882.)

DRAG, by Brewster & Co., New York. (F. Rogers: *A Manual of Coaches,* 1899.)

This dead-axle DRAY was called a MACHINERY TRUCK by the builders, The Haywood Wagon Co., of Newark, New York, though the former term is more appropriate. Having a capacity of 7 tons, it was built about 1910.

A heavy dead-axle DRAY, believed to have been photographed in Lowell, Massachusetts, about 1900, showing an unusual and interesting manner of suspending the driver's seat. (Smithsonian photo.)

Two-wheel DRAY mounted on truck springs. (R. H. Allen Co., New York, 1883.)

Several two-wheel DRAYS can be seen in this interesting cut that decorated a New York billhead in 1846. (Original in Smithsonian collections.)

Russian DROSHKY. Wheels, 26½-inch and 33½-inch; body and gearing, black with no striping; trimming, green cloth. (*The Carriage Monthly,* November 1884.)

DUC. Wheels, 26¾-inch and 38½-inch; body, maroon; seat panels, yellow; body striped carmine; gearing, carmine, striped black. (*Le Guide du Carrossier,* April 1887.)

This general purpose DUMP CART was popular among farmers. On 54-inch wheels, it dumped when the hooks in front were released. (Studebaker Bros. Mfg. Co., South Bend, Indiana, 1903.)

of bottom-dump WAGON called a SPREADING WAGON was so constructed that the driver could accurately control the flow of material from the body, for such purposes as road building. Another type of WAGON intended for hauling materials such as coal had side chutes underneath both sides of the body, but this could not accurately be called a dump wagon since the coal needed to be shoveled towards the chutes. Some WAGONS had mechanisms for raising the entire body before dumping, to facilitate dumping the load down a chute. Numerous patents were granted on dump wagons beginning about 1850. (O)

DUOBUS — A London CAB, very much like a BOULNOIS CAB, except that it was larger; the term was, however, sometimes applied to the BOULNOIS CAB. The duobus was invented by a Mr. Harvey. (O)

DUQUESA — (From the Spanish, meaning duchess.) A type of PHAETON, somewhat resembling a VICTORIA, that was used in Spain. (H)

DUQUESITA — (From the Spanish, meaning little duchess.) A Spanish carriage resembling the DUQUESA, but somewhat lighter, used with one horse, and with room for but one person on the driving seat. (H)

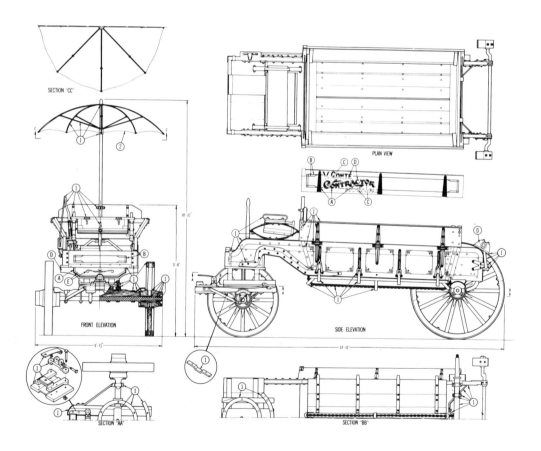

DUMP WAGON of about 1910. (Smithsonian drawing from the original vehicle at the Horse'n' Buggy Museum, Gettysburg, Pennsylvania.)

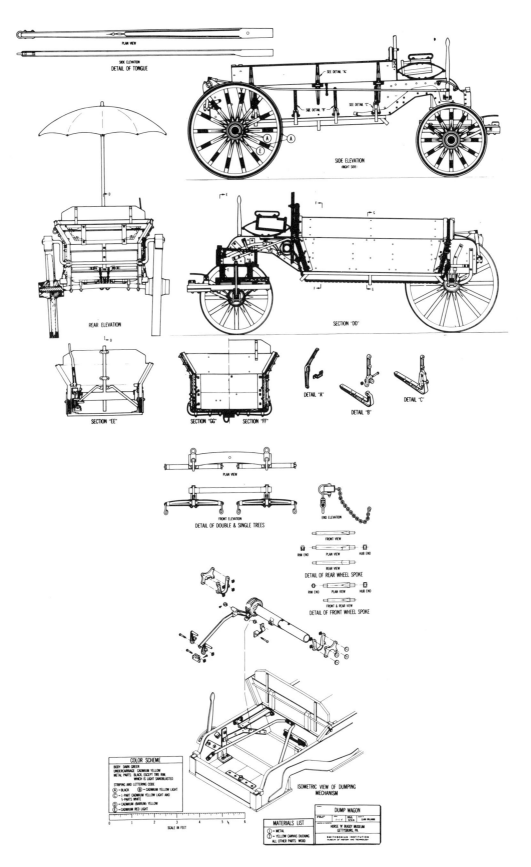

DUMP WAGON.

DUQUESA with rumble. Wheels, 32-inch and 39-inch; body panels, dark blue; moldings, black striped with fine carmine; gearing, black striped with blue and white, close together; trimming, blue cloth and blue goat skin. (*The Hub,* April 1882.)

DUQUESITA. Wheels, 32-inch and 40-inch; body panels, dark green, moldings, black; gearing, dark green, with two broad stripes separated by a fine line of carmine; trimming, green goat skin. (*The Hub,* April 1882.)

E

ECCENTRIC CARRIAGE — A carriage so constructed that the front axle, when turning, moves off its center, the object being to shorten the carriage-part, and yet allow the front wheel to pass under the body without coming in contact with it. (C)

EILWAGEN — An Austrian term that is synonymous with DILIGENCE. This carriage was also known as a MALLEPOSTE. Stratton states that the term is synonymous with BRITZKA CHARIOT. (H)

EKKA — A primitive, two-wheel, one-horse, one-passenger vehicle used in India. (O)

ELEVATING GRADER — See EXCAVATOR.

ELYSIAN CHAPEL-CART — A late nineteenth-century variety of pleasure CART built by James Henderson & Company of Glasgow, Scotland. (H)

EMBALMING WAGON — A variety of UNDERTAKER'S WAGON, being a type of business WAGON or BUGGY. It has a two-passenger seat as does a BUGGY, generally a top (either standing or falling), and a large boot in which the paraphernalia of the undertaking profession could be carried to a home where a body was to be prepared for burial. (O)

EMIGRANT WAGON — A general term applied to any covered WAGON used by emigrants moving to the American West during the latter half of the nineteenth century. (O)

EPIRHEDUM — An ancient Roman vehicle of unknown description. (H)

EQUIBUS — A proposed variety of two-wheel vehicle, intended to carry four passengers besides the driver. It was suggested in the *Scientific American* in 1878, and the vehicle was to straddle the back of the horse, with passengers over the animal's back, and the driver nearly over its neck. It is not known if any of these oddities were ever built. (O)

EQUIPAGE — (From the old French *equiper* or *esquiper*, to equip, to arm.) It is strictly applicable to a traveling retinue in general, such as that of a prince, including horses, carriages, equerries, footmen, etc. It is often applied specifically to an aristocratic carriage, together with horses, harness, and other appointments. (H)

EQUIROTAL — (From Latin, indicating wheels of equal size.) The equirotal carriage was the invention of William Bridges Adams, and is described in detail in his book, *English Pleasure Carriages*. Adams believed that the front and rear wheels of a carriage should be of one size, and designed a carriage that was pivoted near the center of the perch, rather than at the front axle, thereby eliminating the fifth-wheel. This caused the body to be made in two separate sections, and both front and rear sections turned with the respective pair of wheels, since each axle was fixed permanently to its portion of the body. This enabled front wheels as large as those in the rear, for unlike the wheels of a conventional carriage, these did not move toward the body in turning. Adams claimed that this construction provided easier draft and a shorter turning radius, and, since the driver's seat pivoted with the front axle always keeping the driver squarely behind the horses, also gave him better control over them. In addition, the equal-size wheels permitted the springs to be mounted on the same level, which Adams believed would allow a more uniform motion of the body. Adams was a practical and enthusiastic carriage builder, and abundant means allowed him to experiment with his theories to an extent not often permitted other inventors. He applied his theory to many types of carriages, as well as to ordinary WAGONS, and believed that his invention was a success. Posterity, however, has not confirmed his claims, for the invention did not outlive the inventor. (H & A) See also PHAETON, EQUIROTAL.

ESCORT WAGON — The all-purpose freighting WAGON of the army, replacing the six-mule WAGON in 1878, though both were used for some years after 1878. Drawn by four animals, it was slightly smaller than the six-mule type, and had a spring seat inside for the driver. This wagon continued in use to the end of the horse-drawn era, surviving an unsuccessful attempt during World War I to replace it with a type known as

An EKKA of India. (M. M. Kirkman: *Classical Portfolio of Primitive Carriers,* 1895.)

    the COMBAT WAGON. The standard load of the escort wagon was 3000 pounds, but loads of 5000 pounds were frequently carried. (O)

ESSEDA — A form of light war CHARIOT, with or without scythes, used by the ancient Britons. (H)

EXCAVATOR — A machine that scoops up earth as it is drawn forward, having a conveyor geared to the wheels, which in turn lifts the earth either to the interior of the excavator, or to another WAGON alongside. Also called a WAGON LOADER, SELF-LOADING WAGON, and ELEVATING GRADER. (O)

EXCURSION WAGON — A PASSENGER WAGON somewhat similar to a MOUNTAIN WAGON. One variety had a body more like a light FARM or DELIVERY WAGON, with seats on elliptic springs, while the gear was a regular farm gear with bolster springs. A standing top often was provided. (S226)

EXPRESS WAGON — A freight-carrying WAGON in the medium-weight class, intended for carrying larger parcels and boxes to homes or business houses, and for carrying trunks and baggage to and from the depot. The driver's seat frequently was higher than that on the DELIVERY WAGON, and the body was provided usually with flare-boards to assist in holding bulky loads. Occasionally the EXPRESS WAGON was hung on three elliptic springs, but more often it was on platform springs. The body was either open, or with a roof supported by four corner posts, but it rarely had a paneled body like the DELIVERY WAGON. It was sometimes called a BAGGAGE WAGON. (S226)

Side and rear views of an EQUIBUS. (*Scientific American*, April 27, 1878.)

EMBALMING WAGON equipped with tail gate. Wheels, 41½-inch and 44½-inch; body and gear, black, gear striped with yellow; trimming, leather; lamps and rail, silver plated. (*The Hub,* May 1897.)

Slusser's EXCAVATOR (self-loading and self-dumping WAGON) of 1874 reportedly could load 100 cubic yards per day. (Built by G. G. Haslup & Bro., Sidney, Ohio.)

EXCURSION WAGON with body on bolster springs and seats on elliptics. Wheels, 44-inch and 52-inch; body, vermilion; gearing, yellow; top and curtains, white duck. (Studebaker Bros. Mfg. Co., South Bend, Indiana, 1903.)

The Eberhard Mfg. Co. called this a PLATFORM SPRING WAGON, which, because it is a *Wagon* on platform springs, is technically correct. It is not, however, what those in the trade generally meant by the term; in reality, this is an EXPRESS WAGON. (Eberhard Mfg. Co., Cleveland, Ohio, 1915.)

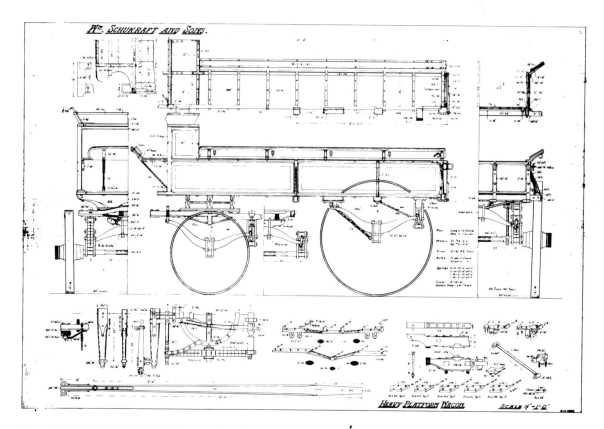

Heavy PLATFORM WAGON built in 1898 by William Schukraft and Sons, Chicago. Called a *Platform Wagon* because of the platform springs, many builders would have called this an *Express Wagon*. Wheels, 38-inch and 56-inch; dark green body on red or yellow gear, neatly striped. (Drawing from Smithsonian collections.)

F

FARM WAGON — A general type of all-purpose farm vehicle, featuring a square-box body. It is found in a wide variety of styles and sizes, often with sideboards, and spring seats, and sometimes with flare-boards and turn-under wheels, though this last feature is unusual on a farm wagon. The body is commonly mounted directly on the carriage, without the use of springs, but occasionally bolster springs may be found. (S226)

FIACRE — A general term in France, applied to a HACKNEY-COACH. The word was in use from about 1650 until recent times. The name is derived from the Celtic saint, Fiachra, but there is some difference of opinion among historians concerning the circumstances under which the carriage acquired his name. Some believe that these carriages first stood, about 1650, in front of a Paris inn named the St. Fiacre. Another theory is that a likeness of the saint was painted on the doors of the coaches, since he had, nearly a thousand years earlier, established a hospital in France for travelers and the poor, and used wheeled vehicles to carry the sick to his hospital. His carriages supposedly came to be known as FIACRES, and the term was eventually applied to public carriages. The seventeenth-century Fiacre resembled a COACH, carrying six passengers, two on each of the transverse, vis-a-vis seats, and one by each door, facing outward. (H & O)

FIAKR — A Bohemian term, obviously borrowed from the French, *Fiacre*, for a two-horse public carriage. (O)

FLEMING CARRIAGE — A variety of ROCKAWAY, developed by R. J. Fleming, of Harrisburg, Pennsylvania, during the 1850s. Like the GERMANTOWN, it had no wheel-house, and like the ROCKAWAY, it omitted the front pillars from the projecting roof. (O)

FLOAT — (1.) A four-wheel WAGON or TRUCK, surmounted by a platform, prepared for the purpose of display in public processions or parades. (2.) A term applied in London and vicinity to a heavy DRAY. The Studebaker Company also used the term FLOAT for a type of heavy dray. (3.) In certain rural areas of England, the term was applied to a deep CART, having large wheels and a cranked axle, used to carry livestock or coal. This vehicle was also called a FLOATER. (H, O, & S226)

FLOATER — See FLOAT, meaning 3.

FLY — An English expression, being a contraction of FLY-BY-NIGHT, as SEDAN CHAIRS on wheels were known during the Regency. The term was also applied generally to two-wheel, horse-drawn CABS, and to four-wheel public vehicles such as COACHES, DILIGENCES, and particularly to HACKNEY-CARRIAGES that were used in villages. (H & O)

FLYING HOSPITAL — See AMBULANCE.

FLYING MACHINE — Synonymous with FLYING COACH. This term was frequently applied in the American colonies to any public stage whose proprietors were boasting of the speed of their vehicles. (O)

FORGE CART — A two-wheel army vehicle, being a portable blacksmith's forge, equipped with the necessary tools of the trade. It is also called a TRAVELING FORGE, though it differs from the vehicle described under that entry by being a two-wheeler. (O)

FOURGON — An old form of French luggage vehicle, afterward domesticated in England. It was employed in the pre-railway era as an attendant to precede a traveling COACH and convey the courier and baggage. In form it usually resembled a CABRIOLET body attached to a huge trunk. The term came to be applied to numerous types of BAGGAGE WAGONS, DELIVERY WAGONS, general service ARMY WAGONS, AMMUNITION WAGONS, and the small, hand-drawn BAGGAGE WAGONS used around railway terminals. (H)

FREIGHT WAGON — Generally, any WAGON used for hauling freight, such as a CONESTOGA. Specifically, the term was applied to the high, square-box, western, freight-carrying WAGON, with sides from three to five feet high. Several of these were frequently coupled together and drawn by a large number of animals. (S226)

FURNITURE WAGON — A WAGON, similar to either a light DELIVERY WAGON or to an EXPRESS WAGON, used for furniture delivery. (S226)

The Columbian FARM WAGON, on steel axles; wheels, 43-inch and 50-inch. (Columbia Wagon & Body Co., Columbia, Pennsylvania, ca. 1908.)

The FLEMING CARRIAGE. Wheels, 38-inch and 50-inch. (*The American Coach-Makers' Illustrated Monthly Magazine,* July 1855.)

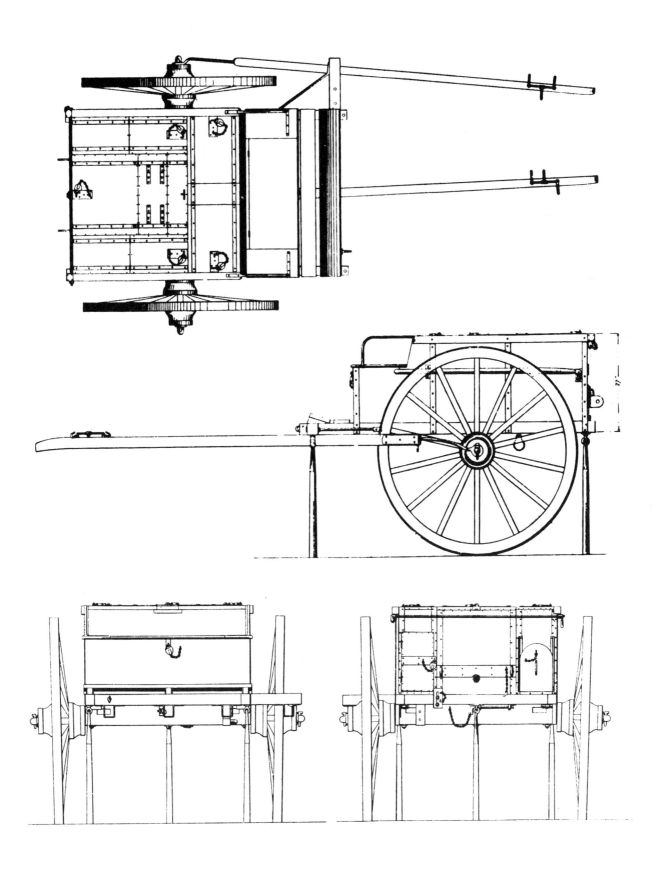

Cavalry FORGE CART. (From the *Annual Report of the Chief of Ordnance*, Washington, 1874.)

A variety of heavy FREIGHT WAGON that was popular in the West. Wheels, 44-inch and 54-inch; body, 12 feet long, and available from 36 inches to 52 inches deep. (Studebaker Bros. Mfg. Co., South Bend, Indiana, 1903.)

Unusually large FREIGHT WAGON used by the Fortuna Mining Co., of Yuma County, Arizona. The wheels are 5 feet and 8 feet in diameter, and the body is 20 feet long, 5 feet deep, and 3 feet 8 inches wide. It weighs 6,515 pounds, and has a carrying capacity of 12 tons. (*The Carriage Monthly*, November 1899.)

G

GADABOUT — An English variety of pleasure CART, adopted in America during the 1880s. It could carry four passengers. (H & O)

GALLYMANDER — A vehicle equipped with massive rear wheels, and a boom and winch, used for moving huge pieces of granite during quarrying operations in New England. (O)

GAMBO — A type of CART used in Wales and western England for harvesting crops. In use well into the twentieth century, the gambo is an open CART, having ends, or end stakes, and racks on the sides to prevent hay from catching in the wheels. Most were equipped with shafts for use with one horse, but the gambo of the early nineteenth century is said to have sometimes been constructed with a pole, and often drawn by oxen. (O)

GARBAGE CART (or WAGON) — A two- or four-wheel vehicle for carrying garbage. In the later period the body usually was made of iron or steel, so as to be leakproof. Most of these were dumping vehicles, dumping over backwards rather than from the bottom, so as to retain the leakproof feature. (O)

GENTLEMAN'S WAGON — See COAL-BOX BUGGY.

GEORGE IV PHAETON — England's George IV, formerly the exponent of the dangerous PERCH-HIGH PHAETON, was showing signs of age in 1824, at which time he requested his coachmaker to design a light, low-hung PHAETON. The result was a pony PHAETON weighing just over 300 pounds, supported by four elliptic springs, and having a very low CABRIOLET body and extremely small wheels, 21 and 30 inches in diameter. This badly proportioned vehicle was improved gradually during the next decade, until it became one of the most popular carriages for ladies' driving. It was provided with a rumble, and a graceful, flaring dash, the sides of which protruded beyond the body and continued downward so as to form the front fenders. The CABRIOLET body sometimes had caned sides, with a skeleton bottom side. Frequently drawn by two horses, it could also be used with one. Later, this PHAETON contributed some of its features to the design of the VICTORIA. (O)

GERMANTOWN (or GERMANTOWN ROCKAWAY) — A style of carriage that is believed to have been first built in 1816, by C. J. Junkurth, of Germantown (now a part of Philadelphia). A very marked influence of the COACHEE, also a product of the Philadelphia area, can be seen in the Germantown, which, in fact, appears to be the successor to the COACHEE. The body displays nearly identical lines, having side doors, and pillars supporting a roof that provides protection to passengers and driver alike. Seating for six passengers differs slightly from the COACHEE, the seats being arranged as in an extension-front ROCKAWAY, with the center seat facing toward the rear. Some of these carriages carried only four persons, the driver's seat then being located farther back inside the body, rather than at the front, or projecting slightly beyond as in the larger vehicles. It is not known just what suspension the first Germantowns employed, but subsequent ones used two elliptic springs. About 1830, a Jamaica, Long Island, carriage builder constructed the first primitive ROCKAWAY. As this vehicle evolved into a more sophisticated form it showed a notable resemblance to the Germantown. The latter then often borrowed the name of the Long Island carriage, and became known as a GERMANTOWN ROCKAWAY, but it appears likely that the ROCKAWAY, being a later development, actually acquired some of its characteristics from the Germantown. The features distinguishing a Germantown from a Rockaway were that the former had roof pillars at the front edge of the roof, while in the Rockaway, this pair was omitted; the ROCKAWAY generally had a wheel-house, while the Germantown did not; and the Germantown usually had an OGEE back, while the ROCKAWAY seldom displayed this feature except in some of the earlier versions. The Germantown frequently had a storm hood attached to the front of the roof to assist in protecting those on the front seat, since the seat was often set slightly beyond the body. Both vehicles could be used with either one or two horses. (C, H, & O)

GADABOUT. Wheels, 46-inch; body panels and dash, cherry varnished in the natural color; gearing, natural finish; iron parts, black; trimming, green York "terry." (*The Hub,* April 1882.)

GALLYMANDER at Vinalhaven, Maine, used in granite quarrying. (Photo from John R. Danley.)

Five-hundred-gallon wet GARBAGE WAGON built by The Haywood Wagon Co. of Newark, New York, about 1910.

Four-seat GERMANTOWN ROCKAWAY. Wheels, 44-inch and 50-inch. (Lawrence, Bradley & Pardee, New Haven, Connecticut, 1862.)

GERMAN WAGON — The name applied in England to the original form of BAROUCHE, when that carriage was first introduced into England from Germany around 1760. See BAROUCHE. (H)

GHARRY — A four-wheel carriage employed in India, particularly Calcutta, as a HACKNEY-CARRIAGE or type of OMNIBUS. It is essentially a rectangular SEDAN CHAIR on wheels, with paneled sides and sliding doors. Air and light are admitted by Venetian blinds set in the ends or sides. The gharry has a double roof, with an inch or an inch and a half space between, to better insulate occupants from the heat of the sun. (O)

GIG — The derivation of this word is unknown, but it is believed to imply a light and rapid motion; possibly applied to a two-wheel vehicle because of the ease with which it could be turned around. The gig was simply a better suspended and more sophisticated form of CHAISE, according to William Felton, who defined the vehicle in his 1796 *Treatise on Carriages*. He wrote, "Gigs are one-horse chaises, of various patterns, devised according to the fancy of the occupier; but, more generally, means those that hang by braces from the springs; the mode of hanging is what principally constitutes the name of Gig, which is only a one-horse chaise of the most fashionable make; Curricles being now the most fashionable sort of two-wheeled carriages, it is usual, in building a Gig, to imitate them, particularly in the mode of hanging. The Gig mostly hangs from the middle of the hind pillars, and is built as light and easy as possible; all one-horse chaises, that are neat and fancifully constructed, are named Gigs, and called by the name that the body is distinguished by; such as a step-piece, a tub-bottom, or a chair-back Gig &c."

In the nineteenth century the term became almost synonymous with CHAISE, particularly when the more modern forms of springs were employed. The DENNETT, TILBURY, and STANHOPE were modifications of the earlier gig. When a ventilated locker for carrying dogs was added to the gig, the vehicle became a DOG CART. See also DENNETT, TILBURY, and STANHOPE. (H, C, & O)

GIG CURRICLE — A GIG that is so constructed that it may be used with either one or two horses. Frequently this construction enables the shafts to be removed from their sockets, and placed together into another socket, to form a pole. The weight is kept comparatively light, similar to that of an ordinary GIG, so that it is not too heavy when used with a single horse. (F)

GIG, FANTAILED — A two-wheel vehicle popular in America during the first quarter of the nineteenth century, so named because of the fantail extension at the rear of the body. (H)

GIG, GORST — A peculiar style of GIG invented by a Mr. Gorst, of Liverpool, England. (H)

GIG, SUICIDE — A type of GIG used in Ireland, so called because of its excessive height. (O)

GIG, TUB-BODIED — A contemporary of the FANTAILED GIG, so named because of the shape of the body. It was popular in America during the first quarter of the nineteenth century. (H)

GILL — (1.) A pair of wheels supporting a framework on which timber is conveyed; it is the English equivalent to the American LOGGING WHEELS. In a few instances, the English applied the term to four-wheel vehicles that were used for the same purpose. (2.) A painter's TRUCK. (C, H, & O)

GINNY CARRIAGE — (1.) A small, strong carriage of wood or iron, used to convey the tools and materials used by railway workers. (2.) The term was also applied at one time to a variety of low-wheel BASKET-PHAETON. (H & C)

GLADSTONE — A four-wheel carriage, hung very low, having two inside seats, a driver's seat, and usually a rumble, with a calash-top over the back seat. (C & O)

GLASS COACH — Name originally given to private COACHES to distinguish them from the HACKNEY COACH. (C)

GOABOUT — A light driving WAGON, of simplified construction, equipped with a folding rumble. It was built by the Henry Hooker Company of New Haven Connecticut. (O)

GO-CART — (1.) A variety of VILLAGE CART. (2.) A framework on casters, used to support a child who is learning to walk. (3.) A type of CABRIOLET used in London during the early part of the nineteenth century, alluded to as follows in W. B. Adams' *English Pleasure Carriages*, "Old Chariot bodies were cut down and numberless transformations made . . . the truth is, they all more or less bear a strong resemblance to the vehicles called 'Go-Carts' which ply for hire, as a sort of two-wheeled stage, in the neighborhood of Lambeth, the deep-cranked axle being the principal distinction." (H)

GO-DEVIL — A United States term applied to a rough SLED used for dragging logs. One end of the log rested on the SLED, and the other end dragged behind. (O)

GOING-TO-COVER CART — A term sometimes applied to a DOG CART, because of the CART shown in the famed print by C. C. Henderson, entitled *Going-to-cover*.

GOLF WAGON — A modified square-box WAGON without top, having two or three seats, and used as a sporting vehicle. (O)

An Indian GHARRY. (*The Hub*, October 1896.)

GIG, on C-springs and braces, of about 1810. (Smithsonian collection.)

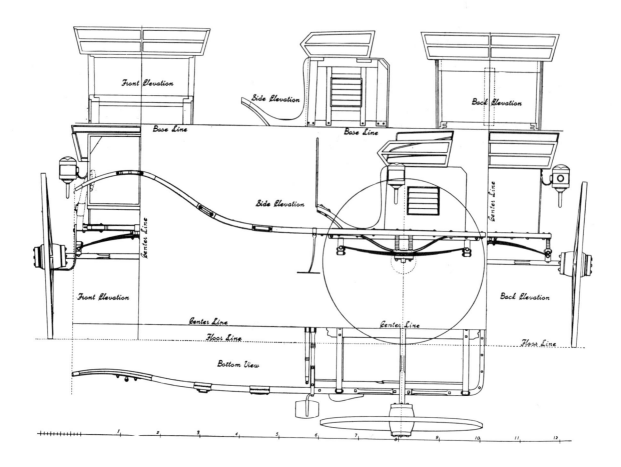

GIG. Wheels, 48-inch; panel, white; slat blinds, red striped with black; gearing, red striped with black; trimming, Bedford cord. (*The Carriage Monthly,* July 1898.)

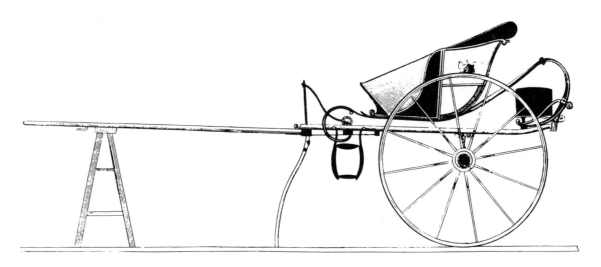

GIG CURRICLE with the convertible shafts that can be made into a pole. Since the pole needs to be longer than the shafts, the shafts are made extra long, and slide into three sockets on the gearing; when put together to form a pole, only two pole-sockets hold the pole at points B and C, thus extending the pole a greater distance forward. (W. Felton: *Treatise on Carriages,* 1796.)

GLADSTONE, with 40-inch and 48-inch wheels. (Lawrence, Bradley & Pardee, New Haven, Connecticut, 1862.)

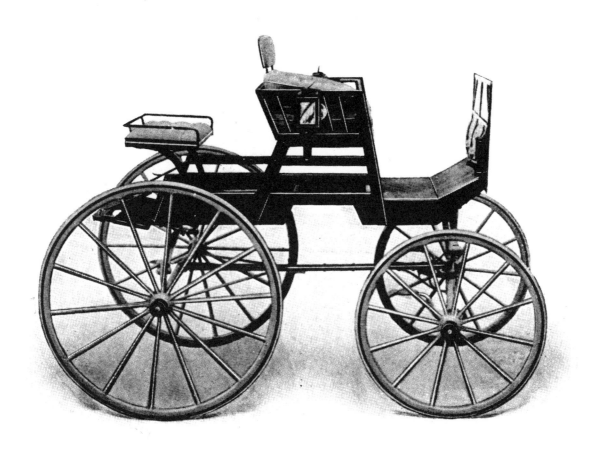

A cut-under GOABOUT with disappearing rumble. (Henry Hooker & Co., New Haven, Connecticut, ca. 1905.)

GOLF WAGON. Wheels, 42-inch and 46-inch; body, black; moldings, same color as gearing; gearing, coaching red or primrose; trimming, russet leather. (Cortland Wagon Co., Cortland, New York, 1902.)

GONDOLA – (From the Italian, signifying a boat.) A public carriage, so named because of its long, boat-shaped body. The term was applied in France, during the early part of the eighteenth century, to a type of COACH that was characterized by its unusual size. Roubo says, in his *L'Art du Menuisier Carrossier* (1771): "Next to the *Coches* [sic] the largest vehicles are the Gondolas, which are sometimes even more capacious than the former, at least as far as the body is concerned, being capable of seating a dozen persons." Straus, in his *Carriages and Coaches*, stated that the passengers sat on two longitudinal benches, facing one another, and two more sat on single seats at each end. This vehicle apparently was the predecessor of the OMNIBUS. (H & O)

GONDOLA-LANDAU – Same as CANOE-LANDAU. (H)

GONDOLA OF THE STREETS (or OF LONDON) – London slang for a HANSOM CAB. (H)

GOOLD CUTTER – See ALBANY CUTTER.

GOSPEL-WAGON – A large open WAGON used for open-air religious services. It had a roof supported by pillars, and longitudinal benches inside for the choir and speakers. A section of the side was hinged, and could be left down even with the floor of the WAGON so that the speaker stood outside the body of the vehicle, where he was protected from the sun by an awning. This WAGON, built in 1887 by Pearce & Lawton, of Washington, D. C., also accommodated an organ. (O)

GOVERNESS CART – A type of light, two-wheel WAGONETTE, developed in England late in the horse-drawn era. It has two longitudinal seats like a WAGONETTE, and entry is gained through a small door in the rear. The body is frequently of basketwork. (H)

GRASSHOPPER-CHAISE – An old form of CHAISE, or WHISKEY, so named because it employed the grasshopper spring, which Felton described in his *Treatise on Carriages* (1796): "All the framings form an agreeably connected line; it is exactly on the same principle as the Whiskey, which was built from them, having the springs, in the same way, fixed to the axletree, and the body united with the carriage, but only different in its shape; the framings of the body, being much wider, shows more panel, which extends to the shafts at the corners, and are arched up in an agreeable form, between the bearings; they have a more solid appearance than the Whiskey, and are, on that account, preferred by some persons, and, in particular, by those called Quakers, and for that reason are by some called Quaker's Chaises, and, by others, Serpentine, or Sweeped-bottom Chaises." (H)

GREEN MACHINE — The nickname applied to a vehicle that is claimed by some to have been the first BROUGHAM, so named because of its color. (H)

GROCER'S BOB-SLED — A grocer's delivery SLED composed of the body of the regular grocer's WAGON, mounted on bob-runners. (H)

GROWLER — A slang expression applied in England to a four-wheel CAB. (H)

GUAGA — The public omnibus of Cuba. As described by a correspondent of the Chicago *Daily News*, "in appearance it much resembles an antiquated street-car, or a very old-fashioned Daguerreotype gallery on wagon wheels. . . . Four little Cuban ponies are attached to it." (H)

GURNEY — Contraction of GURNEY-CAB, a variety of four-passenger public CAB invented and patented in 1882 by J. T. Gurney, of Boston, Massachusetts. The seats of this rear-entrance CAB are longitudinal as in an OMNIBUS, and the rear and side glasses are made to drop. *The Carriage Monthly* reported in 1904 that a modification of this CAB was being used rather extensively in the United States, and on the Pacific Coast the name was applied somewhat inaccurately to a variety of public CABS. (H & M)

GYPSY-WAGON — A WAGON intended to approximate a dwelling house on wheels, including conveniences for sleeping and preparing food, as used by gypsies, surveyors, traveling photographers, and other migratory parties. Many of these wagons were ingeniously and efficiently constructed with many small, hidden compartments, in order to afford the maximum possible storage space in a relatively compact vehicle. An attractive fireplace, complete with a hearth and decorative mantel, provided heat for cooking and bodily comfort, while still other wagons might, instead, be equipped with a small stove. Those used by the gypsies were usually elaborately decorated with gold leaf, and bright colored designs. Also known as a WARDO or VARDO. (H & O)

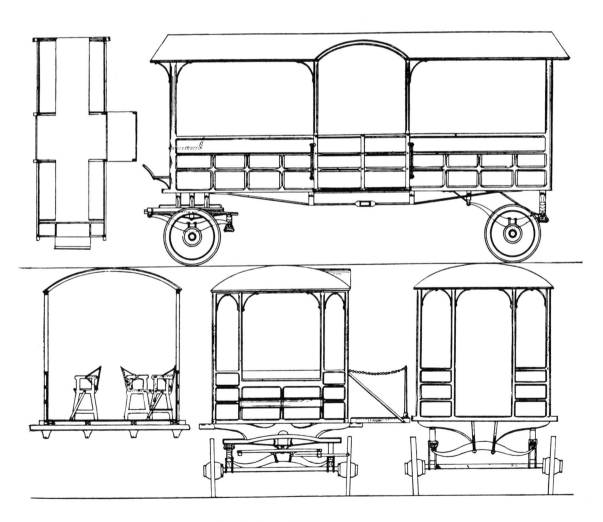

GOSPEL-WAGON. Wheels, 36-inch. (*The Hub,* August 1887.)

GOVERNESS CART. Wheel size, optional, 34-, 36-, 38- or 40-inch. Lower panels, black or dark green; basketwork, natural finish; gearing, carmine, yellow, or green with black striping; trimming, Bedford or whipcord. (J. A. Lancaster & Co., Merrimac, Massachusetts, ca. 1910.)

GRASSHOPPER, or three-quarter panel **CHAISE**; it is much like the *Whiskey* but shows more panel, and thus has a more solid appearance. (W. Felton: *Treatise on Carriages,* 1796.)

GURNEY CAB. Wheels, 52-inch; lower panels, olive; upper panels, black; imitation canework, white; fine gold and carmine stripes on body; gearing, black striped with two medium lines of orange, glazed with carmine; trimmings, maroon plush, with head-lining of terry. (*The Hub,* February 1884.)

H

HACK — An abbreviated form of HACKNEY, having the same meanings. It was also, on occasion, applied to the driver of a HACKNEY-CARRIAGE, and to a wornout horse. (H, C, & O)

HACKERY — Derivation unknown. (1.) A type of bullock CART used in India to transport goods. (2.) The term was also applied in western India, Ceylon, and Bengal, to a lighter vehicle that was used for carrying passengers. (3.) One of Frith's photographs, taken in Bombay, shows a two-passenger CART, with a light and highly decorated body, paneled to the arm-rail, and supported by platform springs on two heavy spoked wheels. To the pole are attached two small oxen. The corner pillars extend to a convenient height, and support a canvas top and roll-up canvas sides; a further canvas screen extends from the top, well forward over the oxen, and is supported in front by a bamboo rod resting on the pole. The driver is seated astride the pole. A coat of arms on the side panel, showing the British lion and unicorn, suggests that this is probably a modernized pattern. (H & O)

HACKNEY (or HACKNEY-COACH or -CARRIAGE) — (The derivation of this word has been disputed by etymologists, and its actual origin is unknown. The most likely source seems to be the French *haquenee*, or *hacquenee*, meaning an ambling horse. The word is also attributed to the Welsh, *hacknai*.) It is said that the first COACHES that ran for the conveyance of casual passengers started from Hackney (a London suburb) carrying fares into London. Pepys, in his Diary of 1662, writes of riding his *hacquenee* to Woolwich, evidently referring to a hired saddle horse. The term was also used to refer to an ordinary saddle horse, as distinguished from a draught horse or military horse. Eventually the term came to be applied to the public COACHES that had begun service in London around 1605. These first hackney-coaches were the second-hand COACHES of the gentry, and frequently retained the coats of arms of the original owners. The driver usually rode the near horse. Eventually, toward the end of the eighteenth century, some hackney-carriages became smaller and shorter, carrying only two passengers inside, with a third sharing the seat with the driver, though some drivers still preferred to ride the near horse. During the last days of the hackney-carriage — about 1825-1860 — the vehicles and service were left to deteriorate immensely, and filthy and dilapidated carriages were driven about the streets by dishonest and indifferent drivers. Hackneys were then succeeded by various forms of CAB. The term lingered for many years, however, being generally applied, particularly in the abbreviated form, HACK, to any vehicle kept for public hire.

The term was applied in a general way to horses, as stated above, but more specifically the Hackney Horse is a popular breed that originally developed in the eighteenth century. A harness horse, it is characterized by compactness, power, and speed. (H, C, & O)

HACQUET WAGON — A term applied to the four-wheel WAGON used in the military service of some nations to carry pontoons. The underframe of this vehicle is built similar to the crane-neck carriage, by which means it can be turned sharply without difficulty. In French, the term means DRAY. (W & O)

HALF-PLATFORM WAGON — See PLATFORM WAGON.

HAMMOCK CARRIAGE — A primitive form of wheeled bed or litter used by the Anglo-Saxons about the twelfth century, the passenger being carried in a hammock slung between two posts attached to the axles. This is apparently the first instance of a completely suspended body. (H)

HANDBARROW — A wheelless barrow borne by hand, as distinguished from a wheelbarrow. A STRETCHER and a BIER are forms of handbarrows. (H)

HAND-BUGGY — Another name for JINRIKISHA. (H)

HAND-CART — A small CART drawn or pushed by hand, often used around warehouses. (C)

HANDY WAGON, FARMER'S — This term was applied by some manufacturers to a farm-type running gear with

An Anglo-Saxon HAMMOCK CARRIAGE of the twelfth century. (Ezra M. Stratton: *The World on Wheels,* New York, 1878.)

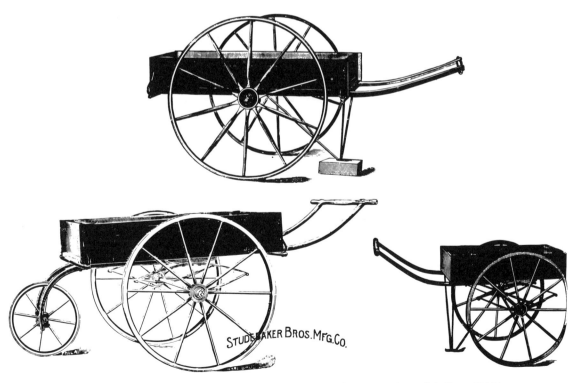

A variety of HAND-CARTS, all on 36-inch wheels. (Studebaker Bros. Mfg. Co., South Bend, Indiana, 1903.)

adjustable coupling pole. It was used by farmers for hauling logs or lumber, or with various types and sizes of platforms and boxes. An all-purpose farm gear. (O)

HANSON (or HANSOM CAB) — See CAB, HANSOM.

HANSOM, BOON — A patented variety of public CAB, with the driver's seat at the rear, but not centered, and entrance at either front or rear, originated by Messrs. Boon & Ries, of London, England. (H)

HANSOM, BROUGHAM — See BROUGHAM-HANSOM.

HARMA — A term which was, according to Stratton in *The World on Wheels*, applied both to the war CHARIOTS and peasant vehicles of ancient Persia. (O)

HARMAMAXA (or HARMAXEN) — A four-wheel carriage of ancient Persia. The body was long enough to enable one or two passengers to recline at full length. The top was supported by a number of pillars, and the sides were covered with awnings which could be opened or closed at the pleasure of the occupants. Some were enclosed with richly decorated hangings. These carriages were most frequently employed by women. (H & O)

HAY WAGON — An exceptionally long WAGON, 16 or 18 feet in length, with either a flat body or a shallow, boat-shaped body having flaring sides. A high frame, such as an "A" frame, is fixed at both ends, and these assist in holding and tying down large loads of hay. Also called a HAY-RIG. (H & O)

HEARSE — (From the French, *herse*, meaning harrow [twelfth century]; through the Italian, *erpice*; from the Latin, *hirpex*, meaning a large rake, used as a harrow.) The term was next applied to a triangular frame, similar in shape to a harrow, that was used to carry candles; then, to a framework that held candles over a bier at a funeral; then, to the framework supporting a pall over the bier; next, to the bier itself; and finally, to the carriage carrying the body to the grave. In both England and America the latter application occurred in the seventeenth century, possibly earlier in England. The American hearse undoubtedly evolved from a simple, flat-bed WAGON, ornamented with the draperies of mourning, eventually acquiring a flat roof supported by pillars. About mid-nineteenth century it began to have more sophisticated lines, the body became enclosed, generally with glass in the sides and ends, and the driver's seat was detached from the body and moved forward like that of a COACH, and was often ornamented with a hammercloth. Spring suspension was provided, usually by ellipticals in front and platform springs in the rear. The bodies were heavily ornamented with carvings and moldings, and the tops carried plated railings, urns, pompons, or other decorative devices. Lamps were large and ornate. The hearse was customarily black, though a smaller size, intended for children, was white. Inside were hung the draperies of mourning. (H, C, & O)

HEAVY CARRIAGES — The general name by which all standing top, or carriages having outside driver's seats, are designated. (C)

HECCA — A two-wheel CART used in India. Built without springs, it had low wheels and a wooden axle, and was generally provided with a canopy top. The shafts met just above the horse's shoulders, where they were attached to the pommel of the saddle. (H)

HELL-CART (or HELCART) — A slang expression used in mid-seventeenth-century England, referring to the much despised public COACHES then coming into use. The term first appeared in the work entitled *The World Runnes on Wheels: or Oddes betwixt Carts and Coaches* (London, 1623), by John Taylor, the "Water Poet," wherein he says: "An olde Coach is good for nothing but to cousen and deceiue people, as of the olde rotten Leather they make Vampires for high Shooes, for honest Country Plowmen, or Belts for Souldiers, or inner lynings for Girdles, Doggeschollers for Mastiffes, indeede, the Boxe if it were bored thorow, would be fittest for a close stoole, and the body would (perhaps) serue for a Sow to pigge in. If the curses of people that are wrong'd by them might haue preuailed, sure I thinke the most past of them had been at the deuill many yeeres agoe. Butchers cannot passes with their cattel for them: Market folkes which bring prouision of uictualls to the Citie, are stop'd, stay'd and hindred. Carts or Waynes with their necessary lading are debard and letted, the Milke-maydes ware is often spilt in the dirt, and peoples guts like to be crushed out being crowded and shrowded vp against stalls, and stoopes, whilst Mistris Siluerpin with her Pander, and a paire of cram'd Pullets, ride grinning and deriding in their Hel-cart, at their miseries who go on foote." Evelyn, in his *Character of England*, published in 1659, advises that Londoners still called coaches HEL-CARTS, the name thus applied by Taylor, thirty-six years earlier. See also, TORTOISES, FOUR-WHEELED. (H)

HERDIC — A peculiar pattern of OMNIBUS named after the inventor and patentee, Peter Herdic of Williamsport, Pennsylvania. It was used to some extent after 1880 in Philadelphia, and in several other large cities. The original Herdic was a four-wheel vehicle, with longitudinal seats and rear entrance like the conventional

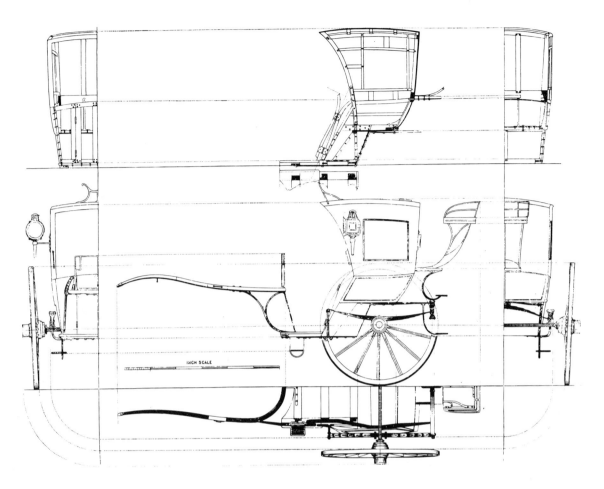

HANSOM CAB. Wheels, 55-inch. (*The Hub,* January 1889.)

HANSOM CAB, built by Brewster & Co., of Broome Street, New York. Wheels, 56-inch; body, blue, with black moldings striped light blue; gearing, blue striped black; trimming, blue cloth. (*The Hub,* August 1892.)

HANSOM CAB photographed in London in 1895. (Smithsonian photo.)

Persian HARMAMAXA. (Ezra M. Stratton: *The World on Wheels*, New York, 1878.)

166

Pennsylvania-style HAY WAGON (Lancaster and Lebanon Counties), available with either 16-foot or 18-foot body; 43-inch and 50-inch wheels. (Columbia Wagon & Body Co., Columbia, Pennsylvania, ca. 1908.)

OMNIBUS, but featured both a fixed front axle equipped with steering knuckles, and a novel method of suspension. Cranked axles were employed to lower the body, which was suspended on regular platform springs in the rear, and on two pairs of single-elbow springs in front. The first Herdics carried eight passengers, and were very compact and light in weight, reportedly less than half the weight of an ordinary OMNIBUS, so that they could be readily drawn by one horse.

    The term Herdic was also applied to a two-wheel, four-passenger vehicle that was similar to the one described above, and built by the same inventor. (H, M, & O)

HOBBY HORSE — Synonymous with DRAISINE.

HOODLUM WAGON — An extra WAGON that sometimes accompanied a CHUCK WAGON, used to carry water barrels and extra provisions, as well as the bedrolls and other duffel of the round-up hands. (O)

HOOK-AND-LADDER TRUCK — See TRUCK.

HOOPOE — Derivation unknown. This term was applied to a form of gypsy traveling WAGON that was used by anglers, artists, etc. (H)

HORSE-VELOCIPEDE — A one-wheel SULKY or ROAD CART built about 1881, probably by Joseph Haslip of Maryland. It is most unlikely that many were built. In 1894 an English carriage dealer also developed a variety of one-wheel ROAD CART in which he employed the wheel of a high-wheel bicycle, with the saddle over the wheel, and short shafts rigidly attached to the girth. Again, there is no evidence of the vehicle gaining any popularity. Numerous other fruitless efforts were also made at other times to design a one-wheeler. (O)

HOSE-CARRIAGE — A piece of fire apparatus, mounting a large reel on which fire hose is carried. Like many pieces of fire equipment, these vehicles often are highly ornamented. If mounted on two wheels, the apparatus is called a HOSE-CART. On four wheels it might be called a HOSE-CARRIAGE, HOSE-REEL, or HOSE-TRUCK. The term HOSE-REEL is applied to both two- and four-wheel types. (H & O)

HOSPITAL VAN — A large vehicle resembling an OMNIBUS, built in 1888 by the Studebaker Brothers Manufacturing Company, for the use of soldier invalids at the National Military Home, Ft. Leavenworth, Kansas. It

HEARSE. Wheels, 37-inch and 46-inch. (*The American Coach-Makers' Illustrated Monthly Magazine*, August 1856.)

HEARSE. Wheels, 39-inch and 48-inch; all black; carriage part striped with a 3/8-inch stripe of deep blue; trimming, black cloth; rails, hinges, and door handle, silver plated. (*The Carriage Monthly,* June 1891.)

is equipped with both rear and side doors, and has ten reclining chairs inside, of a type similar to those used in railway cars. It is not known whether more than one of these was built. (O)

HOTEL COACH — A public COACH employed expressly in cities, where it carried passengers from the railroad depot to the various hotels. Abbot, Downing's CONCORD COACHES were most often used for this purpose, the size being generally smaller than those used on the road, and lacking the leather covering to the rear baggage rack. The colors were less likely to be the usual red body on a yellow carriage, but were frequently green, orange, white, blue, olive, black, maroon, etc. (O)

HOURLY — A local term applied to a public COACH, descriptive of the schedule which it kept. Story (1815-1895), the sculptor, in an interview in the New York *Evening Sun*, of March 19, 1887, said: "I was going to catch the 'Hourly,' as the coach was called, that ran in those days every hour, between Boston and Cambridge, for it was long before the time of the omnibus and horse-car." (H)

HOWDAH (or HOUDAH) — (From Hindustani, *haudah*; Arabic, *haudaj*.) A litter or platform, usually equipped with a railing and canopy, that is fastened to the back of an elephant or camel, for conveying passengers. (H)

HOWELL GIG — A form of GIG PHAETON first designed by C. M. Britton, of New York, in 1872, and so named because the first vehicle of this type was built to the order of Howell, the celebrated New York photographer of that day. Its characteristics are the adaptation of a GIG body to a PHAETON gearing, and a general lightening of all the parts, to render it suitable for road use. (H)

HUB-RUNNER SLEIGH — A wheeled vehicle on which hub-runners have replaced the wheels for winter use.

HUG-ME-TIGHT — A slang expression applied in the United States to a BUGGY with a narrow seat upon which the two passengers were forced to sit snugly. This type of carriage was popular as a courting vehicle. (O)

HURDLE — A type of rude SLEDGE, once used in England to convey convicted traitors to the place of execution. (O)

HURDLE-CART — An English variety of DOG CART used for sporting purposes. (H)

HEARSE exhibited by Crane & Breed, of Cincinnati, at the World's Columbian Exposition in Chicago, Illinois. *The Carriage Monthly* evidently felt it was too elegant to carry the ordinary name, "Hearse," for they called it a *Massive Funeral Car*. Wheels, 38-inch and 50-inch; finish, black; hammercloth embroidered with gold work; mountings and lamps, gold plated. (*The Carriage Monthly*, September 1893.)

An Indian HECCA of the 1870s. (Ezra M. Stratton: *The World on Wheels*, New York, 1878.)

A two-wheeled HERDIC, with 60-inch wheels. Body above glass-frames, and all moldings, black; below glass-frames, Tuscan red glazed with carmine; glass-frames, natural finish; gearing, yellow striped with a 5/8-inch black line; trimming, red plush. (*The Carriage Monthly*, May 1881.)

An improved type of HERDIC, equipped with Ackermann steering. Wheels, 45-inch and 56-inch. (*The Hub*, September 1881.)

This larger size HERDIC was used in Washington, D. C., and apparently carried about fourteen persons, necessitating a pair of horses. It appears from the photograph that this model does not employ either the front suspension or the steering of the original *Herdic,* and is perhaps a *Herdic* in body style only. (Photo courtesy of Robert H. Renneberger, The Carriage Shop, Frederick, Maryland.)

This 4-wheel HOSE CARRIAGE was drawn by two horses and carried 1400 to 1800 feet of hose. (Amoskeag, Manchester Locomotive Works, New Hampshire, 1899.)

This two-wheel HOSE CART carried 500 to 700 feet of hose, and could either be hand-drawn, or attached to a steam fire engine. (Amoskeag, Manchester Locomotive Works, Manchester, New Hampshire, 1899.)

HOSPITAL VAN. Wheels, 43-inch and 55-inch; panels, wine color striped with gold; boot and moldings, black; gearing, wine color, striped black and gold; coat of arms of the United States on sides of boot. (*The Hub,* October 1888.)

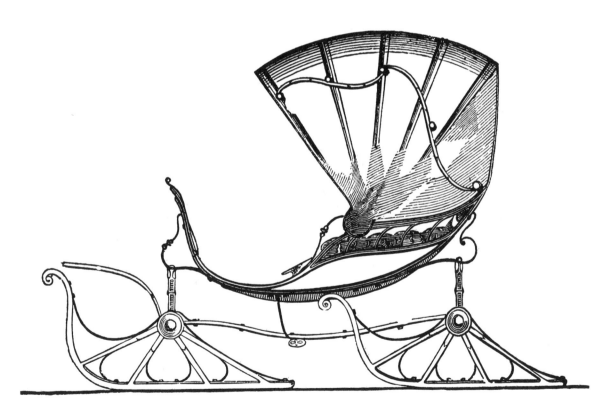

Hub-runners replace the wheels on this PHAETON to make the latter a HUB-RUNNER SLEIGH. (*The American Coach-Makers' Illustrated Monthly Magazine,* September 1856.)

I

ICEBOAT (or ICE YACHT) — A boat-shaped vehicle equipped with runners, and propelled over the ice by means of sails. (O)

ICE-CART — A CART that carries ice. This term is often misused for ICE WAGON, but it should be remembered that a CART has two wheels, instead of the WAGON's four. (H)

ICE-WAGON — Generally, any WAGON that carries ice, but more specifically, a WAGON especially designed for this purpose. It is usually of medium size, mounted on platform springs, and with a covered driver's seat. The body is enclosed with high sides and a roof which projects beyond both ends to prevent the sun's rays from reaching the ice. It is frequently equipped with a weigh-scale. The painting of ice wagons was often very colorful and decorative; many had immense oil paintings covering the sides, generally depicting polar scenes. (S226)

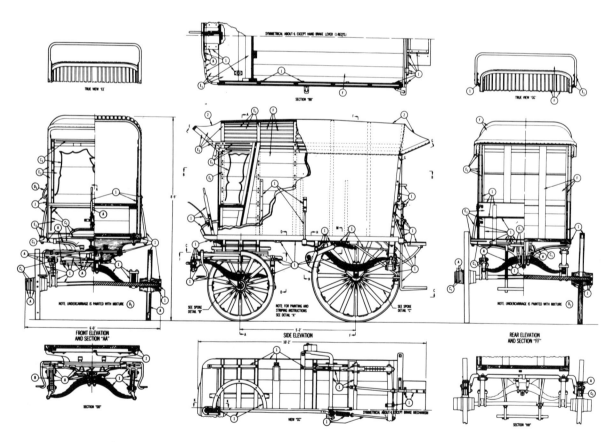

ICE WAGON of about 1910. (Smithsonian drawing from original vehicle at the Horse 'n' Buggy Museum, Gettysburg, Pennsylvania.)

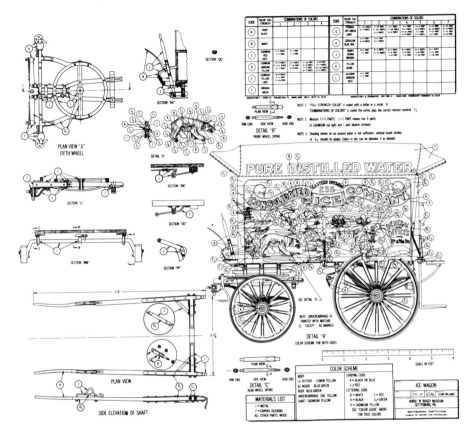

ICE WAGON.

J

JACK WAGON — An unidentified type of WAGON built by Amos Stiles, a carriage builder of Moorestown, New Jersey, 1812-1821, though there is no reason to believe that he was the only builder. The notations in Stiles's Day Book indicate that there were variations of this type, the main feature being that it was mounted on braces suspended from upright jacks, and it had a standing top, with curtains. Other known features, which may not have been characteristic of all jack wagons, were rear doors, sometimes with glass, three seats, inside linings, and a choice of shafts or pole. Probably there was much similarity between these and BOULSTER WAGONS, JERSEY WAGONS, and COACHEES, and possibly differed from the BOULSTER WAGON only by having jacks and braces. (O)

JAGGER WAGON — This term originated in the New York area, and is supposedly taken from the name of the maker. Originally it was applied to a square-box BUGGY or a light business WAGON having a body set on bolsters with imitation bolster-stakes and side braces. Lacking external springs, the seat was set on two side bars, or wooden springs, inside the body. By the 1860s the vehicle had acquired regular spring suspension, with either elliptics or some other form of light spring, and by 1900 the term became almost synonymous with SPRING WAGON in some shops. Canopy tops and curtains were often fitted to these WAGONS, while the seat, or seats, were sometimes so constructed that they could either be relocated or completely removed. (C, H, & O)

JAUNTING-CAR — A light, two-wheel open CART that developed in Ireland from the TROTTLE-CAR about 1815. The body is mounted on long double-elbow springs, and the two seats are arranged lengthwise over the wheels, so that the four passengers (sometimes six) sit back to back, with their feet resting on boards outside the wheels. A small driver's seat at the front is situated above the well that lies between the backs of the passenger-seats. Small parcels can be carried in this well. (C, H, & O)

JAUNTING-WAGON — A late American innovation on four wheels, but employing the seating arrangement of the Irish JAUNTING-CAR. (O)

JAUNTY (or JAUNTY-CAR) — An abbreviated form of JAUNTING-CAR.

JENNY — A small two or four-wheel carriage without a top, made to be drawn by ponies. (C)

JENNY LIND — A variety of BUGGY having a canopy top. Named after the popular Swedish singer, it was introduced around the middle of the nineteenth century, and enjoyed great popularity for many years. (O)

JERKER — See MAIL JERKER.

JERSEY WAGON — A type of traveling WAGON used in America during the late eighteenth and early nineteenth centuries, the term apparently being almost synonymous with the earlier form of COACHEE. Most common in the New Jersey-Philadelphia area, though it later migrated to more distant parts, it was described by Robert Sutcliff, who traveled the United States during 1804-1806. Referring to an illustration in his *Travels in Some Parts of North America*, he wrote: "The open carriages described in this plate are called wagons, and the best of them *Jersey Wagons*. They are made very light, hung on springs with leather braces, and travel very pleasantly. They are covered at the top with canvas painted. On the sides, there are three rows of curtains and those in the outer rows are likewise of canvas painted. Those in the middle row are of linen, and the inside curtains are green baize. The season and the weather regulate the use of them." Numerous other references indicate that this was a popular carriage.

By mid-nineteenth century the term was applied to a square-box carriage much like a CARRYALL, or JUMP-SEAT WAGON, and in this sense the term continued in use until nearly the end of the horse-drawn era. (H & O)

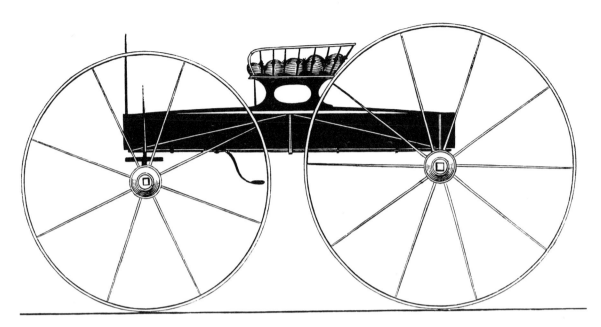

An example of JAGGER WAGON that adheres closely to the original pattern. The side-bar springs supporting the seat can be clearly seen above the body. Wheels, 46-inch and 50-inch. (*The New York Coach-Maker's Magazine*, July 1858.)

JAGGER WAGON with removable top, and seats that could be moved about or entirely removed. Note, too, the elliptic end springs. Wheels, 41-inch and 48-inch. (Lawrence, Bradley & Pardee, New Haven, Connecticut, 1862.)

JAUNTING CAR. (From Francis T. Underhill: *Driving for Pleasure*, 1897.)

JIGGER — A slang expression that was applied in New York City to a small, one-horse, street-railway car. It was operated without a conductor, having only a driver, who also handled fares and gave change. (H & O)

JINRIKISHA (or JINRIKSHA) — (From the Japanese, *Jin*, man; *riki*, power; *sha*, carriage.) A light, two-wheel, man-drawn vehicle, with a one- or two-passenger CABRIOLET-like body, used in Japan and other Asian countries, as well as in South Africa. One man was generally sufficient to pull the carriage, but when necessary another could be added in front, tandem fashion, and one or two could push from behind. Its prototype is said to have been invented about 1871 by Reverend Jonathan Goble, an American missionary living in Yokohama. About a decade later, an American carriage builder, James H. Birch, of Burlington, New Jersey, began to manufacture jinrikishas, and soon became one of their leading manufacturers. Birch offered a variety of styles, some resembling the PHAETON-BODY ROAD CARTS used in America. Nearly all were equipped with a falling top, and many had pneumatic tires mounted on wire-spoke wheels. Since the duties on carriages entering many foreign countries were much higher on new vehicles than on used ones, Birch saved this added cost for his customers by shipping many of his jinrikishas in mud-splattered, "used" condition. These vehicles survived World War II, but were finally banned in most countries during the following decade. (H & O)

JOB-WAGON — A New England term, applied to a light EXPRESS WAGON kept on hire for jobbing purposes. (H)

JOGGING CART — A late American variety of VILLAGE CART, used as such, or for exercising trotters. The name seems unfortunate, as it means literally, *jolting-cart*, implying a quality that is just the reverse of that which has given this CART its popularity. (H & O)

JOLT WAGON — A primitive, western, FARM WAGON, sometimes referred to as a CROCK-WAGON. (H)

JOSS — An old New England term for a high-back, primitive form of SLEIGH. (H)

JUGGERNAUT CAR — (From the Hindi *Jagannath*, meaning Lord of the World.) The sacred pyramidal vehicle of a Hindu idol representing the god Vishnu; pilgrims once sacrificed themselves under the moving wheels of this vehicle. (H)

JUMPER — A crude SLED or SLEIGH, having runners made of saplings, the ends of which turned upward and continued forward so as to also serve as shafts. The runners were held in position by several cross-pieces. The term was also used in New England to refer to a variety of light, two-passenger SLEIGHS. (C, H, & O)

JUMP-SEAT CARRIAGE — A carriage with one or more adjustable seats, so arranged that when not in use they can be hidden from view, by which means it can be converted from a two-passenger to a four-passenger vehicle, or vice versa. It is also called a SHIFTING-SEAT CARRIAGE. (H)

JUMP-SEAT WAGON — Any WAGON with the rear seat-board so constructed that it can be turned over. (H)

The American JAUNTING-WAGON built by the Parsons Vehicle Co., Columbus, Ohio. Wheels, 36-inch and 44-inch; track, 4 feet 8 inches; body, black with carmine toe boards striped black; gearing, carmine with bands, clips, etc., black, but no striping; trimming, Bedford cord. (*The Hub,* November 1901.)

JENNY LIND. Wheels, 45-inch and 47-inch; body, black; gearing, dark green striped with vermilion; trimming, dark green cloth. (*The Carriage Monthly,* February 1882.)

Robert Sutcliff's plate showing OPEN CARRIAGES, "the best of them Jersey Wagons." (*Travels in Some Parts of North America, 1804-1806,* 2nd edition, 1813.)

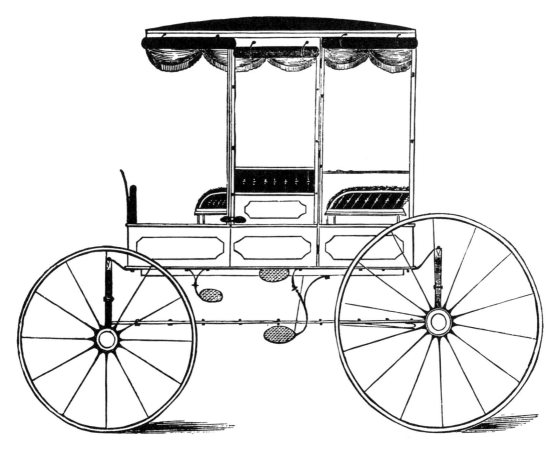

A late type of JERSEY WAGON, closely related to the *Carryall* and early *Depot Wagon*. Wheels, 42-inch and 49-inch; main panels, umber; smaller center panels, light lake; moldings, black. (*The American Coach-Makers' Illustrated Monthly Magazine,* May 1855.)

JAPANESE JINRIKISHA. (Smithsonian photo.)

A more sophisticated form of JUMPER. (*The Hub*, July 1882.)

KAGO — A form of PALANQUIN used by the Japanese.

KARIOL — The Danish spelling for CARIOLE.

KIBITKA — A rude four-wheel vehicle, commonly employed in Russia as a posting-wagon. In the absence of springs, seats are often formed by winding rope back and forth from one side of the top framework to the other, and sometimes cushions or hay are placed on these rope seats. Wooden bows support a cloth cover. (C, H, & O)

KIMBALL CUTTER — See PORTLAND CUTTER.

KITTEREEN (also KITTERINE or KITTAREEN) — Several derivations are offered for this word, but none can be substantiated. One attributes it to the town of Kettering, in Northamptonshire, England; and another to public vehicles bearing the name, operated by Christopher Treen (Kit Treen). The term was applied to several different two- and four-wheel carriages; unfortunately, few details of any of them are known. On the island of Jamaica, and in America, references have been found as early as 1737 until about 1800, applying the term to a variety of one-horse CHAISE, equipped with a top. Early in the nineteenth century, usage in western England referred to a four-wheel car or van that served as a public vehicle. Again, no details are found. In 1885, the term again was used in Jamaica, but this time the application was to a light, four-wheel BUGGY of American origin, provided with a movable hood. (H & O)

KNIFEBLADE — A slang expression applied to a BICYCLE. (H)

L

LANCE TRUCK — One of the vehicles making up a United States Army field telegraph train (ca. 1876). It was a dead-axle vehicle which carried from three hundred to five hundred lances (on which the wire was strung), and all the tools and insulators necessary to string ten miles of line. See BATTERY WAGON. (O)

LANDAU —(Name variously said to have derived from the Bavarian town of Landau, where a carriage of this type supposedly was built, or from Landow, the English designer of a later version of such a carriage.) The precise date of the Landau's origin is unknown, but it appears to have developed in Germany in the late sixteenth century; the name, however, was not applied until the eighteenth century. It migrated to England about the 1740s, but apparently did not rapidly follow the BERLIN into France, for Roubo, in his very complete work of 1771, did not mention a Landau or describe any vehicle with a let-down top. Felton, in 1796, described a Landau in detail, and gave an accurate plan of one; he complained that they were "heavier and more expensive than the common Coaches" (these circumstances were later overcome), and said that "the upper parts are covered with a black grain-leather which cannot be japanned, and of course does not look so well as fixed roofs," but he admitted that "they are the most convenient carriages of any."

A member of the COACH family, the Landau is essentially a COACH with a falling top. The body is like that of a COACH, from the belt-rail downward, but the top is made of two separate folding tops which, when up, lock together in the middle. The door is low, but arranged in the same manner as the lower part of a COACH door; a patent window fastener supports the glass frame when it is up, and does away with the extra frame originally used. The patented supports, hinged so as to be folded down upon the door when the glass is lowered, were perfected by Frederick Wood, of Bridgeport, Connecticut, without which improvement the Landau would not have become so popular.

Two principal types of Landau are the ENGLISH-QUARTER LANDAU, having a drop-center and angular lines, and the CANOE-LANDAU, the bottom line of which presents a continuous curve, resembling a canoe in profile. In England the former style was sometimes known as a SHELBURNE LANDAU, after the Earl of Shelburne, while the CANOE-LANDAU was known as the SEFTON LANDAU, after the Earl of Sefton. A variety of ENGLISH-QUARTER LANDAU, the BARKER-LINE LANDAU, is characterized by the Barker quarter, which has curved corners rather than angular, and was named after the London coachbuilding house of Barker, who first popularized the style. The upper quarters of a Landau are either of leather, when it is known as a LEATHER-QUARTER LANDAU, or of wood and glass forward of the hinge pillars, when it is known variously as a GLASS-FRONT, GLASS-QUARTER, or FIVE-GLASS LANDAU, according to the number and arrangement of the lights. Numerous devices have been adopted for supporting the wood and glass tops, the two most common styles being the Lohner, or parallel system, and the Kellner, or French system, named after the inventors, Jacob Lohner of Austria, and M. Kellner, of France. The parts of the framework of the Lohner system lay parallel with one another when folded down, while the parts of the French system lay at varying angles. The front glass drops before the top is lowered, and the quarter-glasses slide backward and drop into the door recesses, along with the door glasses. In many instances the weight of the tops is balanced by springs, to assist in lowering and raising them, and these are known as automatic tops.

The Landau retained its popularity until around 1900, and was one of the last heavy carriages to pass from use. It was one of the most difficult carriages to build, and countless devices were invented to support and secure tops. Originally designed for paired-draught, the demand for lighter carriages late in the nineteenth century caused some manufacturers to build extremely light Landaus that were suitable for use with one horse. (H, C, & O)

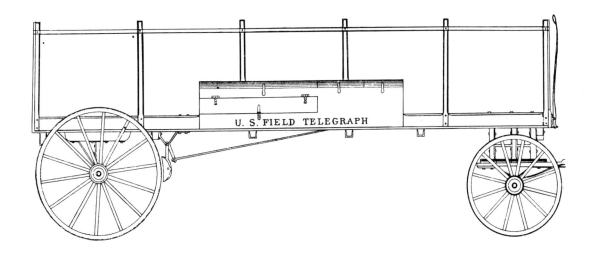

LANCE TRUCK of the United States Army field telegraph system. This vehicle carried 17-foot lances. (*Report of the Board of the U. S. International Exposition, 1876*, Washington, 1884.)

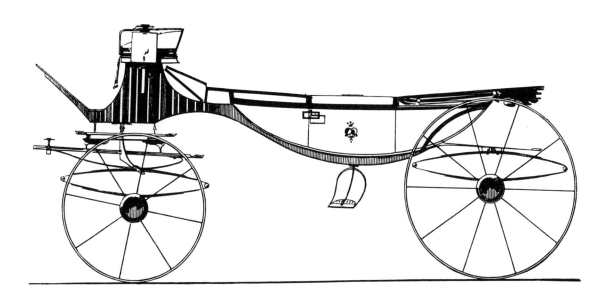

This glass-front LANDAU, with canoe-shaped body, was of French design, and was not nearly so common in the United States as the English style with the drop-center. Platform springs were much more commonly used on vehicles of this class, than the elliptic springs shown here. Wheels, 37½-inch and 46-inch; body, dark green; moldings, black with fine line of carmine; gearing, green with two fine lines and medium center line of carmine; trimming, green skins and cloth. (*The Hub,* March 1881.)

LANDAU made on the Lohner system. Wheels, 38-inch and 48-inch; body and gearing black, the latter striped with a 1/8-inch line of vermilion with a fine line of same on each side, glazed with carmine; trimming, green cloth. (*The Carriage Monthly*, November 1881.)

LANDAU built on the Kellner, or French plan. Wheels, 36-inch and 44-inch; lower panels, dark green; upper body and moldings, black; gearing, dark green striped with a ½-inch line of light green with two fine lines on each side; trimming, back, cushions, falls, and doors, green morocco or goat skin, and the balance of green cloth. (*The Carriage Monthly,* November 1882.)

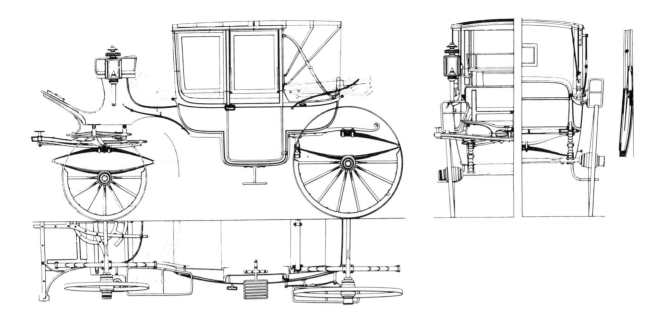

Five-glass LANDAU. Wheels, 36-inch and 46-inch; body panels, Nile green; moldings, black; gearing, green with carmine stripes; trimming, dark green cloth. (*The Hub,* August 1891.)

LANDAU exhibited by Brewster & Co. at the World's Columbian Exposition in Chicago. Wheels, 37-inch and 48-inch; body, ultramarine blue; moldings and boot, black with 1/8-inch pale blue strip on molding; gearing, ultramarine blue striped with pale blue; hub bands, black; trimming, blue morocco for lower, and blue cloth for upper. (*The Carriage Monthly,* July 1893.)

LANDAU, DEMI — Synonymous with LANDAULET. (H)

LANDAU-ROCKAWAY — See ROCKAWAY-LANDAU.

LANDAULET — A carriage bearing the same relationship to the LANDAU that the Chariot or Coupe does to the full Coach. It is simply a Coupe with a falling top; or, more properly, a Landau *coup'd* (cut in half), seating two instead of four passengers. A full Calash-top covers the rear seat, while the front portion is a framework of wood, having glasses made to drop. The front folds down, by various means such as the French system, or it may be entirely removed and the Calash-top used alone. Like the Landau, the Landaulet is built in different sizes, some being sufficiently light so that they can be drawn by one horse. The terms *Landaulet*, and *Demi-Landau* are used synonymously. In 1796 Felton described one of these vehicles, using the two terms interchangeably, and his detailed account shows that they were essentially the same carriage. (H, C, & M)

LANDAULET-ROCKAWAY — See ROCKAWAY-LANDAULET.

LAND-RAFT — Used synonymously with SLEDGE by William Bridges Adams, in his *English Pleasure Carriages*. (H)

LAND-SHIP — See BARCO DE TIERRA. (H)

LAWRENCE WAGON — A form of BUGGY devised by James W. Lawrence, of the firm of Brewster & Company, of Broome Street, New York City. (H)

LEAMINGTON CART — A variety of DOG CART. (H)

LECTICA EQUESTRE — A horse-litter. (H)

LIGHT CARRIAGES — A general term for carriages, with or without top, but not paneled up, that are suitable for use with one horse. (C)

LIMBER — The detachable forepart of a gun carriage (or other two-wheel military vehicle such as a BATTERY WAGON, COMBAT WAGON, TRAVELING FORGE, etc.), the addition of which converts the carriage into a four-wheel vehicle, for draught purposes. It consists of an axle, two wheels, and a pole or shafts. One or two ammunition chests are often mounted on the limber, providing seats for several artillerymen. The lunette or ring on the trail, or stock, of the other carriages fits over the pintle or hook of the limber.

Limber is also an obsolete English term for the shaft of a wagon or carriage. In this sense it was usually used in the plural. (H & O)

LIMOUSINE — Originally the name given to a caped cloak worn by the natives of the French province of Limousin. Later the term was applied to an automobile body having a closed compartment behind that carried several passengers, with a driver's compartment that was open but under roof. In 1902 the editors of the French trade journal, *Le Guide du Carrossier*, suggested that since automobile designers borrowed ideas from the carriage trade, the idea should also be reversed, with the carriage builders borrowing automobile designs. Accordingly, a two-wheel, horse-drawn LIMOUSINE was designed, having a body just like the auto, but minus the hood, and having its door in the rear. No evidence has been found to indicate that the idea gained acceptance. (O)

LINEIKA — A Russian term, defined in Tolstoy's *Anna Karenina* as a four-passenger DROSHKY. (H)

LITTER — (From the old French, *litiere*; low Latin, *lectaria*; Latin, *lectus*, meaning a bed.) A wheelless conveyance that generally carried but one person. First borne by human carriers, they were introduced into Rome from the East, and were most frequently used by women and royalty. Ralph Strauss, in his work, *Carriages and Coaches*, stated, "Julius Caesar restricted their numbers, and in the reign of Claudius permission to use them was granted only as a particular mark of royal favor." In Rome a larger type of litter soon developed, known as a BASTERNA, that was carried on long poles between two mules or horses in tandem. Frequently these were covered by a sort of canopy, with openings in the sides, and the passenger rode in a semireclining position. The use of litters spread throughout the Continent during the Middle Ages, and finally to England about the year 1200. During these years the litter often served in place of a wheeled carriage as the vehicle of state.

At a later date litters were intended primarily for the sick or wounded, again being carried either by men or horses. Litters were then devised that could be suspended from the sides of a single animal, though the two-horse type also remained in limited use. See also CACOLET. (H, R, & O)

LIVERY HACK — A term sometimes applied to a MOUNTAIN WAGON. (O)

LOG (or LOGGING) SLED — A stout SLED, of simple construction, with fittings for lashing and dragging logs. (H)

LOG (or LOGGING) WAGON — A heavy wagon gear used for hauling logs. The logs are placed on bunks, or false bolsters, attached over the regular bolsters. The bunks usually raise the load higher than the tops of the

LANDAULET built by William Dixon of Toronto, Canada. Wheels, 36-inch and 46-inch; body panels, ultramarine blue over Prussian blue; moldings, black striped with carmine; gearing, like panels, striped with carmine; trimming, blue satin. (*The Hub,* September 1882.)

LANDAULET with a straight front. Wheels, 36-inch and 44-inch; door panels and rear quarter, green striped with fine lines of carmine; gearing, lighter green striped with one broad and two fine lines of black; trimming, green cloth. (*The Hub,* January 1889.)

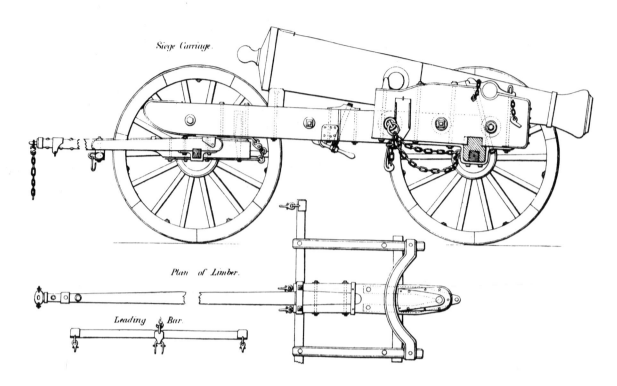

LIMBER with the trail of a seige carriage attached, in the traveling position. (From John Gibbon: *The Artillerist's Manual*, D. Van Nostrand. New York, 1863.)

The horse-drawn LIMOUSINE as shown in a 1902 number of *Le Guide du Carrossier*. Wheels, 49-inch; lower panels, black; middle and upper panels, maroon; spindles, red; gearing, red; striping, red on body and black on gearing.

French HORSE-LITTER of about the fourteenth century. (Ezra M. Stratton: *The World on Wheels,* New York, 1878.)

British MULE-LITTER used in Crimes. (*Medical & Surgical History of the War of the Rebellion,* Part III, vol. 2, Surgical History, Washington, Government Printing Office, 1883.)

wheels, which are generally low, and in some cases extend out over the wheels to facilitate rolling the logs off the side of the WAGON. (S226)

LOG (or LOGGING) WHEELS — A large pair of wheels, 8 feet or more in diameter, with axle and pole, used in logging operations. The wheels are placed astraddle the log, near its center, and with the pole raised to a vertical position. The log, or logs are chained to the axle or bolster at a point off the centerline of the axle, so that when the pole is pulled down to a horizontal position, the logs are raised off the ground; if supported slightly ahead of center, the rear ends of the logs will trail on the ground. (O)

LONG-WAGON — An early form of English traveling carriage, used during the fourteenth century. When used by persons of high rank, these WAGONS were richly decorated and comfortably lined with cushions and other trimmings. A number of curved bows supported the top covering. The term is believed to be synonymous with WHIRLICOTE. (H & O)

LORRY — An English term, commonly applied to vehicles in the classification of TRUCKS and DRAYS. In the midland counties of England they are called DRAYS, but in the north the term LORRY is most often used. They are heavy vehicles, with 5- or 6-inch tires, often carring 7 or 8 tons. Generally they resemble a DRAY, but in some cases they are lighter, with removable sides and back, and then resemble a TRUCK. Also called LURRY.

Vehicles built in the United States were sometimes given this name also; for example, one known as a RAILROAD LORRY, which resembled English LORRIES, with very low sides, and ends inclining gently outward. (O)

MOUNTAIN WAGON also known as a LIVERY HACK. Wheels, 42-inch and 48-inch; body, black; gearing, red or yellow striped black; trimming, black leather. (*The Carriage Monthly,* April 1891.)

LOG WAGON with Bernhard rear brake. Wheels, 41-inch front and rear, the usual extra height of the rear wheels being dispensed with so that the logs could be rolled on and off over the wheels. (Studebaker Bros. Mfg. Co., South Bend, Indiana, 1903.)

Two pairs of LOGGING WHEELS, shown about 1897 in front of the Macclenny, Florida, blacksmith shop of J. M. Isenberg (third from left).

English LONG-WAGON of the fourteenth century. (Ezra M. Stratton: *The World on Wheels*, New York, 1878.)

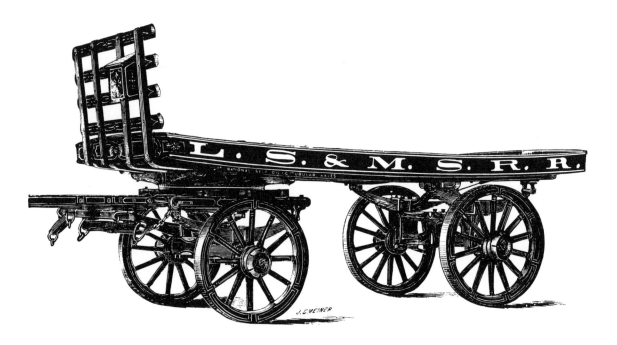

Railroad LORRY, borrowing both the name and some of the style from England. (E. Chope & Sons, Detroit, Michigan, 1886.)

LUMBER-WAGON — A heavily built, springless WAGON, intended for hauling timber or lumber, and for general farm use. A long reach is secured to the hind axle by a bolt or pin rendering it easy to lengthen or shorten the distance between the two axles. For hauling lumber it is frequently used without a body, often with bunks or false bolsters put on over the regular bolsters. (H & C)

LUNCH-WAGON — A traveling WAGON used particularly in the larger cities late in the horse-drawn era to serve food and drink to persons in areas where regular lunchrooms were not available. Some served sandwiches, milk, coffee, etc., while others served only milk or coffee with muffins or donuts. They were equipped with a small kitchen containing closets, shelves, oil stove, coffee urn, and ice box, together with an eating area with counter and stools. Some lacked this latter feature, and served only through a window to persons on the sidewalk.

There were also special purpose WAGONS such as the ICED-MILK WAGONS, fitted with hand-operated milk shake machines, and WAFFLE WAGONS, equipped with stoves and waffle irons. Lunch-wagons were frequently painted white, and many were ornately striped and decorated with landscape scenes. (O)

LURRY — Same as LORRY.

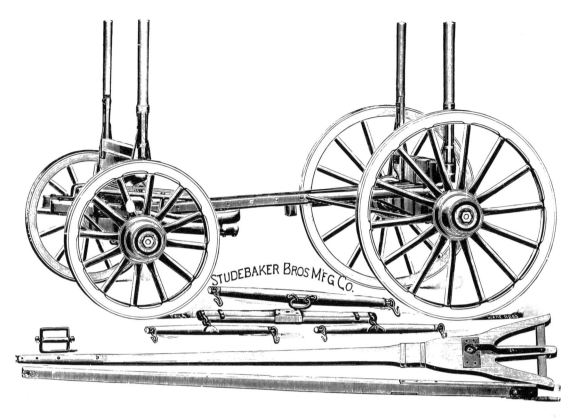

City LUMBER-WAGON, on 38-inch and 52-inch wheels. (Studebaker Bros. Mfg. Co., South Bend, Indiana, 1903.)

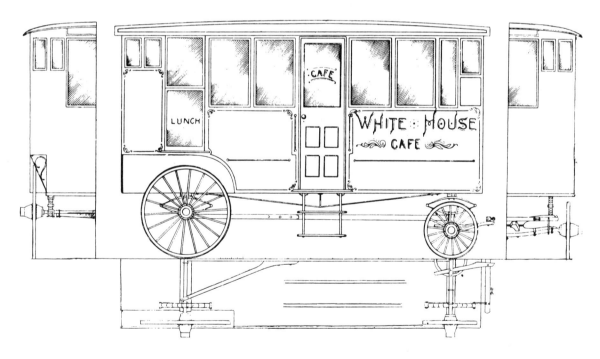

New England LUNCH-WAGON. Wheels, 29-inch and 48-inch; body and gearing, white with gold and red stripes and lettering; interior, stained to imitate cherry. The letters shown on the floor plan represent: A, stools; B, counter; C, closets; D, trap shelf; E, coffee urn; F, drawers at each end; G, case with 7 drawers. 39 inches high; H, shelf, 19 inches from floor; J, oil stove; K, shelf, 39 inches from floor; L, shelves for pots and pans; M, trap. (*The Hub,* June 1894.)

M

MADEIRA BULLOCK SLEDGE — An old ox-sledge peculiar to the Island of Madeira, the body of which is made of willow, stoutly braced, placed on runners, and used on dry ground. *St. Nicholas Magazine* for March 1887, states: "Everything is drawn on runners in Madeira. At that time there was but one wheeled carriage on the Island. The greater part of the people walk; a few ride in carts, a few in hammocks borne on men's shoulders, and for long distances they ride horseback. Merchandise is drawn on sledges. . . . The oxen were small, but handsome and well cared for. Occasionally the boy would stop for the cart, and allow first one runner and then the other to pass over a little bag of grease which he carried in his hand. In this way the runners are greased so that they may glide along easily, and this is what makes the streets so slippery." (H)

MAGNETIC CAR — A four-wheel vehicle used by the ancient Chinese, described as follows in Goodrich's *Columbia*, page 31: "In the year 2700 B. C., the Emporer Wang-ti placed a magnetic figure with an extended arm on the front of carriages, the arm always turning and pointing to the South, which the Chinese regarded as the principal pole." (H)

MAIL JERKER — A light, open PASSENGER WAGON hung on thoroughbraces, made by the Abbot, Downing Company. The body is similar to that of a SPRING WAGON, having two seats, to carry four passengers, including the driver. It is equipped with a dash, a standing top, and a luggage rack behind. (O)

MAIL WAGON — (1.) A small, enclosed square-box WAGON with sliding doors, used for rural delivery of mail. (2.) A closed, paneled DELIVERY WAGON used for carrying mails. (3.) A type of screened BAGGAGE or EXPRESS WAGON, generally used for carrying parcels, or bulk mail. (4.) Generally, any WAGON that carries mail. (5.) A variety of PASSENGER WAGON that was sometimes known as a WESTERN PASSENGER WAGON. (O)

MALLEPOSTE — An Austrian term, synonymous with DILIGENCE. It is also called an EILWAGEN. (H)

MALT WAGON — A water-tight WAGON used by brewers for conveying malt. (H)

MALVERN CART — A variety of DOG CART. (H)

MANGONEL — An ancient war-engine for throwing stones or javelins, sometimes, but not always, mounted on wheels. (H)

MANTELET — An ancient war-engine, being a shelter set on wheels, used in sieges by being driven before pioneers to protect them from the small shot of the beseiged. (H)

MANURE SPREADER — A specially designed WAGON used to haul and spread manure on the fields. A rotating mechanical device on the rear of the WAGON spreads the manure evenly on the ground as the WAGON is drawn across the field. (O)

MEDICAL CART — An ordinary CART, equipped with a low cover, used to carry medical supplies for the army in the field. (O)

MENAGERIE WAGON — A WAGON that is essentially an animal cage on wheels, used by circuses and often highly decorated. (H)

MEXICAN-QUARTER COACH — An Americanism introduced about the year 1850, by Wood Brothers, of New York, and applied to the quarter panel of a COACH, where two curves intersect. (H)

MILK WAGON — A type of light DELIVERY WAGON, usually on three elliptic springs, used to deliver milk to homes. Also used for the delivery of bakery products, this type of WAGON is generally characterized by a closed, paneled body and sliding doors. It follows much the same construction as a MAIL WAGON, though the latter is considerably smaller. (S226)

MILL WAGON — See BUILDER'S WAGON. (S226)

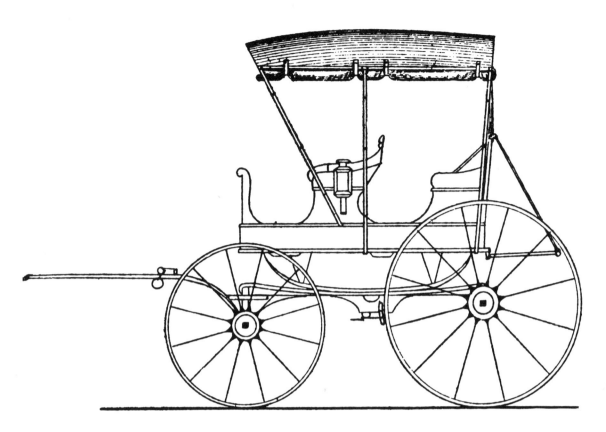

MAIL JERKER. (Abbot, Downing Co., Concord, New Hampshire, ca. 1885.)

Rural MAIL DELIVERY WAGON. The lower panels, and front and rear panels are of wood, while the upper side panels are of duck. The front light is hinged at top and swings inward; the rear curtain rolls up, and the doors slide. Inside is fitted with drawer and pigeon holes as shown by inset, and there is a drawer under the seat, also shown. Wheels, 40-inch and 44-inch; body, blue below and white above; gearing, red. (Studebaker Bros. Mfg. Co., South Bend, Indiana, 1903.)

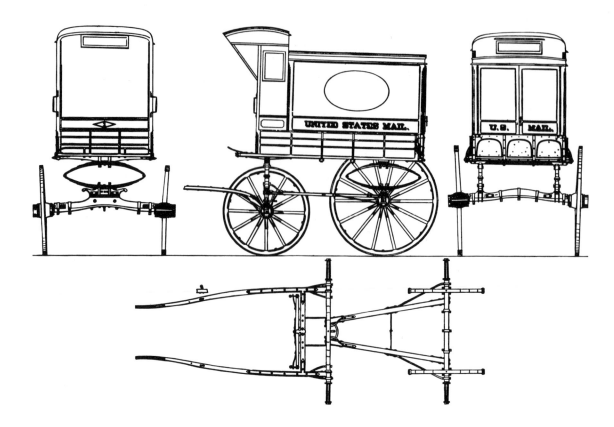

One-horse MAIL WAGON. Wheels, 40-inch and 50-inch; gearing, light cream color striped with one broad line of vermilion and two fine lines of blue; lower body, including upper rail, blue, finely striped with red and white; lettering panels on sides and rear, white; main side panels and doors, red; lettering, gold leaf with double carmine shades; oval contained crest showing post-boy, train, steamboat, and telegraph. (*The Hub,* May 1880.)

MILORD (or MYLORD) — A term applied on the Continent to the CAB-PHAETON. Also see CAB-PHAETON and CABRIOLET. (O)

MINIBUS — See BOULNOIS CAB.

MOCK CALECHE — A COACH body, the upper front quarters, front end, and upper portion of door being removable as those on a BAROUCHE or CALECHE. The back end is paneled instead of having a falling top. (C)

MONACHUS — A very light, two-wheel Roman vehicle somewhat similar to a CISIUM. (O)

MONALOS — A variety of CHAISE devised by a resident of Boston, Massachusetts, in 1867, described as follows in Ezra Stratton's *The World on Wheels*, page 461: "During the year (1867) a distinguished Boston surgeon invented and ordered built for his own use, a vehicle which a friend describes as a sort of Chaise, with wheels five feet in diameter, cranked axle, thoroughbraces, and wooden springs, strapped to the shafts in a novel manner. The body, a sort of Buggy kind, was fitted with a top, having a place for professional instruments made at the back. The learned doctor called it a Monalos, from a Greek word signifying *alone* as it was of the Sulky class intended for one passenger only." (H)

MONOCYCLE — A one-wheel vehicle patented on April 26, 1832, by Charles Hamond, an English civil engineer. The term is also somewhat synonymous with UNICYCLE, though it is not often used this way. (H)

MOORE'S HIGH-WHEEL CARRIAGES — In 1771 Francis Moore succeeded in showing, to the wonderment of Londoners, that large wheels enable vehicles to roll with comparative ease. The journals of that year contain many such announcements as the following: "On Saturday evening, Mr. Moore's new constructed coach, which is very large and roomy, and is drawn by one horse carried six persons and the driver, with amazing ease, from Cheapside to the top of Highgate Hill. It came back at the rate of ten miles an hour, passing

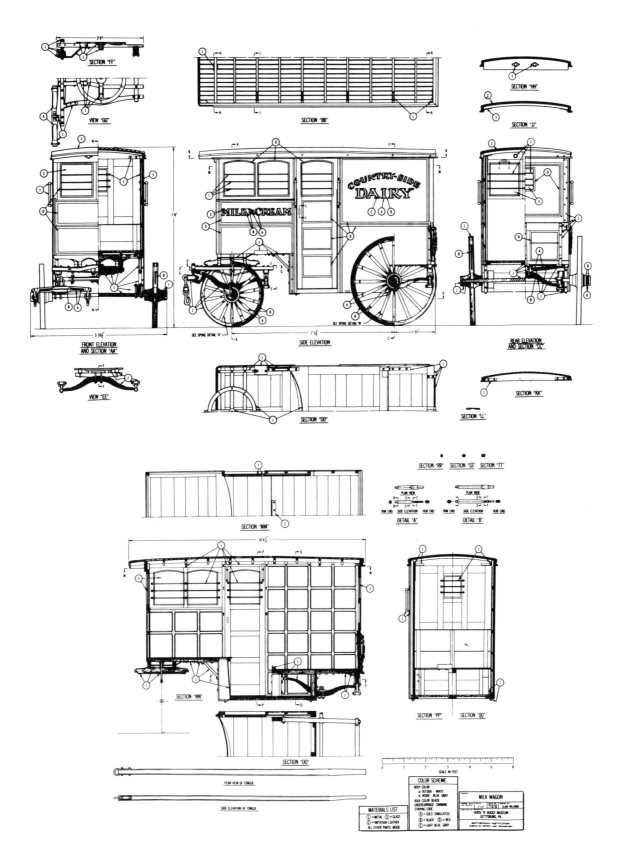

MILK WAGON of about 1915. (Smithsonian drawing from original vehicle at the Horse 'n' Buggy Museum, Gettysburg, Pennsylvania.)

coaches-and-four and all other carriages it came near on the road." Another account contains the following description of the vehicle: "Mr. Moore has hung the body, which is like that of a common coach reversed, between two large wheels 9½ feet in diameter, and draws it with a horse in shafts. The passengers sit sideways within, and the driver is placed upon the top of the coach." Mr. Moore was presented to George III, who is said to have "passed great commendations on the vehicle." Moore's COAL CARRIAGE, similarly characterized by the large wheels, is also frequently noticed in the same year, 1771.

It is believed by some that Hansom's 1834 vehicle was influenced to some extent by these designs of Moore, who obtained patents on later versions of his carriages in 1786 and 1790. (H & O)

MORAY CAR — A low-hung, four-passenger, dos-a-dos, two-wheel vehicle introduced in England by G. A. Thrupp about the middle of the nineteenth century. (O)

MORAY PHAETON — A variety of PARK PHAETON popular in England about 1850. (H)

MOREG — An ancient threshing machine. (H)

MORTAR-WAGON — A vehicle designed for the transportation of seige-mortars and their beds, or of guns or large shot, and shells. A LIMBER similar to the one for seige-gun carriages is used with it. The body consists of a platform of rails and transoms resting on an axletree. The stock is formed by prolonging the two middle-rails. The side-rails projecting to the rear form supports for the pivots of a windlass roller. This roller is used to load guns and mortars on the WAGON by drawing them up the stock. A muzzle-bolster on the stock near the LIMBER, and a breech-hurter near the hind part of the WAGON, are provided and used when long pieces are transported on it. Mortars are usually carried mounted on their beds. (W)

MOUNTAIN WAGON — A PASSENGER WAGON used extensively in the western states during the post-Civil War period. They were equipped with two, three, or four seats, and were built much like a SPRING WAGON, but were generally somewhat heavier. Suspension was by thoroughbraces in some instances, though many of the late WAGONS employed elliptics, side springs, or a combination of the two. Most were provided with a braking mechanism. Some were open, while others had standing tops, and many had a baggage rack on the rear. (H, M, & S226)

MUD WAGON — A term frequently applied to the various types of PASSENGER WAGONS, being a somewhat inferior class of STAGECOACH. Since they were less expensive than the true COACHES, they were often used on routes where the service was more severe, and were generally mud-splattered — thus the name. This term was in use at least as early as the 1830s. See PASSENGER WAGON. (O)

MURPHY WAGON — A heavy western FREIGHT WAGON built by the wagon-building establishment of Joseph Murphy, St. Louis, Missouri, which operated from 1826 to 1894. The WAGONS for which he was most famous were built largely during the several decades around the middle of this period, and were designed to carry two or three tons. (O)

MILORD. Wheels, 35-inch and 42½-inch; body, black striped light green; upper seat panel and gearing, light green; gearing, striped black. (*Le Guide du Carrossier*, 1898.)

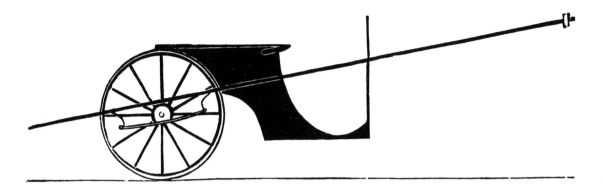

Hamond's MONOCYCLE of 1832. (Ezra M. Stratton: *The World on Wheels*, New York, 1878.)

One of Francis Moore's PUBLIC VEHICLES. (H. C. Moore: *Omnibuses and Cabs*, 1902.)

Four-spring MOUNTAIN WAGON employing both side-springs and elliptics. Wheels, 42-inch and 48-inch; body, dark green; gearing, red or yellow; trimming, leather. (Studebaker Bros. Mfg. Co., South Bend, Indiana, 1903.)

N

NIB (or NEB) — An English term for an apparatus similar to LOGGING WHEELS, and used for the same purpose. Sometimes called a PAIR OF WHEELS or a TIMBER BOB. The term was also applied in England to the shaft or pole of a WAGON, which possibly explains the first-named application, for the logging apparatus was but a pair of wheels, an axle, and a pole, or *nib*. (U & O)

NIGHT-BINER — A slang expression for a public carriage engaged in night service. (H)

NIGHT-HAWK — A New York slang expression for a public CAB engaged in night service. (H)

NIGHTSOIL-CART — A water-tight CART for conveying nightsoil, or the contents of necessaries, so named because its collection and removal in cities was done at night. (H)

NODDY — A public vehicle that was popular in Ireland and Scotland during the last half of the eighteenth and the early nineteenth centuries. Descriptions of the vehicle vary so widely over the years that it appears that the term was rather loosely applied to many types of public vehicles used in these countries. While a few four-wheel carriages seem to have been called noddies, most sources agree that the true noddy was two-wheel — a sort of CHAISE — drawn by one horse. An 1825 source, however, states that one type was drawn by two horses abreast, one of which carried the driver. The one-horse variety carried two or three passengers and had a stool for the driver in front. Early noddies had bodies suspended on braces, which gave them a nodding motion, thence the name. A later reference shows that the term was still in use in 1889, and describes this later noddy as a cumbrous box on two wheels, with a rear door, and driver's seat in front. (H & O)

NORFOLK SHOOTING CART — A variety of DOG CART. (H)

NORIMON — A native Japanese term, meaning PALANQUIN. (H)

NOTTINGHAM CART — A variety of DOG CART.

Japanese NORIMON with door closed and the additional insulating sun roof in place. (Smithsonian collections.)

Japanese NORIMON with door open, sun roof removed, and a portion of the roof thrown back. (Smithsonian collections.)

O

OBOZE — A Russian FREIGHT WAGON. (H)

OGEE ROCKAWAY — A ROCKAWAY body in which the lower ends of the back corner pillars have an ogee shape. (C)

OIL-GEAR — See PIPE-LINE GEAR. (O)

OIL-TANK WAGON — A WAGON, generally on platform springs or truck springs, mounting a cylindrical tank with a capacity between 300 and 1000 gallons. It was used to transport petroleum products such as kerosene from the distributor to retail outlets. (O)

OMNIBUS — (From the dative plural of the Latin *omnis*, meaning all.) A public street vehicle intended to carry a large number of persons. It is in the WAGONETTE class, having a door in the rear, longitudinal seats facing one another, paneled sides, and drop lights in the sides and ends. This vehicle, the outgrowth of the earlier GONDOLA, was developed in Paris in 1819, where it was introduced by Jacques Laffitte, later the Minister of Louis Philippe. Laffitte did not name the carriage, however, this being done about 1828 by a bath-house proprietor named Baudry, who ran one, which he called *L'Omnibus*, for the convenience of his patrons. Laffitte's vehicles were built by George Shillibeer, an English coach builder living in Paris. Shillibeer decided to return to London to begin omnibus service there, this service commencing on July 4, 1829. In the United States the first rear-entrance omnibus type, known as a SOCIABLE, was put into service in New York City in 1829 by Abraham Brower, who had preceded it two years earlier with his ACCOMMODATION, a kind of

OIL-TANK WAGON built by Shaw, Backus & Co., of Indianapolis, Indiana. Wheels, 44-inch front and rear. (*The Hub*, August 1881.)

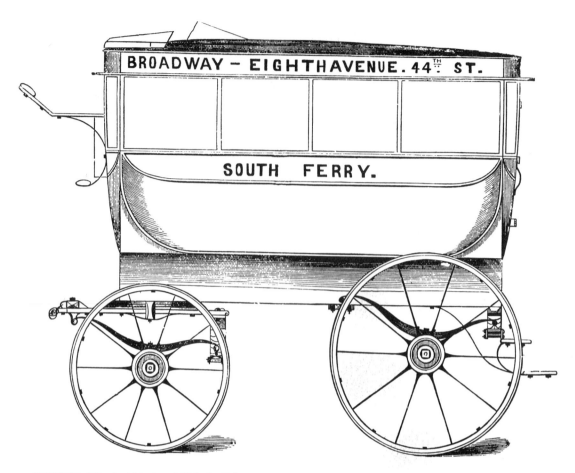

OMNIBUS. Wheels, 46-inch and 55-inch. (*The American Coach-Makers' Illustrated Monthly Magazine*, April 1855.)

"The Pride of the Nation," a famous OMNIBUS built in 1875 by John Stephenson, of New York City. Thirty-six feet long, it was reported by *The Hub* in 1876 to have employed ten horses and carried a hundred and twenty passengers. Even allowing for standees, it is hard to imagine how it could carry so many. (Smithsonian photo.)

This typical OMNIBUS of the late nineteenth century was built by John Stephenson of New York City. Featuring a clerestory roof that is fitted with luggage rails, the bus carried twelve to fourteen seated passengers, plus standees. (Smithsonian photo.)

open, side-entrance public COACH. In 1831 the prototype of the American omnibus was built in New York by John Stephenson, who was later to become the nation's foremost omnibus builder. Stephenson's first omnibus was hung on elliptical springs, which proved inadequate; consequently, later vehicles were mounted on platform springs. Most omnibuses carried twelve to fourteen passengers, yet a number of larger ones were built. They continued in use until the second decade of the twentieth century, when they were replaced by motor buses.

In addition to the mechanical features already described, some omnibuses carried passengers on the roof, either on knifeboard-seats, running longitudinally, or garden-seats, running transversely. Knifeboard-seats came into use in England about 1850; garden-seats were introduced there about 1880, though they were reportedly in use on the Continent many years earlier. Both types were used to some extent in the United States, but double-decked omnibuses were not nearly so common as they were in London. Access to the roof was first by means of a ladder; later, compact stairways were employed. Around 1847 the clerestory roof was introduced, this feature providing additional headroom, light, and ventilation, and if so constructed, it could form either the seat or back of a knifeboard-seat. Other omnibuses featured bombe (or bombay) roofs, curving up in the center, and some were provided with luggage rails on top. A small platform, sometimes called a monkey-board, was installed on the rear of many buses for the conductor, yet a great number of vehicles were operated only by a driver, eliminating the need for this platform. For a complete history of the omnibus, see Henry C. Moore, and Charles E. Lee (refer to bibliography). (C, H, & O)

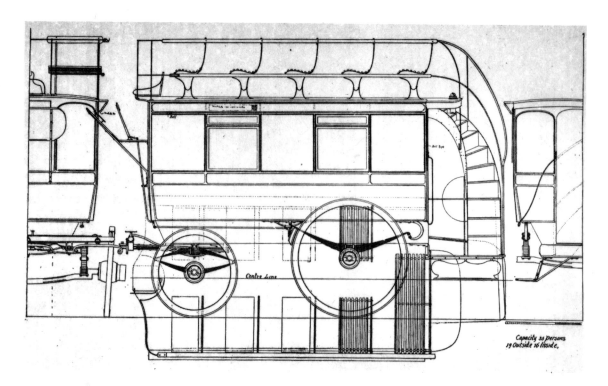

DOUBLE-DECK OMNIBUS with garden-seats on the roof, used by the Fifth Avenue Transportation Co., of New York City. Wheels, 42-inch and 56-inch. (*The Hub,* April 1888.)

OPERA BUS built in two sizes to seat four or six persons. (Geneva Wagon Co., Geneva, New York, 1906.)

ORE WAGON. (Studebaker Bros. Mfg. Co., South Bend, Indiana, 1903.)

An OVERLAND WAGON such as this one by Abbot, Downing & Co. cost $600 in the nine-passenger size, and $525 for the six-passenger size. This illustration is from their catalog of the late 1880s.

This posed hold-up scene shows an OVERLAND WAGON somewhere in Oregon, probably in the 1890s. (Smithsonian photo.)

OMNIBUS, COACH — See COACH-OMNIBUS. (H)

OMNIBUS, WAGONETTE — See WAGONETTE-OMNIBUS. (H)

OPERA BUS — A carriage that is related to the OMNIBUS, and sometimes called a PRIVATE OMNIBUS, that developed about 1870. It has an elevated driver's seat, and a short passenger-compartment behind with seats running lengthwise, facing each other. The upper portion is enclosed in glass. These were used as private carriages. (O)

ORE WAGON — An open WAGON for carrying ores and other heavy bulk materials. These WAGONS were built considerably heavier than general-purpose WAGONS of the same size. (S226)

OVAL HEARSE — A HEARSE having a large oval-shaped glass mounted in each side. (C)

OVERLAND WAGON — A variety of thoroughbrace PASSENGER WAGON. The famed Abbot, Downing Company built large numbers of these. (O)

OWL (or NIGHT OWL) — Synonymous with NIGHT-HAWK and BUCKER, as applied to a street CAB seeking patronage after dark. (H)

OXFORD DOG CART — A local English term applied to one of the earliest types of DOG CART, used especially for tandem driving, and characterized by extremely high wheels known as OXFORD BOUNDERS. (H)

P

PALANQUIN — A covered conveyance, corresponding to the SEDAN CHAIR, used in eastern Asia for transporting a single passenger. It is supported by one or two long poles carried by two men, one being at each end. In Japan it is known as a NORIMON. (C & H)

PANNEL — A military carriage upon which mortars and their beds are conveyed upon a march. (W)

PARK WAGON — A name sometimes applied to a SPRING WAGON. (S226)

PASSENGER WAGON — A type of STAGE WAGON (in the later sense of this term's usage), generally on thoroughbraces but occasionally on springs, varying in size to carry from four to twelve persons. The bodies were most often square-box in shape, many having no doors, but open sides, with curtains to roll down in bad weather. Nearly all had luggage racks on back, but only the heavier ones had racks on top. The heavier varieties were frequently called MUD WAGONS, and were very popular in the West. The cost of these WAGONS was generally only about half that of a regular MAIL COACH, and in some instances was considerably less. The running gear differed from that of a MAIL COACH in the manner of suspension, for the braces did not hang from elevated jacks, but rather from shackles mounted lower on the gear. To compensate for the lowered braces, the body was raised by the addition of a pair of longitudinal members which were mounted some distance below the floor of the WAGON by means of a number of iron supports. Many variations of PASSENGER WAGONS can be found, among them the AUSTRALIAN PASSENGER WAGON, the CALIFORNIA, the POWEL, the WESTERN, the FLORIDA, and the HACK PASSENGER WAGON. These WAGONS were developed during the middle third of the nineteenth century, obviously successors, along with the MAIL COACHES, of the earlier STAGE WAGONS. (O)

PATROL WAGON — A WAGON designed for the use of the police department for patrol duty. According to *The Wagon-Maker*, Chicago, July 1, 1886: "To Chicago belongs the credit of devising and introducing the Patrol Wagon, that modern adjunct of the police system, which increases the serviceability and lessens the cost of police departments of large cities. The first police Patrol Wagon was introduced in the fall of 1879, by Wm. J. McGarigle, then General Superintendent of the Chicago Police Department. A plan of getting the police where they were needed at the earliest possible moment had been talked of for some years by Chief McGarigle and others. . . . The telephone was introduced while these measures were being considered, and it was immediately adopted as a means of sending alarms to the stations, and wagons were decided upon as the means of conveyance, and corner patrol-boxes were built. Mayor Harrison warmly espoused the project. The result was that, before Superintendent McGarigle resigned, in 1882, there were seventeen wagons in use."

Eventually this term became synonymous with PRISON VAN, the vehicle being then used for the conveyance of prisoners, and constructed so as to insure their security. (H)

PEDDLER'S WAGON — A type of WAGON used by peddlers, having a closed body of medium height, containing numerous drawers and compartments, accessible from the sides, in which their wares were carried. Frequently the top had a railing so that boxes could be carried, and the hind portion sometimes had a rack similar to the luggage rack on a CONCORD COACH. The body was generally mounted on platform springs. (O)

PEDESTRIAN CURRICLE — Synonymous with DRAISINE.

PEELERS' CELL — A nickname applied to what is claimed by some to have been the first English BROUGHAM. "Peeler" was the nickname for Sir Robert Peel's police, new at the time. (H)

PERAMBULATOR — (From the Latin verb, meaning to walk through or over.) Originally this was an instrument for measuring distances (Phillip's *New World of Words*, 1706 edition), but later the term came to be applied to a child's carriage. (H)

PERCH CARRIAGE — A vehicle made with a perch, or reach, as distinguished from those wherein this connection is dispensed with, such as those carriages with platform suspension. (H)

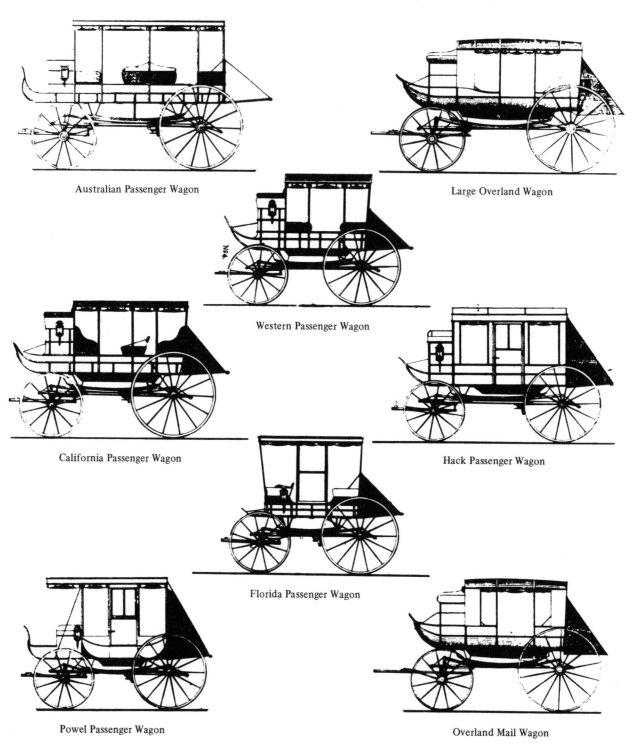

A variety of PASSENGER WAGONS offered on an Abbot, Downing & Co. advertising sheet of about 1885.

PATROL WAGON. Wheels, 36-inch and 50-inch; body, dark green; gearing, carmine striped black; trimming, enameled leather or duck. (*The Carriage Monthly,* September 1885.)

PERITHRON — An English WAGONETTE in which the driving seat was bisected down the center, to permit a passenger entering from the back to reach the front seat. (O)

PETORRITUM — An ancient Roman vehicle of unknown description. (H)

PHAETON — (Derivation: from Phaethon, son of Helios in Greek mythology, who drove the Chariot of the Sun with such recklessness that Zeus struck him down with a thunderbolt, lest he set the earth on fire.) The name is believed to have first been applied to the carriage by the French, during the rage for classic pseudonyms that was prevalent during Voltaire's time, the first known reference being in 1735. Usage of the term spread to England by 1747 or earlier, and to America by the 1760s. The type of vehicle that carried the name was actually in use prior to the application of the term, but was known as a four-wheel CHAISE, or in France, as a four-wheel CABRIOLET. Phaetons rapidly found favor, and were much used as sporting vehicles. They had the advantage of being lighter than other four-wheel carriages, and more comfortable and safer than two-wheelers. Generally the body was single, like a CHAISE body, for the accommodation of two passengers, but double phaetons were also built to carry four persons. One of the principal features of the Phaeton is that, having no driving seat, it is owner-driven, allowing the owner the satisfaction of handling his own carriage and horses, this factor contributing immensely to its popularity as a sporting vehicle. The phaeton usually had a calash-top over the seat, a feature commonly denied when the driver was a servant. The phaeton not only retained its popularity until the end of the carriage era, but grew to be the largest, most varied group of vehicles. The name was applied to large and small types, drawn by one or more horses; some had relatively straight lines, as a WAGON, while others had graceful, curving lines. While most frequently built for two passengers, many phaetons could accommodate four, or occasionally even six, passengers. Seats, where there were several, were often on the same level, yet the front seat was sometimes slightly elevated. A calash-top sometimes covered only the rear seat, or in other instances, only the front, while other phaetons had an extension top over all. Some were not equipped with tops. A rumble for either one or two servants was added to many vehicles of this class. Usually the sides were open, but a few had low doors. Many lighter phaetons had drop-fronts in order to give ample room while retaining a light appearance, yet the heavier phaetons often had a cut-under body to allow the wheels to turn more sharply. Suspension,

PEDDLER'S WAGON. Wheels, 42-inch and 50-inch; body, Chinese vermilion; gearing, straw color striped black. (*The New York Coach-Maker's Magazine,* November 1859.)

    likewise, varied greatly. Usage of the term became so general and loose toward the end of the era, that often there was little distinction between certain phaetons and a carriage bearing some other name. See other main types following this entry. (H, C, & O)

PHAETON, BASKET — A light PHAETON, the body of which is made with an iron frame secured to wooden rockers, and filled in with willow basketwork. Of English origin, it became very popular in America as a park or beach carriage. (C & H)

PHAETON CART (or PHAETON-BODY ROAD CART) — A two-passenger ROAD CART having a PHAETON body with a trimmed seat, and sometimes a falling top. (O)

PHAETON, CRANE-NECK — A contemporary of the PERCH-HIGH PHAETON, but having crane-neck perches to permit sharper turning. Because of this construction, it was somewhat heavier than a PERCH-HIGH PHAETON of the same size. (F)

PHAETON, DEMI-MAIL — A modified and slightly smaller form of MAIL PHAETON, drawn by two horses. It is generally hung on four elliptic springs without a perch. (H)

PHAETON, EQUIROTAL — A three-wheel PHAETON invented in 1856 by C. W. Saladee, editor of *The American Coach-Maker's Illustrated Monthly Magazine.* It was entirely unlike equirotal vehicles of William Bridges Adams, and since the rear wheels were larger than the single front wheel, the term EQUIROTAL seems inappropriate. Saladee's aims were to eliminate the twisting of the vehicle as it passed over uneven ground; to enable a shorter turning radius; and to provide easier access to passengers by removing the conventional front wheels from their path. Saladee reported most satisfactory performance, and claimed to have a number of orders, but there is no evidence that more than these few were built. (H & O) See also EQUIROTAL.

PHAETONETTE — A combination of two carriages, the PHAETON and the WAGONETTE. Used as a family carriage, the front is built like a light PHAETON, and carries two passengers, while the back is built like a WAGONETTE and carries four passengers. This carriage could be equipped with a canopy top. (O)

PHAETON, GENTLEMAN'S — The larger type of driving phaeton, intended for gentlemen's use. They are both higher and heavier than Ladies' phaetons, and are usually characterized by angular lines. (O)

PHAETON, LADIES' — A small, comparatively light phaeton intended for ladies' use. They are generally low to afford easy access, and are most frequently characterized by graceful, curving lines. (O)

PHAETON, MAIL — The PERCH-HIGH PHAETON, one of the earliest sporting vehicles of the Regency, was first modified, in an effort to reduce its ridiculous height, into a gentleman's driving vehicle known as the CABRIOLET-PHAETON. This carriage also failed to fulfill the desires of the sporting gentry, so that about 1830, the mail or telegraph-springs developed somewhat earlier for use on the MAIL-COACH were adapted to a new style of phaeton by English carriage builders, thus carrying over the *mail* term to the new phaeton. This carriage had a heavy square-box body, with a front seat that was protected by an apron and a falling top, and a grooms' seat behind that was on a level, or nearly so, with the front seat. These two seats were movable, so that in inclement weather the hooded front seat might be moved backward, while the groom's seat was moved forward, so that a servant might drive. Construction was most substantial, and the body was suspended on two sets of the mail-springs. Always driven with a pair of horses, the versatile mail-phaeton quickly gained and retained a considerable popularity, and was used not only as a gentleman's driving carriage and a sporting vehicle, but was also commonly employed as a traveling or posting carriage. Later in the century the mail-phaeton became somewhat lighter in construction, often having a cut-under body, while four elliptic springs were frequently substituted for the mail-springs. (H & O)

Sliding-seat PHAETON, in which the main seat slides backward, and a second seat turns up out of the body, creating a four-passenger vehicle. Wheels, 40-inch and 47-inch. (*The American Coach-Makers' Illustrated Monthly Magazine,* January 1856.)

BASKET PHAETON of about 1900. (Smithsonian collection.)

A large crane-neck PHAETON. Body panels striped with green, yellow, and black; doors and sword case, black; silver-plated moldings; knee boot and top, brown leather; gearing, medium green striped with orange and black; carvings, highlighted with yellow; front boot and luggage boot, brown leather, the front boot with silver moldings; hub bands, silver-plated. (W. Felton: *Treatise on Carriages*, 1796.)

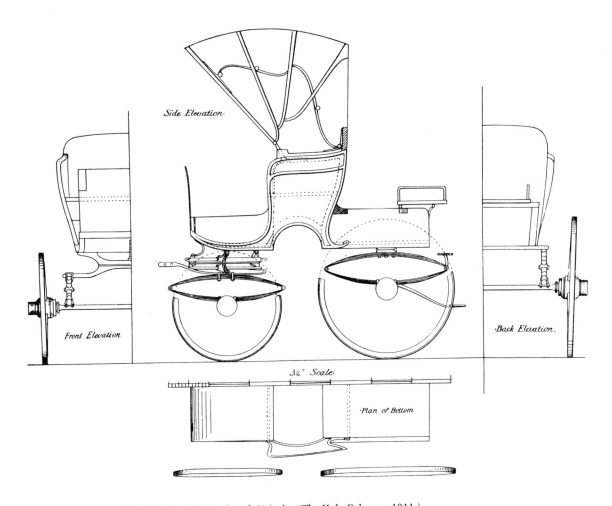

DEMI-MAIL PHAETON. Wheels, 32-inch and 41-inch. (*The Hub*, February 1911.)

PHAETON, MALVERN — A variety of PARK PHAETON that was popular in England about 1850. A similar vehicle used in the United States was sometimes called a four-wheel DOG CART. (H & O)

PHAETON, PARISIAN — A late term applied to a class of LADIES' PHAETONS characterized by Parisian grace of outline and by extreme lightness. It has four low wheels, properly seats only two persons, and generally has no driving seat, being driven either by its lady occupant, or from the rumble which is usually added. The Parisian phaeton usually has angular lines. (H)

PHAETON, PARK — A general term applied to that class of PHAETONS adapted for park driving. Some were four-passenger carriages, much like a four-wheel CABRIOLET, while others were more like a QUEEN'S PHAETON, for two passengers, occasionally with a rumble for one servant behind. (H & O)

PHAETON, PERCH-HIGH — One of the earlier forms of the PHAETON, used in the late eighteenth and early nineteenth centuries. Described and illustrated in Felton (1796), the body of this two-passenger carriage was hung very high above the front axle, this feature being somewhat dangerous, but at the same time probably contributed to its popularity as a sporting vehicle. Suspension was on S-springs and braces in the rear, and on single-elbow springs in front, or on some similar combination. A large boot was fixed in front between the body and the perch carriage. At one time it was a most fashionable driving vehicle, and a favorite of the Prince of Wales, who later became George IV. (H & F)

PHAETON, PONY — A name given to light, low, two-passenger PHAETONS. They may be drawn by one horse, or by a pair of ponies. Also see DUC. (C & O)

PHAETON, QUEEN'S (or QUEEN'S BODY) — A variety of LADIES' PHAETON, characterized by a curved bottom line. (H)

W. B. Adam's EQUIROTAL PHAETON, also equipped with Adam's regulating bow-spring. (W. B. Adams: *English Pleasure Carriages*, 1837.)

Saladee's three-wheeled EQUIROTAL PHAETON. (*The American Coach-Makers' Illustrated Monthly Magazine*, October 1856.)

PHAETONET. Wheels, 36-inch and 42-inch; body and moldings, black; moldings striped with fine line of carmine; seat panel, lake; gearing, carmine striped with two fine lines of black; trimming, maroon cloth. (*The Hub*, June 1889.)

GENTLEMAN'S DRIVING PHAETON. Wheels, 33½-inch and 47-inch; seat cane, yellow; moldings on seat and Stanhope pillar striped with a fine line of yellow; gearing, chrome yellow striped with a ½-inch line of black and two fine lines of same on each side; trimming, green cloth. (*The Carriage Monthly*, April 1883.)

LADIES' PHAETON with rumble seat. Wheels, 30-inch and 41-inch; body, blue; moldings, black, with straw-color stripe; carriage, blue with straw-color striping; trimming, blue colth.

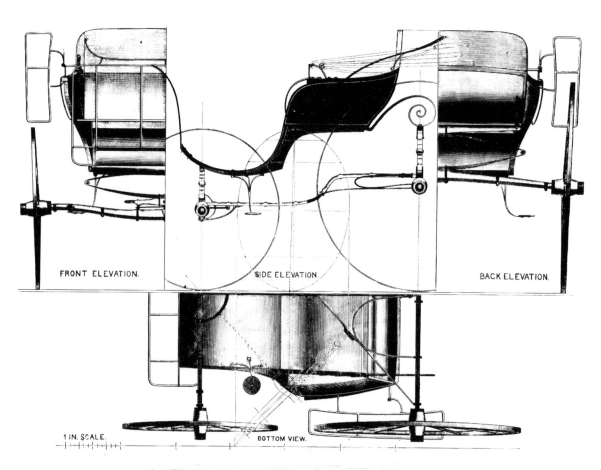

LADIES' PHAETON. Wheels, 34-inch and 44-inch; main panels, deep green; moldings and rockers, black; moldings of body and drop-front striped with fine yellow; gearing, deep green striped with two 1/8-inch lines of yellow; trimming, green cloth. (*The Carriage Monthly,* April 1890.)

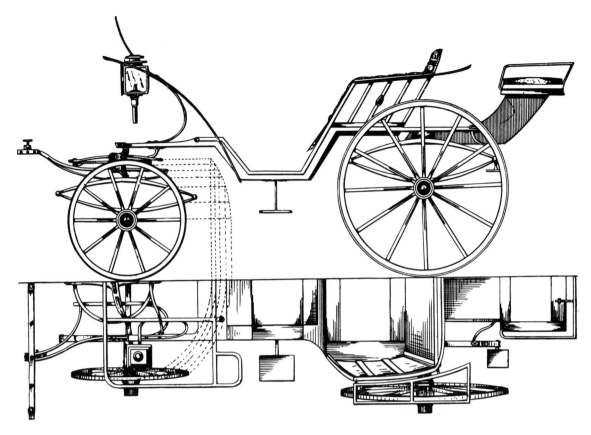

LADIES' PHAETON with paneled rumble. Wheels, 25½-inch and 37-inch; body, burgundy claret with black moldings, and fine strip of vermilion on moldings; gearing, lake, striped vermilion, glazed with carmine; trimming, maroon cloth. (*The Hub,* December, 1882.)

PHAETON, SPIDER – A gentleman's PHAETON, consisting of a Tilbury body on four wheels, sometimes having a spindle-seat and spindle-sides in place of panels. A skeleton rumble accommodates a groom. An American contribution that was introduced about 1861, the spider phaeton became most popular during the last decade of the nineteenth century. (H & O)

PHAETON, STANHOPE – A gentleman's driving PHAETON, much like the MAIL PHAETON, but of lighter build. It has a square-box body, a curved-panel front seat with Stanhope pillar (from which comes the name), and a railed seat behind, on the same level as the front seat, for one or two servants. The front seat is commonly equipped with a falling top, while the rear seat is open. Suspension is generally on elliptic springs, but might also be on mail or platform springs. It can be drawn by one or two horses; customarily it was used with one horse in England, but with two on the Continent. By the mid-1890s the STANHOPE BUGGY was often called a STANHOPE PHAETON, though this usage confuses the BUGGY with the true Stanhope phaeton. (H, C, & O)

PHOONGYES' CHARIOT – An ecclesiastical vehicle used in Burma and India. (H)

PHOREION – An ancient Greek Litter, carried by eight bearers. (H)

PICK-UP CART – A pair of wheels with an arched axle, used in the same manner as LOGGING WHEELS. These CARTS generally have much smaller wheels than the LOGGING WHEELS, and they are used frequently by contractors, farmers, etc. (O)

PILENTUM – (1.) An early Roman carriage described as being light in weight, very comfortable, and consequently often employed for traveling long distances. Some historians believe that the body may have had some form of suspension, but this is purely conjectural. It had a roof supported by pillars, but was open at the sides. Historians do not agree on whether this carriage had two or four wheels. Since illustrations showing both types are offered as being representative of a pilentum, possibly they were constructed with either two or four wheels, with the body style being the principal feature. These vehicles were often trimmed and fin-

PHYSICIAN'S PHAETON. Wheels, 42-inch and 47-inch; body black; gearing deep green striped with double lines of black; trimming green cloth. (*The Carriage Monthly*, June 1883.)

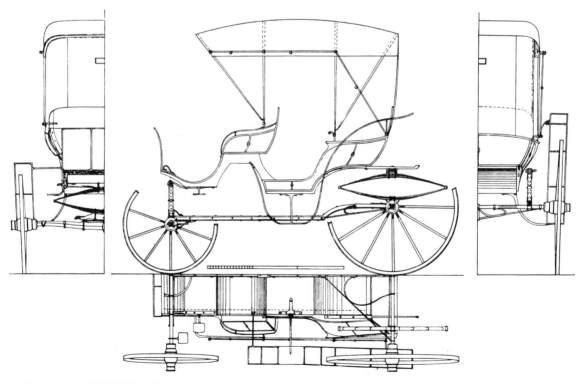

Extension-top PHAETON. Wheels, 34-inch and 46-inch; body black, with dark green panels; gearing dark green striped with two fine lines of orange; trimming green cloth. (*The Hub*, April 1889.)

Four-passenger extension-top PHAETON. Wheels, 44-inch and 51-inch; body, black with gold mountings; gearing, black striped with gold; trimming deep blue cloth. (*The New York Coach-Maker's Magazine,* December, 1870.)

MAIL PHAETON. Wheels, 37-inch and 46-inch; body, black with Stanhope pillar and driving-seat panel dark green; moldings fine-lined with carmine; gearing, dark green with two broad lines and fine center line of carmine. (*The Hub,* December, 1881.)

Malvern PHAETON, popularly used as a pleasure-carriage in England around 1860. (Ezra M. Stratton: *The World on Wheels,* New York, 1878.)

Extension-top PARK PHAETON. Wheels, 34-inch and 46-inch; body, black with dark green panels; gearing, dark green with carmine stripe; trimming, cloth, leather, or whipcord. (Studebaker Bros. Mfg. Co., South Bend, Indiana, 1903.)

Perch high PHAETON. Body, yellow with black sham doors and sword case; top and knee boot, brown leather; gearing, yellow striped black and green; trunk boot, brown leather; moldings, silver-plated. (W. Felton: *Treatise on Carriages,* 1796.)

QUEEN'S-BODY LADIES' PHAETON, with skeleton construction under the seat. This area could also be enclosed. Wheels, 39-inch and 44-inch; panels, dark green; moldings, black; gearing, green, a shade lighter than the body, striped with two medium lines of black; trimming, green morocco for back and cushion, and balance of green cloth. (*The Hub,* August 1885.)

SPIDER PHAETON with rumble. Wheels, 37-inch and 46-inch; panels, dark green; moldings, black edged with a fine line of light green; gearing, green, a shade lighter than the body, striped with one line of black and two fine lines of light green; trimming, green morocco. (*The Hub,* October 1883.)

STANHOPE PHAETON. Wheels, 32-inch and 42-inch; body, black; driver's seat above main panel, chrome yellow striped with a ½-inch line of black; gearing, chrome yellow with a ½-inch black stripe; trimming, hogskin. (*The Carriage Monthly,* October 1887.)

PICK-UP CART with 60-inch wheels. (Columbia Wagon & Body Co., Columbia, Pennsylvania, ca. 1908.)

ished in an elaborate manner, and they were said to have been employed for religious and festive occasions. Roman ladies were permitted to use these carriages when attending games and public festivals, and were thus able to show off their rich garments to good advantage through the open sides of the pilentum. (2.) This term was later applied to a four-wheel carriage that was introduced in England during the second quarter of the nineteenth century. Bridges Adams, in his *English Pleasure Carriages*, describes it as being a "Droitzschka with a curved bottom line instead of a straight one." (C, H, & O)

PILL-BOX — A slang expression applied in London, in the early part of the nineteenth century, to a small one-horse CHARIOT commonly used by physicians. Lord Brougham is said to have copied the compactness and convenience of this vehicle in ordering the carriage that afterward took his name. The term continued to be applied in later years to a physician's BROUGHAM. (H & h)

PILL-BOX PHAETON — An English term, not much used in the United States, for a PHAETON used as a physician's conveyance or a PEDDLER'S WAGON. (M)

PIPE-LINE GEAR — A running-gear without a body, designed for carrying lengths of pipe for oilfield service, or any work using long lengths of pipe. It is somewhat similar to a lumber WAGON, and is sometimes referred to as an OIL-GEAR. (S226)

PLATFORM WAGON — (1.) A square-box WAGON similar to the SQUARE-BOX BUGGY, but larger, with from two to four movable seat-boards on the same level, accommodating two persons on each seat. The size having two seat-boards was by far the most common. This was a most versatile and popular WAGON, much used by farmers because of its adaptability as either a passenger or general-purpose WAGON. When used for carrying goods, all but the front seat were generally removed, and the rear end-gate was usually hinged to facilitate loading. Though frequently an open WAGON, the platform wagon could also have a canopy or extension top. It received its name because the body was hung on platform springs. The terms PLATFORM SPRING WAGON and SPRING WAGON were also applied. Properly, the term PLATFORM WAGON is most applicable if platform springs are used both front and rear; if platform springs are used in the rear and elliptic in front, it can be called a HALF-PLATFORM WAGON, and if it is entirely on elliptics, or some

ROMAN PILENTUM. (Ezra M. Stratton: *The World on Wheels,* New York, 1878.)

PIPE-LINE GEAR with a swivel-reach made of 3½-inch pipe. Wheels, 40-inch and 44-inch. (Kentucky Wagon Mfg. Co., Louisville, Kentucky, ca. 1915.)

patented springs, it can be called a SPRING WAGON. These terms are also synonymous with DEMOCRAT WAGON. A heavier version, usually equipped with brakes, is called a MOUNTAIN WAGON.   (2.) Because of the use of platform springs, many freight-carrying vehicles such as EXPRESS WAGONS, were often called platform wagons.   (3.) This term was also applied by the army in the late nineteenth century to a sort of WAGON used for transporting heavy ordnance. (O & W)

PLAUSTRUM — A primitive form of Roman vehicle, having either two or four wheels, used mainly for agricultural purposes. It consisted simply of a platform on wheels, the latter being of the disc type. Some of these vehicles were later equipped with a large basket, which was placed on the platform to assist in carrying the load, and still others had sides enclosed with the hides of oxen. These vehicles were sometimes used for the conveyance of passengers. (H & O)

PLEASURE CARRIAGE — A general term for any light, showy vehicle. (C)

PLEASURE WAGON — A light New England carriage developed during the early nineteenth century from the light work-wagon. It had a modified square-box body with raved-side construction, and a slightly curving bottom line. In some instances the body was mounted directly on the running gear, but many were suspended on either thoroughbraces or springs. The seat, usually untrimmed, was mounted on wooden cantilevered supports which provided most of the riding comfort that this carriage offered. The entire seat-unit could be readily lifted out, so that the WAGON could be used to carry produce or other items. These carriages were generally painted with bright colors, and panels were frequently decorated with floral designs. From this WAGON the CONCORD WAGON gradually evolved, which in turn contributed to the design of the BUGGY, or ROAD WAGON. Because of the removable seat and the carrying capacity, the pleasure wagon might also be considered an ancestor of the SPRING WAGON. (O)

PLOSTELLUM (or PLAUSTELLUM) — Diminutive of PLAUSTRUM. An ancient Roman CART or WAGON of small size, frequently used by children. (H)

POLE-CART — A variety of pair-horse ROAD CART made popular by Calvin Toomey, of Canal Dover, Ohio. (H)

POLE-CHAIR — A term sometimes applied in America to the CURRICLE, because of the fact that it was equipped with a pole, for use with two horses, instead of being equipped with shafts. (O)

POLO-CART — An English variety of light ROAD CART. (H)

PONTOON WAGON — (Military) Part of the equippage of a pontoon train. The flat-bottom pontoon, or float, used in a pontoon bridge was carried by this wagon. CHESS WAGONS and TRESTLE WAGONS were also used in the pontoon train. (O)

POPCORN WAGON — A light vending Wagon, in which popcorn was roasted and sold on the streets. (O)

PORTE-CHAISE — Contraction of CHAISE A PORTEURS. (H)

PORTLAND CUTTER — Probably the most popular of all American SLEIGHS, this light, square-body CUTTER was developed early in the nineteenth century at Portland, Maine. Attributed to Peter Kimball, of Bryant Pond, Maine, it would appear more likely that his sons, James M., who operated in Portland, and Charles P., who later worked in Chicago, were responsible for the fully developed design. Copied by builders throughout the country, the Portland was also built as a four-passenger SLEIGH. It was also known by the name of its designers — KIMBALL CUTTER or SLEIGH. (C & H)

PO' SHAY — Contraction of POST-CHAISE, used by Thackeray. (H)

POST-CHAISE — (From the Latin *postis*, a post [such as a doorpost], an upright timber.) Upright posts were found to be convenient places to "post" news, or public notices. Other derivatives came to be applied to the riders who carried messages between posts or stations along a route, later to the communications they carried, and finally to the entire system for carrying the mails. Late in the seventeenth century the French were employing two-wheel carriages that not only carried the mail, but also accommodated one or two paying passengers. The door of the French CHAISE-DE-POSTE was hung on the front of the carriage, and was hinged at the bottom so that it fell forward between the shafts. The body was hung by braces extending from whip-springs in the rear, and was supported by elbow springs in front. This carriage was driven by a postilion.

In 1743 this system of post-traveling was introduced into England by John Trull, an artillery officer, who obtained a patent for renting out traveling carriages. At first the post-chaise used in England was very similar to the French vehicle, but within a short time they became four wheel carriages — in reality, CHARIOTS minus the coach-box, and driven by a postilion. For a time the four-wheel post-chaise retained an undesirable feature of its two-wheel predecessor, being equipped with shafts for the off-horse, while the near-horse was harnessed to an outrigger singletree. This arrangement was shortly replaced with a single pole and the conventional paired-draught. Either two or four horses were employed, as occasion demanded. Post-chaises

This three-seat PLATFORM WAGON or PLATFORM-SPRING WAGON was called a *Drummers'*, or *Summer Resort Wagon* by the manufacturer, giving some indication of the uses to which it was put. It was offered with a black body on a red or dark green gear, and imitation leather trimming, for $77. (Elkhart Carriage & Harness Mfg. Co., Elkhart, Indiana, 1907.)

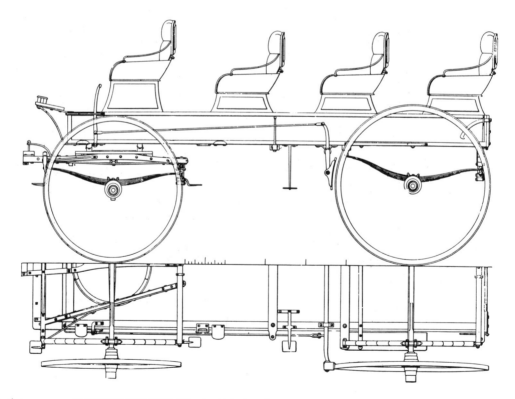

Twelve-passenger PLATFORM WAGON. Wheels, 42-inch and 46-inch. (*The Hub*, July 1900.)

ROMAN HAY-PLAUSTRUM. (Ezra M. Stratton: *The World on Wheels,* New York, 1878.)

New England PLEASURE WAGON with body on thoroughbraces and seat on wooden cantilever-springs. A small, round child's seat turns out from underneath the seat. Wheels, 36-inch and 48-inch; body and gearing, dark green ornamented with light green panel borders and yellow striping. The cross-bar carries an attractively forged wrought iron whiffletree that is common to this type of vehicle. (Author's collection.)

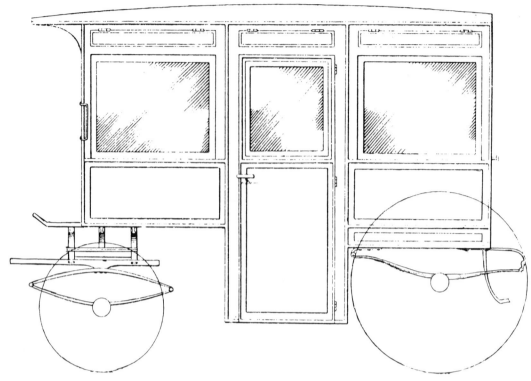

POPCORN WAGON. Wheels, 30-inch and 43-inch; imitation graining on body; gearing, light brown. (*The Hub*, June 1894.)

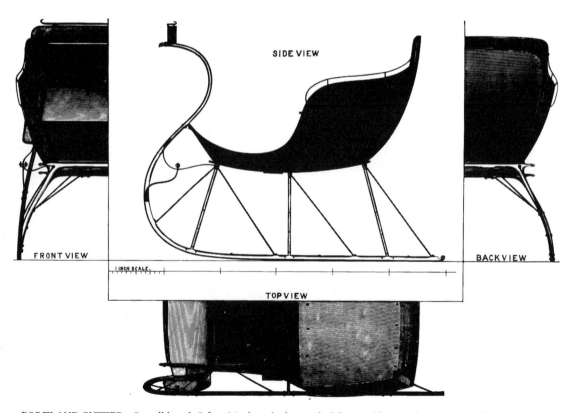

PORTLAND CUTTER. Overall length 5 feet 5 inches; body panels, lake; moldings and other parts of body, black; moldings striped with a fine line of yellow or gold; gearing, lake striped with fine lines of black and yellow or gold; trimming, brown cloth. (*The Carriage Monthly*, August 1887.)

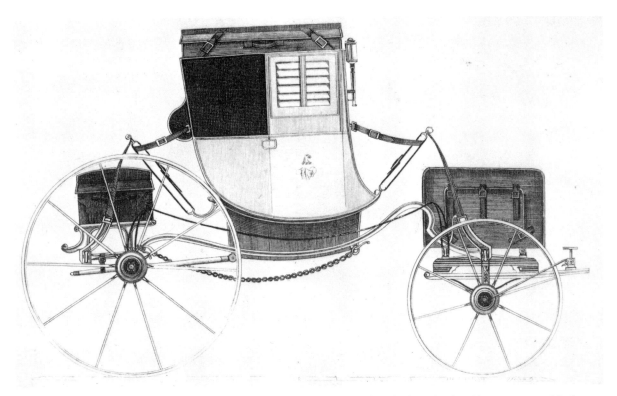

A traveling POST-CHAISE on a CRANE-NECK CARRIAGE. Body, black with silver-plated moldings; Venetian blinds, yellow; gearing, dark green striped with yellow; trunks and imperial, brown. The springs are wrapped with cord, and the equipment includes a drag-staff and chain. (W. Felton: *Treatise on Carriages,* 1796.)

were not only operated as public vehicles, but were also frequently kept as private carriages.

The post-chaise is believed to have arrived in America by mid-eighteenth century, where it became most popular, as in England. It is doubtful, however, that it carried much mail, serving rather in the capacity of a public or private carriage. Post-chaises customarily carried two passengers, yet some accommodated only one, while others carried a third passenger on a movable inside seat. For long distance traveling the carriage usually had a trunk in the rear, a trunk-boot in front, and sometimes an imperial on the roof. Some CHARIOTS were so constructed that they could be converted into post-chaises by the removal of the coach-box. These vehicles were generally termed POST-CHARIOTS. Post-chaises and POST-CHARIOTS continued in use into the middle of the nineteenth century. (H, C, & O)

POST-CHARIOT — A CHARIOT constructed with a removable coach-box, so that it could be used as either a CHARIOT or a POST-CHAISE. (A)

POST-OMNIBUS — A public vehicle of the OMNIBUS type used for the conveyance of the mails as well as passengers. It was in general use in pre-railway times, and was still employed on short routes in the latter part of the nineteenth century. (H)

PRAIRIE SCHOONER — A term applied to various types of western covered wagons from the 1840s onward. The name was suggested since the wagons, like ships, carried both freight and emigrants with their belongings, and were topped by linen or duck covers that were reminiscent of a ship's sails. Prairie schooners differed slightly from eastern freight wagons such as the CONESTOGA, by having nearly straight lines, with vertical or only slightly inclined end-gates, and top bows that were almost vertical. They were, as a rule, medium-size wagons carrying two to three tons, and were frequently equipped with an inside seat. Continuing in use into the twentieth century, they were commonly found making up emigrant trains, as well as carrying freight to points not serviced by water or rail transport. (H & O)

PRAIRIE YACHT — A vehicle invented by a Dr. Wheeler of Grand Forks, North Dakota. Modeled after the ICE YACHT, it was designed to skim over the snow-covered plains by the aid of the wind. (M)

PRINCE ALBERT — A variety of PARK-PHAETON, with a calash top over the seat, and a rumble for a servant in the rear. (O)

Double line of PRAIRIE SCHOONERS drawn up for a rest. Date and location are unknown, though the site bears some resemblance to Cimmaron Crossing on the Santa Fe Trail. (Photo from Smithsonian collection.)

PRINCE ALBERT. Wheels, 35½-inch and 44½-inch; body, black; front seat panels and gearing, tan brown; striping, black. (*Le Guide du Carrossier,* October 1891.)

PRISON VAN — A closely paneled WAGON, with driver's seat in front and a rear portion arranged on the WAGONETTE plan, specifically designed for the safe conveyance of prisoners. This vehicle was popularly known as the BLACK-MARIA. (H)

PRIVATE OMNIBUS — See OPERA BUS.

PULKA (or PULKHA) — The sleigh used by Laplanders for driving reindeer during the late eighteenth and nineteenth centuries. Drawn by a single reindeer, it resembled the front half of a canoe cut transversely, and closed at the back by a perpendicular panel of wood. It had a keel, four to six inches wide, finishing in a point in front. The reindeer was driven with a single line. This sleigh was also called a PULK. (H, C, & O)

PUNG — (From *tom-pung*, a corruption of several similar-sounding words from various American Indian dialects, describing a crude sledge used by these people. The word *Toboggan* comes from the same source.) In New England the term PUNG was applied sometime during the eighteenth century to a simplified box-sleigh, which usage continued throughout the nineteenth century. The pung was usually drawn by a single horse. According to Houghton, the term eventually was applied in New England to any light business vehicle on runners. (H, C, & O)

PUSH-CART — A HAND-CART intended to be pushed instead of drawn. (H)

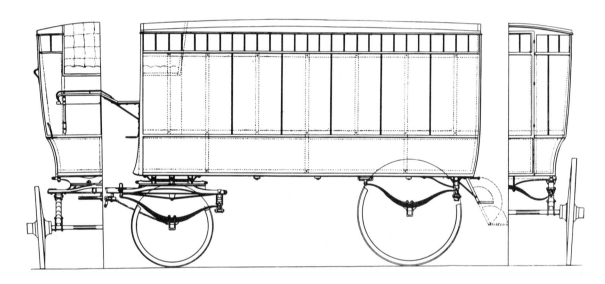

PRISON VAN. Wheels, 38-inch and 50-inch; panels, Ketterer green; moldings, black with carmine striping; gearing, wine color, striped white. (*The Hub,* January 1893.)

LAPLAND PULKA (Otis T. Mason; *Primitive Travel & Transportation,* Smithsonian Institution, Washington, D. C., Government Printing Office, 1896.)

Q

QUADRICYCLE — A four-wheel, pedal-propelled CYCLE used either on rails or ordinary roads. (H)

QUADRIGA — A Roman CHARIOT drawn by four horses abreast. The term was sometimes applied to the horses as well as to the vehicle. (O)

QUAKER CHAISE — See GRASSHOPPER CHAISE. (H)

QUARTER (or QUARTETTE) — This term was suggested by *The Hub* as a class-name for the family of vehicles commonly known as four-passenger extension-top PHAETONS. Apparently the idea never achieved any success. (H)

QUARTOBUS — A variety of four-wheel public CAB, invented by a Mr. Okey, and introduced in London in 1844. It accommodated four passengers, and was characterized by short coupling to promote easy draught. According to H. C. Moore, in his work, *Omnibuses and Cabs*, this carriage was not a success. (H)

Tandem QUADRICYCLE. (Smithsonian photo.)

R

RALLI-CAR or RALLY CART — A four-passenger English CART, brought prominently into public notice by a patent law suit. Equipped with dos-a-dos seats, this vehicle was introduced by C. S. Windover in 1885, and was named after the first purchaser. It was nicknamed the CLOTHESBASKET by the London *Daily Telegraph*. (H & O)

RATION CART — An ordinary CART used for carrying rations to army troops in the field. (O)

RECKLER — A native CART, peculiar to India, somewhat similar in construction to the American SULKY, but provided with a pole, and intended to be drawn by a pair of bulls. (H)

RED RIBBON CUTTER — A CUTTER patented in 1881, and built by Frank B. Miller, of Enon, Ohio, having a body shaped like a reindeer. (O)

RED RIVER CART — A primitive wooden CART much used by the half-breed settlers of the Red River area in both the United States and Canada. This CART, drawn by a single ox, horse, or pony, began to develop around 1800, and was used for over three-quarters of a century. Serving as a freighter, it carried meat, furs, and skins to such places as St. Paul, where the merchandise was sold or traded to the Hudson Bay Company. A load of about 800 pounds was generally carried. A lighter variety of CART, finished somewhat more attractively with red and yellow decorations, was also built for personal use. Some of the latter had skin or oil-cloth covers to protect the occupants. No metal was employed in the construction of Red River carts, but where some reinforcement was required, such as around the hubs, strips of buffalo rawhide were bound around. The harness was also of this rawhide. The wheels had no tires, the unshod felloes coming in direct contact with the ground. Often the felloe joints were reinforced by small pieces of wood that were bound alongside by rawhide. (O)

REEFER — A New England name for the STATION WAGON. (O)

REINDEER SLEDGE — See PULKA. (H)

REPUBLICAN WAGON — Same as DEMOCRAT WAGON. An attempt was made to apply this term by a few builders, apparently in an effort to give "equal time" to the second major party, but with little success. (O)

RHEDA — An ancient Roman vehicle. According to Stratton, the term was applied to a wide variety of vehicles, both two-wheel and four-wheel, open and closed, and with or without cushions or linings. The various types were used for carrying both freight and passengers, for military and state purposes, as a posting vehicle that was available for hire, and as a family carriage. Most frequently drawn by oxen, they appear to have been so general in nature that it would be hard to ascribe any characteristic features to them. (H & O)

RHEDA CABALLARIA — (Literally meaning horse litter.) Bridges Adams, in his *English Pleasure Carriages* (page 28), says, "with the Normans came the horse-litter, a native originally of Bithynia, and from thence introduced into Rome. . . . Malmsbury records that the dead body of Rufus was placed upon a *rheda caballria*, or, as Fabian translates it, a 'horse-litter.' King John, in his last illness, was conveyed from the Abbey of Swinstead in a 'lectica equestre.' These were for several succeeding reigns, the only carriages in use for persons of distinction." (H)

RICKSHAW — The Anglicized abbreviation of JINRIKISHA. (H)

RIDING CHAIR — A simplified one-passenger, two-wheel vehicle of the eighteenth century. See CHAIR. (O)

RIG — (Undoubtedly a contraction from "rigging.") An American slang expression applied to certain vehicles intended for either industrial or agricultural purposes, when provided with special appurtenances or riggings, such as a HAY-RIG. The term is also commonly applied to a carriage with its horse or horses, and frequently, to any light or dilapidated vehicle. (H&M)

ROAD CART — A light two-wheel CART made for either one or two passengers. The most common variety had a

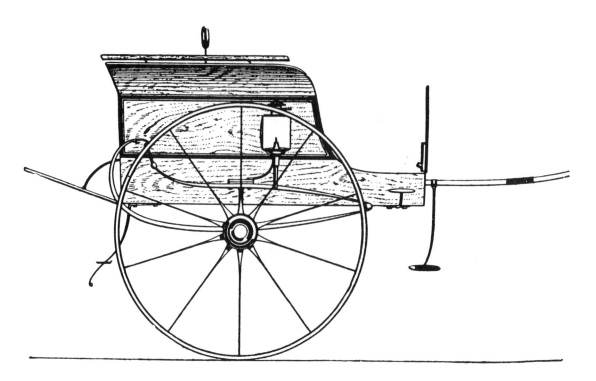

RALLY CART with 54-inch wheels. Natural finish on body with only the upper framework painted. (*The Hub*, August 1892.)

MILLER'S RED RIBBON CUTTER, from an advertisement in *The Hub* in August of 1882.

one-passenger seat mounted over the axle on a skeleton frame. This seat was frequently suspended on a spring, or springs. W. S. Frazier & Co., of Aurora Illinois, claimed to have introduced the name about 1881, though this type of vehicle had been popular for many years previously, having earlier been known as a ROAD SULKY. Road carts were extremely popular, and were most economical, often selling for $10.00 or less after 1900. In addition to widespread general use, many rural mail carriers employed road carts. The term was also applied to more elaborate varieties such as the two-passenger PHAETON-BODY ROAD CART. (S226 & O)

ROAD SCULLER — A vehicle resembling a combined TRICYCLE and parlor rowing machine, propelled by the efforts of the rider being applied to mechanical oars. (H)

ROAD SULKY — See ROAD CART. (C)

ROAD WAGON — Synonymous with the American BUGGY. Some of those engaged in the trade used the term in preference to BUGGY, while others had a tendency to use it only if the vehicle had no top. This term was also loosely applied to larger WAGONS that were employed in the movement of materials or merchandise over the roads. (C & O)

ROBINSON HANSOM CAB — See VICTORIA-HANSOM. (H)

ROCKALET — Synonymous with, and a contraction of ROCKAWAY-LANDAULET. Sometimes the term is also applied to any small ROCKAWAY, but such usage of the term is not recommended. (H & M)

ROCKAWAY — A four-wheel, covered carriage, with either paneled or curtained sides, with or without perch, always on elliptic or platform springs, having a driver's seat that is included in the body proper (and on a level with the other seat, or seats), and a common roof that projects over the driver's seat. In a few instances the driver's seat did not have the protection of a roof, but this was comparatively uncommon. The carriage was drawn by either one or two horses, depending on the vehicle's weight.

The rockaway is a distinctive American style of carriage which is believed to have descended from the late eighteenth-century COACHEE, through the GERMANTOWN of a slightly later era. About 1830 a carriage builder of Jamaica, Long Island, constructed a WAGON that was apparently similar to a New England PLEASURE WAGON. Later, elliptic springs were added, as well as several wooden bows to support a cloth top. Several residents of New York saw the vehicle, and realizing its potential, persuaded a New York carriage-dealer to offer the Jamaica builder's WAGONS in his warerooms. The vehicles attracted much interest, and when the dealer was asked who made them, he attempted to keep the WAGONS an exclusive item at his repository by deliberately misleading the questioners by saying that they were made at Rockaway, where there was, in fact, no carriage builder. By the following season the truth of the WAGONS' origin was known, and demands were heavy on the Jamaica carriage builders, yet the name ROCKAWAY persisted. Gradual improvements, such as the addition of curtains, panels, doors, windows, and better suspension, resulted about mid-century in the fully developed rockaway, though continued styling changes frequently altered its appearance thereafter. It is not clear to what extent the GERMANTOWN may have contributed to the design of the rockaway, or later, perhaps, the rockaway to the GERMANTOWN, but similarities seem to indicate that there was an exchange of ideas. From about 1870 onward the carriage known as the DEPOT WAGON was often nearly identical to the rockaway, and after 1890 the term STATION WAGON was frequently applied to the former carriage. It is difficult to determine any rule that might apply to the usage of these various terms, but it would appear that STATION or DEPOT WAGONS have bottomsides that in the rear are straight, or nearly so, which construction allows the quarter-lights to drop.

Rockaways were built in a profusion of styles, and a variety of names was applied to them. Several distinctive types appear to have been predominant, and the application of specific names to these types was fairly consistent. The first of these was the COUPE ROCKAWAY, carrying four passengers and completely enclosed with glass windows. A second popular four-passenger variety was enclosed by a combination of glass and curtains, the glass being in doors and front, and was known as a CURTAIN ROCKAWAY. If entirely enclosed with curtains, it was known as a LIGHT ROCKAWAY, yet some builders did not use the term "light" but applied the name CURTAIN ROCKAWAY to either of the last two types. Of six-passenger rockaways there was the EXTENSION-FRONT COUPE ROCKAWAY, glass enclosed, with a third, rear-facing seat in the extended front; and last, the EXTENSION-FRONT CURTAIN ROCKAWAY, enclosed by glass and curtains.

The rockaway was considered to be representative of a democratic people, because of the protection that was given the driver. It was a most popular family carriage, frequently being owner-driven, but equally adaptable to being driven by a servant. It retained its popularity until the end of the carriage era, and then finally contributed its body to the LIMOUSINE. Also see GERMANTOWN, STATION WAGON, and

RED RIVER CART of 1882. (Smithsonian collection.)

ROAD CART on 48-inch wheels. The body has a natural wood finish, while the gearing is red. (Studebaker Bros. Mfg. Co., South Bend, Indiana, 1903.)

Phaeton-body ROAD CART. Wheels, 46-inch; body, black; gearing, Brewster green or carmine; trimming, leather or corduroy. (Cash-Buyer's Union, Chicago, ca. 1900.)

DEPOT WAGON. (H, M, C, & O)

ROCKAWAY-LANDAU (or LANDAU-ROCKAWAY) — A vehicle combining the features of a six-passenger ROCKAWAY and a LANDAU. It has a falling top over the rear seat, and a removable standing top over the forward passenger seat the the driver's seat. (H)

ROCKAWAY-LANDAULET (or LANDAULET-ROCKAWAY) — A vehicle combining the features of a ROCKAWAY and a LANDAULET. It has a falling top over the rear seat, and a removable standing top over the driver's seat. It is also called a ROCKALET. (O)

ROLLING BARREL — A hogshead, with an axle fixed through the heads, to which was attached ropes for drawing the barrel along the road. Either man or animal power was used, and these barrels were used mainly in the eighteenth century in the tobacco country, for hauling tobacco to the wharves. (O)

ROLLING BRIDGE — A kind of mobile bridge, composed of two carriages connected by a coupling pole, and placed in creeks that were not fordable. Described in *Practical Instructions for Military Officers*, by E. Hoyt, Greenfield, Massachusetts, 1811. (O)

ROLLING-CART — A primitive form of manure CART used in England in the eighteenth century. It consisted of three pieces of strong elm, two feet in diameter and each eighteen inches long, through which a strong iron axle was passed, so as to protrude a few inches beyond the rollers, allowing an inch between them for convenience in turning around. On the projecting parts of the axle, a fixed frame was placed to support the body. The vehicle could be employed simply as a roller, or for carrying manure, etc., on land where common wheels could not be supported. The men who handled these rolling-carts were called 'high-rollers,' and were honest sons of toil, whereas the American 'high-rollers' of the late nineteenth century were persons whose hands were adept at the manipulation of cards. (H)

ROLLING-CHAIR — The equivalent of CHAISE-ROULANTE. (H)

ROUETTE — A modification of the term ROULETTE. (H)

ROULETTE (or ROULETTE-CHAIR) — A French vehicle of the eighteenth century, also called a BROUETTE or VANAIGRETTE. See BROUETTE. (H)

Six-passenger ROCKAWAY of about 1850. (Smithsonian collection.)

LIGHT ROCKAWAY. Wheels, 36-inch and 46-inch; lower part of body, deep green; moldings and upper part, black; front seat panel, deep green with black moldings; gearing, deep green striped with two fine lines of pale green; trimming, green cloth. (*The Carriage Monthly,* May 1894.)

ROUND-UP WAGON — A type of WAGON, with bows to support a cover, used by cattlemen on trail drives in the West. Also called a CHUCK WAGON, it carried the cook's supplies and utensils. At the rear of the WAGON a high box was constructed, having drawers inside for the smaller and more compact supplies. The rear cover of this box formed the end-gate of the WAGON, and could be secured in a horizontal position to provide a work area for the cook. Inside the WAGON the larger supplies were stored, such as flour, potatoes, beef, or some items of trail equipment. (S1887)

RUNABOUT (or RUNABOUT WAGON) — (1.) A BUGGY or ROAD WAGON made without a top. Many builders used the term somewhat synonymously with DRIVING WAGON. Often a runabout was made very light, sometimes with a stick-seat or a spindle-seat, and sometimes with bike wheels. (2.) Houghton describes the vehicle somewhat differently, saying that it was a two or four-passenger general utility WAGON, first introduced at Syracuse, New York. According to this description the vehicle was apparently the same as a SPRING WAGON, or DEMOCRAT. This is borne out by an 1891 catalog of the Youngstown Carriage and Wagon Company, which shows four-passenger SPRING-WAGONS, with or without top, under the name of RUNABOUT WAGONS. (H, M, & O)

RUTT — An ornate cart, mounted on springs, used by the wealthy gentry of India. It is fitted with rich curtains to shield the occupants, and is drawn by oxen. (O)

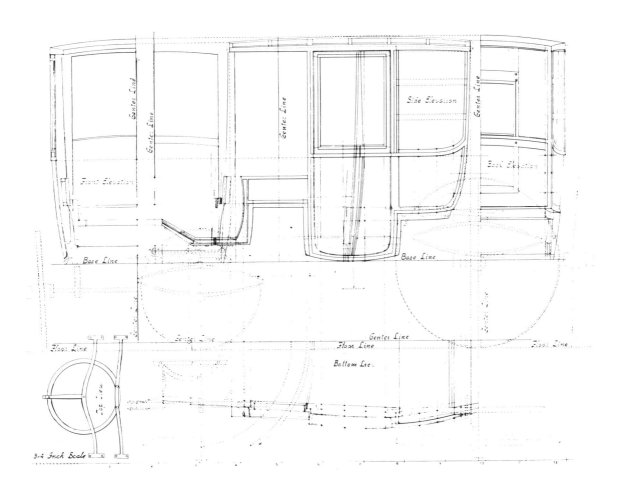

COUPE ROCKAWAY. Wheels, 34-inch and 48-inch; lower panels, deep blue; moldings and upper part of body, black; metal door moldings, and space adjacent to door moldings, blue, so as to carry the blue to the upper part of the body; front seat panels, pillars, and front, blue with black moldings; gearing, deep blue, or slightly lighter; gearing and body moldings striped pale yellow; trimming, blue cloth. (*The Carriage Monthly,* October 1897.)

Henry Hooker & Co. called this a STATION ROCKAWAY, combining the names of two similiar vehicles. This is actually a CURTAIN ROCKAWAY. Hooker also applied the *Rockaway* name to vehicles that were clearly *Station Wagons,* showing the tendency of some builders to use terms somewhat inaccurately. (Henry Hooker & Co., New Haven, Connecticut, ca. 1905.)

ROCKAWAY-LANDAU. Wheels, 36-inch and 48-inch; lower panels, dark olive green; upper part and moldings, black, the latter striped with fine dark green; gearing, dark green striped with two 3/16-inch lines of bluish green with a fine line between them; trimming of cushions and backs, dark green goatskin, balance, dark green cloth. (*The Carriage Monthly,* October 1881.)

ROCKAWAY-LANDAULET, or ROCKALET. Wheels, 36-inch and 47-inch; body, black; gearing, carmine striped with three lines of black; trimming, green cloth or goat-skin. (*The Carriage Monthly,* January 1884.)

ROLLING BARREL, transporting tobacco, with men holding back in the rear. (Smithsonian photo.)

ROUND-UP WAGON. (Studebaker Bros. Mfg. Co., South Bend, Indiana, 1887.)

INDIAN RUTT. (*The Hub,* July 1896.)

S

SAIL-CARRIAGE — See WIND-CARRIAGE.

SAM-BARK — A litter, shaped like a boat, in which the statue of the Egyptian god, Amon, was carried in religious processions. (H)

SANTA FE WAGON — Somewhat synonymous with MURPHY WAGON, but also applied rather generally to any WAGON used on the Santa Fe Trail. (O)

SARRACUM — An ancient Roman vehicle, on two or four wheels, and probably somewhat crude. Used by country folk for conveying produce, it also was a means of personal transportation. During times of plague, it was used to remove bodies from Rome. (H)

SAVANILLA PHAETON — The local name of a variety of PHAETON formerly used in Bangkok, Siam. (H)

SAWDUST CART — A DUMP CART similar to the SLAB CART, but of lighter construction, used for hauling sawdust from sawmills. (H)

SCAVENGER WAGON — A vehicle employed in cities for removing ashes and other waste materials from residences. A GARBAGE WAGON. (H)

SCIRPEA — An ancient Roman vehicle of unknown description. (H)

SCOOP — A form of side-spring BUGGY having a scoop-shaped skeleton body, first built by J. Kelly at Hermon, New York, in 1873, and later popularized by New England builders. (H)

SCORPION — An ancient war engine used for throwing arrows or stones, used principally to defend the walls of a town. (H)

SCRAP-CART — A term applied in London to a vehicle used for gathering scraps of food, which were afterwards prepared in the form of stews, pies, etc., and sold at a nominal price. (H)

SCYTACLAE — A primitive roller-sledge described by Aristotle. (H & O)

SEDAN — See SEDAN CHAIR.

SEDAN CAB — A variety of CAB invented and patented in 1870 by Chauncey Thomas of Boston, Massachusetts. It was so named because of the resemblance of its body to that of the SEDAN CHAIR. (H)

SEDAN CART — A two-wheel substitute for the SEDAN, used in eighteenth-century London. The body was made large enough to accommodate one person, and rested upon shafts, the back ends of which were attached to the axle. It was drawn by one horse, which the driver rode. These vehicles are said to have been used only by persons of middle class. (C & O)

SEDAN CHAIR (or simply, SEDAN) — The derivation of this term is uncertain. It is sometimes said to have taken its name from the French town of Sedan, but this origin is believed unlikely. Some etymologists believe it comes from the Italian, *sede* (Latin, *sedes*), sit; *sedere*, to sit. This last supposition seems more likely since Sedans appear to have been in use in Italy early in the sixteenth century. Sedans were also used in Spain during that century, and late in the century were first seen in England, though they were not widely accepted there until the 1630s. They were apparently also introduced into France early in the seventeenth century. The rise of the use of the Sedan chair in England is credited to Sir Saunders Duncombe, who was granted a patent in 1634 "to put forth and lett for hire" Sedan chairs for a term of fourteen years. It was his intention that the crowded condition of London streets be alleviated by replacing the larger HACKNEY-COACHES with these smaller, man-borne Sedans. Supported between two poles ten to twelve feet in length, and carried by two men, the first Sedans were clumsy conveyances, somewhat resembling dog kennels. They were often made without glasses, while later ones had drop lights in the sides and in the door, which was in front. The body of the public Sedan remained rectangular, but gradually changed from the dog-kennel appearance to a more upright form, approximately five feet tall by thirty to thirty-two inches square. They

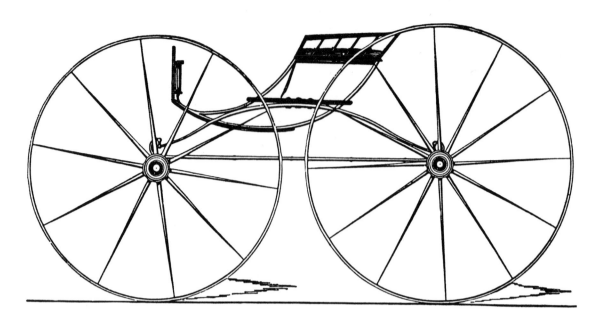

SCOOP. Wheels, 46-inch and 50-inch; body, black; gearing, dark lake; blue cloth trimming. (*The Hub,* March 1882.)

Eighteenth-century ENGLISH SEDAN-CART. (Ezra M. Stratton: *The World on Wheels,* New York, 1878.)

were constructed with a wooden frame, covered with leather. The chair-men, or pole-men who carried the public Sedan chairs were assisted in carrying their burden by shoulder straps, and were, according to Straus, "often drunk, often careless, and nearly always uncivil." By 1821 only about half a dozen public Sedans remained in London, and by 1830, these too had passed from the scene. Some of the Sedans made for private use had bodies shaped more like a carriage, and were elaborate in finish and trimmings. By about 1750 the private Sedan chair had reached the highest point of its development, being made lighter in weight, and richly decorated with moldings, carvings, and paintings. See CHAISE-A-PORTEURS. (H, C, & O)

SEGE — A vehicle peculiar to the Azores Islands, described as follows in Walker's, *The Azores*, page 280: "Now and again, and generally in out of the way places, the old-fashioned ségè is seen — another relic of a by-gone age, a compromise between a modern Hansom cab and an ancient Sedan Chair. Swung high on two wheels, and drawn by a pair of horses, postillion-fashion, the sége had heavy leather curtains in front, completely shutting off the occupants from view when closely drawn." (H)

SEXTET (or SEXTETTE) — A term suggested by *The Hub* as a class name for the family of vehicles now commonly known as six-passenger extension-top PHAETONS. It is not known that this idea ever achieved any success. (H)

SHAY — A colloquial spelling of CHAISE, made famous by Dr. Oliver Wendell Holmes's poem, entitled, *The Deacon's Masterpiece, or The Wonderful One-Hoss Shay*. See CHAISE. (H)

SHEBANG — A slang expression sometimes applied in the western United States to a carriage and horse, just as the term RIG is sometimes used. It was particularly applicable if the entire outfit was rather shabby. (H & M)

SHEEPWAGON — A home on wheels for a sheepherder, used in the Rocky Mountains, or in places where the weather was likely to be unfavorable for tent or outdoor living. The sides of the body are vertical between the wheels, then extend outward above, so as to overhang the wheels. Inside, these overhanging portions serve as benches. Inside is a bed, stove, and table. A box for food supplies is attached to one side, and is accessible from the inside through one of the benches. The top is covered with canvas, and the inside is often lined with blankets, linoleum, or even sheet metal. This type of vehicle supposedly was developed about 1884 by James Candlish of Rawlins, Wyoming, and improved about 1892 by Marshall Buxton for the Schulte Hardware Company, of Casper, Wyoming. The Schulte plans became standardized, and were followed by many other builders. (O)

SHIFTER — A New England term for a one-horse SLEIGH having shafts that can be shifted off-center, so that the near runner will follow the horse, thereby permitting him to use the track of pair-horse SLEIGHS when country roads are only partially broken. Also called a FOLLOW-SLEIGH. (H)

SHIFTING-SEAT CARRIAGE — Same as JUMP-SEAT CARRIAGE.

SHILLIBEER — A term sometimes applied to early OMNIBUSES in London, named after the builder, George Shillibeer.

SHIP-OF-THE-PLAINS — A slang expression for an emigrant WAGON, or PRAIRIE SCHOONER. (H)

SHOFUL — A slang expression (meaning counterfeit) used in London in the late 1830s, referring to a CAB made in imitation of Hansom's PATENT SAFETY CAB. (Moore, page 224)

SHOW CART — A highly finished CART, with wire wheels and pneumatic tires, used for show purposes. (O)

SIDE-BAR WAGON — A variety of ROAD WAGON, characterized by side-bar suspension. (H)

SKELETON WAGON — A four-wheel trotting WAGON, very light in construction (about 75 pounds), having a frame to which the seat supports are attached. They were intended for track use. (H)

SLAB CART — A heavy DUMP CART for hauling slabs and refuse from saw mills, the load being usually dumped into it from a chute overhead. (H)

SLED, SLEDGE, or SLEIGH — (From Middle English *sliden*, and Middle High German, *sliten*, to slide or slip. Equivalent to Middle Dutch and Middle Low German, *slede*.) These terms are applied almost synonymously to vehicles, equipped with runners instead of wheels, that move either passengers or goods over snow and ice. Not always restricted to winter use, the goods-carrying types are sometimes dragged over grass, or even over the earth. While it is impossible to assign specific usages to any of the three terms, SLED is most commonly used in the United States for the heavier, goods-carrying type, while SLEIGH is frequently used for the lighter, passenger-carrying type. The term SLEIGH is restricted largely to the United States and Canada, and is a contraction of the words *sled*, or *sledge*, by the omission of the letter *d*. SLEDGE is more commonly employed in England, but it is also frequently used in the United States.

Historically the sled is believed to be man's earliest vehicle, and probably originated with the branches of trees supporting a load as it was dragged over the earth. Later, more sophisticated sleds, equipped with

SEDAN CHAIR of the French Empire, period of Louis XV, with elaborate gilt moldings and paintings attributed to Watteau. Height without finial, 74 inches. (In Smithsonian collection, on loan from Margaret Garber Blue.)

straight skids or runners, were provided with rollers to facilitate the movement of heavier loads, thus beginning the evolutionary development of the wheel.

In the United States, one of the most popular sleighs was the CUTTER, this being a small two-passenger sleigh having but a single seat-board, usually drawn by one horse. The ALBANY and PORTLAND CUTTERS were among the more common varieties (see these separate entries). Double sleighs, having two seat-boards for the accommodation of four persons, were also popular. Two of the four-passenger styles were patterned after the PORTLAND CUTTER and the ALBANY CUTTER, the former being in effect two bodies in one, arranged in tandem, each having a full seat-board. The four-passenger ALBANY SLEIGH had the continuous curve of the CUTTER body, and two full seat-boards, though the rear seat was more roomy than the front. The ALBANY pattern was also used for a six-passenger sleigh that was sometimes open, or equipped with an extension-top as a half-top.

Many sleighs were also built with bodies identical to carriage bodies, these being identified by compound names, such as BROUGHAM-SLEIGH, VICTORIA-SLEIGH, LANDAU-SLEIGH, COACH-SLEIGH, VIS-A-VIS-SLEIGH, etc.

In the United States sleigh bodies were generally supported by benches and beams (see BOB-SLEIGH gearing nomenclature, page 36), while in Europe, iron scrollwork was more commonly used. Several methods were also used to convert ordinary wheeled carriages into sleighs. One of these employed *bobs* (see Carriage Nomenclature section) and was used largely in New England. The other method retained the carriage axles, with runner attachments substituted thereon in place of wheels.

Also see BOB-SLED. (H, C, M, & O)

SLEIGH, FOLLOW – See SHIFTER. (H)

SHEEPWAGON. Dimensions of bed, 11 feet long, 42 inches wide, 20 inches deep, with 18-inch wings on top of bed; five bows support cover; inside height, 6 feet. (Studebaker Bros. Mfg. Co., South Bend, Indiana, 1903.)

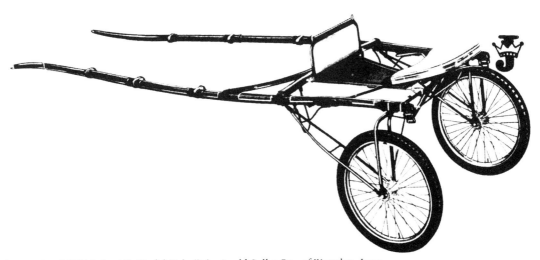

Modern SHOW CART is Jerald's Model G, built by Jerald Sulky Co., of Waterloo, Iowa.

SKELETON WAGON equipped with pneumatic tires. (F. C. Underhill: *Driving for Pleasure*, 1897.)

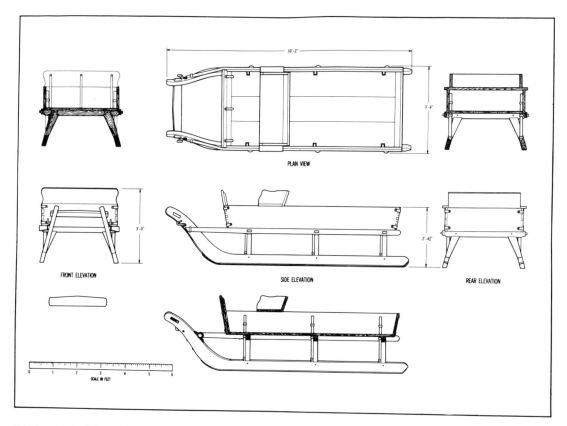

FARM SLED of about 1890. (Smithsonian drawing from original vehicle at the Horse 'n' Buggy Museum, Gettysburg, Pennsylvania.)

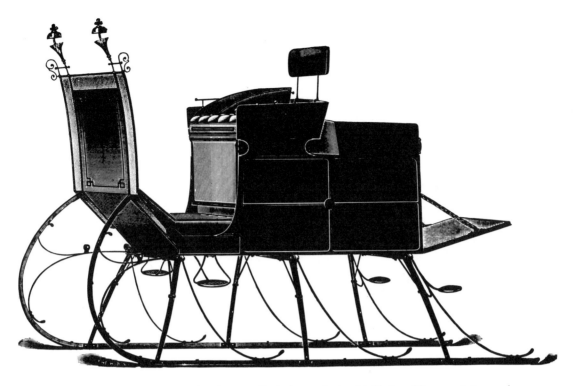

Gentleman's DRIVING SLEIGH. Length 9 feet overall; dark green body with black moldings; seat risers and runners, coaching red; red stripe on moldings; black ironwork; wire screens, rails and gongs, brass; trimming, drab corduroy with black welts; pom-poms, red. (*The Hub*, June 1897.)

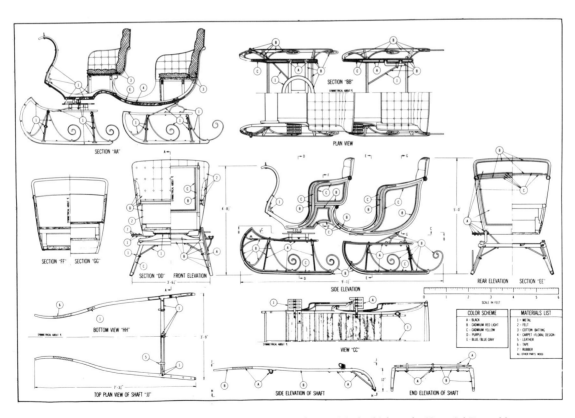

Four-passenger SLEIGH of about 1895. (Smithsonian drawing from original vehicle at the Horse 'n' Buggy Museum, Gettysburg, Pennsylvania.)

SLEIGH, HUB-RUNNER — Any vehicle having its wheels replaced by hub-runners, for winter use. (O & H)

SLEIGH, LONG-RUNNER or SINGLE-RUNNER — Any SLEIGH made with a single fixed runner on either side, as distinguished from a TRAVERSE-RUNNER SLEIGH. (H)

SLEIGH, PONY — A class name applied to any two-passenger SLEIGH, with or without a rumble, usually aristocratic in appointments and finish. Some builders used this term almost synonymously with CUTTER, but it was more commonly applied to the elaborately finished SLEIGHS equipped with a rumble. (H)

SLEIGH, SHIFTER — See SHIFTER. (H)

SLEIGH, TRAVERSE-RUNNER — A SLEIGH body mounted on two sleds, the rear one being fixed, while the forward one is supplied with a king-bolt, and revolves the same as the front gearing of a carriage. (H)

SLICE-OFF-AN-OMNIBUS — See BOULNOIS CAB. (H)

SLIDE-CAR — A primitive vehicle used in Ireland from the eighteenth century through the twentieth. It was a variety of TRAVOIS, shafted, and with a box, basket, or framework on the rear for carrying the load. (O)

SLING-CART — A heavy two-wheel vehicle used in the military service. The wheels were usually six to eight feet in diameter, and the load was suspended underneath the axle by chains. These CARTS were employed to move siege and garrison guns, the carriages of which were not adapted for transport. The lighter types, drawn by men, were called HAND-SLING-CARTS, and were used only for short distances. (W)

SLIPE — A primitive vehicle used in Ireland from a very early period to the present time. It is a type of SLEDGE and is found in many forms. It may have either one or two runners, with a crude framework, platform, basket, or box for carrying the load. It has no pole or shafts, but is simply dragged along the ground. (O)

SNOW ROLLER — A specialized piece of equipment used in New England, where heavy accumulations of snow during the long winter season made the removal of said snow impractical. Instead of attempting to plow the snow aside, the roller simply compacted it, making a firm surface for the animals and vehicles to negotiate. The roller generally consisted of two rollers on one axis, in order that there would be a differential effect which assisted in rounding corners. The rollers were built of planks and were frequently about eight feet in diameter. (O)

SOCIABLE — (1.) A four-wheel pleasure carriage that apparently developed in the early or mid-eighteenth century. While there is evidence that it may have originated in Germany, it was predominantly an English carriage. It was, according to Felton (1796), a "phaeton with a double or treble body, and is so called from the number of persons it is meant to carry at one time." It had either two or three seats (those in the main part of the body facing each other, as in a COACH), carrying two or three persons on a seat, according to the width of the body. The driving seat of this carriage was sometimes omitted. Sociables were made both with and without doors, and had either paneled or caned sides. Occasionally open, they more frequently had a half-top, an umbrella, or a combination of both. In some instances, the owners of these vehicles purchased only the body, and in fair weather substituted this body on the running gear of their COACH or CHARIOT. Its use continued into the nineteenth century, by which time it was very similar to a BAROUCHE or CALECHE. In 1862, Lawrence, Bradley and Pardee, of New Haven, Connecticut, were offering several models which they termed a *Sociable Half-top Caleche.* Shortly thereafter, while the type continued in popularity through the 1880s, the name was gradually supplanted by VIS-A-VIS, an eighteenth-century term that originally referred to an entirely different type of carriage. See VIS-A-VIS. (2) Sociable was also the individual name of the second public street carriage (forerunner of the OMNIBUS) that was introduced in New York City. Built by Wade and Leverich, it was put in service in 1829 by Abraham Brower, as a companion vehicle to the ACCOMMODATION. It was a rear-entrance carriage, with seats running lengthwise, as in an OMNIBUS. (C, H, F, & O)

SOLO — Synonymous with SULKY and RIDING CHAIR in eighteenth- and early nineteenth-century America, the term being applied to single-passenger two-wheelers. They were made both with and without tops. (O)

SPAR WAGON — An unidentified type of WAGON built by Amos Stiles, a carriagemaker of Moorestown, New Jersey, 1812-1821. It was a one-horse vehicle, and presumably was a light PASSENGER WAGON, probably on side spars, as suggested by the name. Possibly this WAGON was nearly identical to Stiles's BOULSTER WAGON, with the exception of the suspension.

According to Houghton, SPAR WAGON was the old name for a SIDE-BAR WAGON, the terms *spar*, or *side-spar*, and *side-bar* being synonymous. (H & O)

SPEEDING WAGON — An extremely light variety of DRIVING WAGON, most frequently made for only one passenger. (O)

SPINDLE WAGON — A ROAD WAGON or BUGGY having a body divided longitudinally, the lower section being

SLEDS and SLEIGHS used in Pennsylvania early in the nineteenth century. (Robert Sutcliff: *Travels in Some Parts of North America, 1804-1806,* 2nd edition, 1813.)

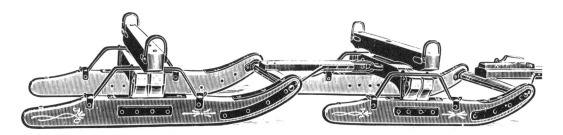

FARM SLOOP SLED with camel-back runners; 44-inch track, center to center, on 2-inch x 2½-inch runners. (The Adams Wagon Co., Ltd., Brantford, Ontario, ca. 1910.)

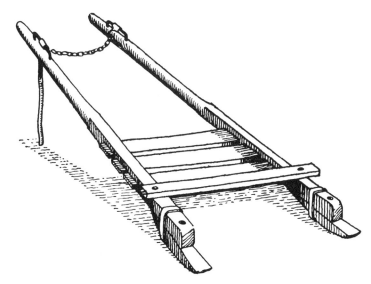

SLIDE-CAR. (G. B. Thompson: *Primitive Land Transport of Ulster,* 1958.)

255

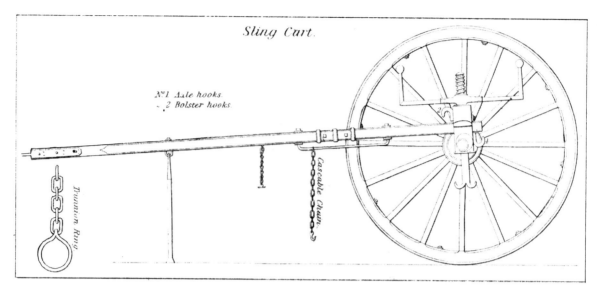

SLING-CART, showing the powerful screw on the axle that was used to lift a gun from the ground. (From John Gibbon: *The Artillerist's Manual,* D. Van Nostrand, New York, 1863.)

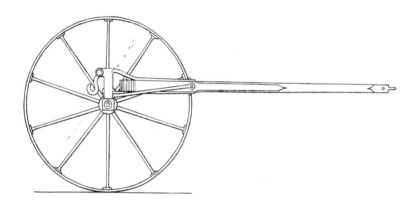

HAND SLING-CART. (From John Gibbon: *The Artillerist's Manual,* D. Van Nostrand, New York, 1863.)

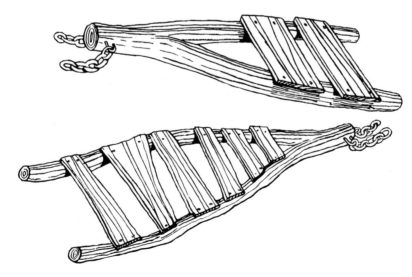

FORKED SLIPES. (G. B. Thompson: *Primitive Land Transport of Ulster,* 1958.)

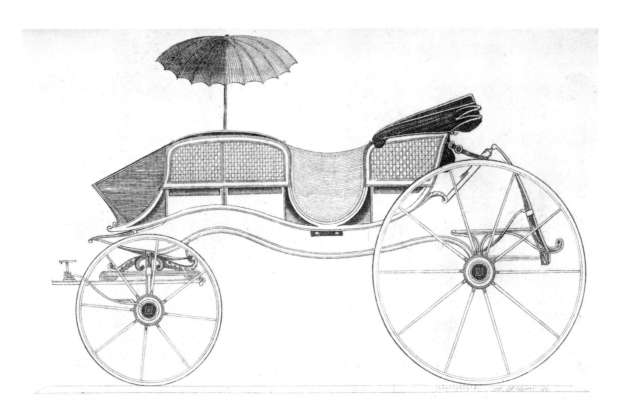

SOCIABLE. Body and gearing, rich medium blue striped with light blue; caned seat panels, sword case, black; knee boot and top, brown leather; umbrella, lavender. (W. Felton: *Treatise on Carriages,* 1796.)

BROWER'S 1829 SOCIABLE. (Ezra M. Stratton: *The World on Wheels,* New York, 1878.)

paneled, and the upper section being filled in with spindles, thus giving a light appearance to the vehicle, which is usually further emphasized by the use of a spindle-seat. (H)

SPRAY AND DISINFECTING CART (or WAGON) — A two- or four-wheel vehicle carrying a tank for liquids, and a force pump for spraying the liquid. Some types were used by farmers for spraying plants, trees, etc. Some were also used by the army and by municipalities for insect control, and treatment of objects or areas which presented a health hazard. (O)

SPREADING WAGON — See DUMP WAGON.

SPRING WAGON — See PLATFORM WAGON.

SPRINKLER — A CART or WAGON, provided with a suitable water-tank and sprinkling device, used for wetting down the dust on public thoroughfares. Some were also equipped with a sweeping apparatus. Sprinklers were likewise used for spraying oil onto roads and highways. (H)

STAGE — A general term applied to any public traveling vehicle. (H)

STAGE-BOAT — A FERRY-BOAT, used in connection with STAGECOACHES or STAGE WAGONS on post routes. (H)

STAGECOACH — A COACH used for public transportation, so-called because the journeys were accomplished by successive stages, after each of which the horses were invariably changed, and occasionally the COACH as well. Public COACHES operated between Edinburgh and Leith, in Scotland, about 1610, and the name STAGECOACH is known to have been used at least by the middle of the century. In America, STAGES ran from Boston to Newport, Rhode Island, as early as 1716, but it was almost mid-century before STAGE service began to increase to any extent. The first STAGES were, in reality, STAGE WAGONS, being simply covered WAGONS with benches inside. By the 1760s some of these WAGON bodies were hung on leather thoroughbraces. As eventually built, the conventional WAGON tops were replaced with comparatively flat roofs that were supported by slender pillars, the driver sitting inside on the front bench. Most of these STAGE WAGONS held about twelve persons who were seated on four benches, only the rear bench being provided with a back. Entry was gained through the open front of the WAGON.

The bodies of these STAGE WAGONS began late in the eighteenth century to resemble COACH bodies. Just prior to 1820 an oval-body stagecoach was developed, having a rounded top, a door in one side, an outside driver's seat, and thoroughbrace suspension on a three-perch running gear. About a decade later the famous American MAIL COACH was developed almost simultaneously by J. S. Abbot, then an employee of Lewis Downing, of Concord, New Hampshire, and by several carriage builders in Troy, New York. Known as CONCORD COACHES and TROY COACHES, the two were almost identical, sizes ranging from six- to sixteen-passenger, exclusive of passengers on roof seats.

After 1800 nearly all of the principal cities as far west as Pittsburgh were connected by STAGE lines, and from 1800 to 1840 the stagecoach provided the only public cross-country transportation that was available to a large percentage of the population. In 1832, 106 STAGE lines ran out of Boston alone. Even after the development of the railroad system, COACHES retained their importance, many thousands serving between and beyond rail lines, particularly in the West where the latter were widely separated. The greater number of these COACHES were also under contract with the government to carry mail. During the nineteenth century the speeds and fares varied according to period and locality, generally running between four and twelve miles per hour at a rate of three to fifteen cents per mile. Production of the famous CONCORD COACHES continued until about 1910, and service in some remote areas was provided for another decade.

In addition to the CONCORDS, built by the Abbot, Downing Company, this firm also turned out an extensive line of less expensive STAGES that are more properly classed as PASSENGER WAGONS, even though many of them have body features that might also qualify them as COACHES. These PASSENGER WAGONS were not only offered by Abbot, Downing, but also by numerous other carriage builders.

Also see CONCORD COACH, TROY COACH, PASSENGER WAGON, and STAGE WAGON. (H, C, & O)

STAGE WAGON — A primitive type of public traveling carriage used in England and America during the eighteenth century, and the early years of the nineteenth. The earliest forms were nothing more than ordinary covered WAGONS, with several transverse benches inside. The benches had no backs or padding, and the bodies were set directly on the running gears without benefit of springs or thoroughbraces, so that riding was most uncomfortable. In America, a gradual improvement took place in the construction of the STAGE WAGON during the last third of the eighteenth century. The seats were sometimes placed on springs, and eventually, the bodies were suspended on thoroughbraces. Bow-supported cloth tops gave way to permanent standing tops supported by eight slender pillars, leaving the sides open, with rolled curtains that could be let down in inclement weather. There were as yet no doors, and passengers had to crawl in, with difficulty,

SPEEDING WAGON on tubular steel running gear, wire wheels, and pneumatic tires. Built by George Werner, Buffalo, New York.

SPRAY AND DISINFECTING CART of about 1917. (National Archives photo.)

STUDEBAKER SPRINKLER WAGON. (Studebaker Bros. Mfg. Co., South Bend, Indiana, 1901.)

This two-seat SPRING WAGON had platform springs in the rear and an elliptic in front. It also came fitted as a *Delivery Wagon* as shown by the inset. With a black body on a red or dark green gear, and imitation leather trimming, it sold for $47. (Elkhart Carriage & Harness Mfg. Co., Elkhart, Indiana, 1907.)

American STAGECOACH of about 1820, showing some of the features incorporated into the design of the *Troy* and *Concord Coaches* that were introduced about 1827. Charles Veazie flattened the roof and added the luggage rail, while Orsamus Eaton added the roof-seat for extra passengers, just behind the driver's seat. (Basil Hall: *Travels in North America in 1827-1828,* 1830.)

STAGE WAGON shown in *The Pennsylvania Gazette,* April 24, 1760.

through the open front of the vehicle, over the driver's seat, the latter being under the same roof and on the same level as the passenger seats. The driver often shared his seat with one or two passengers.

Late in the century the body profile began to depart slightly from the straight lines of the WAGON body, and became somewhat curved, so as to resemble a COACH body. Beginning about 1800, some STAGE WAGONS were built with a door in one or both sides, which in turn caused the foremost passenger seat to face the rear. During the second decade of the century this carriage evolved into an oval-shaped COACH, which in the 1820s resulted in the famed American-style STAGECOACHES of the Troy and Concord types. STAGE WAGONS, according to Isaac Weld (1795), were built in varying sizes to accommodate from four to twelve persons. The term continued in use throughout the nineteenth century, long after the original versions had passed from the scene, and came to be somewhat synonymous with PASSENGER WAGON, particularly in the West. See Weld's description of a COACHEE, and his comparison of it to the STAGE WAGON. (O)

STANHOPE (or STANHOPE GIG) — A member of the CHAISE family, named after its designer, Fitzroy Stanhope. The Stanhope was designed about 1815, in an attempt to make the suspension of two-wheel vehicles easy to both horse and passengers. The body was mounted on two cross-springs, the ends of which were shackled to two side-springs that were in turn mounted on the axle. The ride furnished the passengers is said to have been pleasant, but the attempt to provide greater comfort for the horse reportedly was a failure at first, for the shafts vibrated considerably. Later, improved shafts, known as *lancewood fulcrum shafts*, were applied to the Stanhope, giving the horse its share of comfort. The Stanhope had a distinctive no-top body, that later was applied to a BUGGY and a PHAETON, the Stanhope name carrying over to those types (see STANHOPE BUGGY and STANHOPE PHAETON). The Stanhope was very much like the TILBURY, except in the manner of suspension, and the body of the Stanhope was placed on a locker, making a commodious space in which to carry luggage or parcels. The seat could be paneled, spindled, or caned, and in some lighter versions the locker was omitted, with a drop-seat-box sometimes substituted in its place for the accommodation of small parcels. A gentleman's vehicle, the Stanhope was drawn by one horse, or occasionally two in tandem, and was popular in both England and the United States. While it was most popular during the earlier period of its history, it nonetheless enjoyed a considerable popularity until the end of the carriage era. (H, A, C, & O)

STATION FLY — A fly used for railway service. (H)

STATION WAGON — Synonymous with DEPOT WAGON, except that the term is seldom applied to the earlier Dearborn type of DEPOT WAGON. (S226 & O)

STIVERS WAGON — A name sometimes applied to a PIANO-BOX BUGGY, in honor of the designer of that type of WAGON. (O)

STOLKJAERRE — A two-wheel CART peculiar to Norway, having a front seat for two persons, and a driver's seat in the rear. It is generally without springs. (H & O)

STONE-DRAG — The most primitive of vehicles, this DRAG consists of a plank or platform, the front end of which is shaped like a boot bracket. Stone, or other burdens, are placed on this conveyance, and it is dragged over the ground by man or beast. The ICE-DRAG of Russia approaches it in simplicity. Also called a STONE-SLEDGE, or STONE-BOAT. (H & O)

STREET-SWEEPER — See SWEEPER.

STUDEBAKER WAGON — A style of FARM WAGON widely used in newly settled areas of the West. It was built by the Studebaker Brothers, of South Bend, Indiana. (H)

SUICIDE — An Irish term, now obsolete, applied to one of those eccentricities in vehicles common in Great Britain early in the nineteenth century. It was described by Edgeworth in his *Essay on the Construction of Roads and Carriages*, (London, 1813), page 143: "Perhaps this name for a very high Gig, with a groom mounted on something like a stool, three feet above the driver, has not yet reached London. It was given to this species of carriages in Ireland, the land of appropriate agnomens." (H)

SULKY — The original application of this term was to a vehicle akin to the French DISOBLIGEANTE. Apparently about mid-eighteenth century, it was applied in a general way to a carriage having room for but one passenger, and from this aloofness of the rider, came the application of the adjective *sulky* to the vehicle. Said Felton, in 1796: "A Sulky is a light carriage built exactly in the form of a Post-Chaise, Chariot or Demi-Landau, but, like the Vis-a-vis, is contracted on the seat so that only one person can sit thereon, and is called a Sulky from the proprietor's desire for riding alone."

In America the term soon came to be applied to two-wheelers that were nearly identical to CHAISES,

STAGE WAGON. (Isaac Weld, Jr.: *Travels Through the States of North America and the Provinces of Upper and Lower Canada During the Years 1795, 1796 and 1797,* London, 1800.)

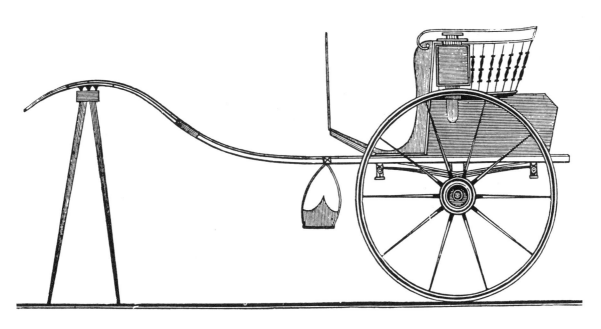

STANHOPE. (W. B. Adams: *English Pleasure Carriages,* 1837.)

STATION WAGON, available with pole and shafts. Wheels, 34-inch and 46-inch; body, dark green; gearing, dark carmine striped black; trimming, cloth, leather, or whipcord. (Studebaker Bros. Mfg. Co., South Bend, Indiana, 1903.)

STUDEBAKER THIMBLE-SKEIN FARM WAGON. Wheels, 44-inch and 54-inch. (Studebaker Bros. Mfg. Co., South Bend, Indiana, 1903.)

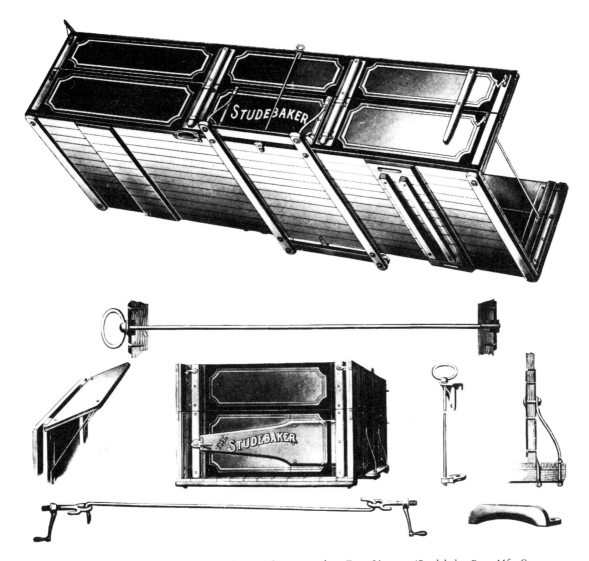

Details of Studebaker's "Twentieth Century" wagon box, as used on *Farm Wagons*. (Studebaker Bros. Mfg. Co., South Bend, Indiana, 1903.)

CHAIRS, or RIDING CHAIRS, except that the vehicles retained the single-passenger characteristics of the English SULKY. Sulkies were used almost exclusively by men, which caused such annoyance to some American ladies that they called the vehicle a SELFISH. Eventually, early in the nineteenth century, the word arrived at the final usage, to which it has since adhered, being applied to a two-wheeler that is reserved for track use. This vehicle consists only of wheels and axle, a skeleton seat for the driver, and a pair of shafts to which are mounted stirrups for the driver's feet. Modern trotting sulkies are very lightly constructed of wood, tubular steel, and aluminum, and are equipped with wire bicycle wheels and pneumatic tires. Also see CHAIR. (H, C, & O)

SULKY SADDLE — A variety of SULKY patented by a Dr. Stillman of New York City, wherein the rider's seat is advanced over the horse's back. (H)

SUMPTER-TRUCK — A term suggested by *The Hub* for a certain type of TRUCK, in reference to its capacity for carrying heavy freight. A Sumpter-horse or Sumpter-mule is one for carrying burdens. The same descriptive word might also be applied to other freight vehicles. (H)

SUNDOWN — A very light vehicle having two or more seats, but no top. (C)

SURREY (or SURREY WAGON) — A popular American four-wheel family carriage that developed comparatively late in the carriage era. The name supposedly came from an English SURREY-CART, the body of which was adapted to four wheels, but since this information appears to be erroneous, the real origin of the term is not understood. Apparently it was derived, through some misconception, from the English county of that name. The vehicle that obviously did lend its body was the English WHITECHAPEL-CART, the first one of which is believed to have entered the United States about 1867. A short time later, in 1872, James B. Brewster & Co., of New York City, introduced the surrey-wagon (by which name the surrey was first known), which consisted of a WHITECHAPEL body on four wheels. The vehicle might have been more properly named a WHITECHAPEL, or WHITECHAPEL-WAGON, after its progenitor, and it was, in fact, occasionally so-called by several builders. The first surrey-wagons had either one or two seat-boards, for the accommodation of two or four persons. The top line of the body was straight in the early types, following the WHITECHAPEL outline, and if the carriage held four persons, the left front seat usually turned over to permit access to the rear seat. Although the first surrey-wagons were most frequently built without a top, occasionally they were equipped with either a falling or standing top. The popularity of the surrey spread rapidly, and by the mid-1880s the sides of the four-passenger types were being cut down in the center to permit access to the rear without disturbing the front seat. Eventually the two-passenger variety became almost unknown. A large variety of styles developed, either with straight bodies or wheel-houses, panel or spindle seats, open, or with canopy, umbrella, or extension tops, and body styles ranging from STANHOPE types to the nearly straight lines of a SPRING WAGON. Popular as long as the horse continued in use, surreys could be purchased early in the twentieth century for prices ranging from $50 to $100. Toward the end of the carriage era, some builders applied the term CABRIOLET to their finer surreys. (H, M, & O)

SURREY-CART — According to Houghton, this was an English form of DOG CART, characterized by a WHITECHAPEL body. It was supposedly named after the county where it was first built. There is no evidence, however, that such a vehicle existed, and the information, therefore, is believed to be in error. (H & O)

SWEEPER — An apparatus mounting a large cylindrical brush, used for sweeping refuse from the streets. The brush was usually driven by gears or chains, or a combination of both, from the axle. (O)

SWELL-BODY (or SWELL-SIDE CUTTER) — See ALBANY CUTTER.

This modern SULKY is made by the Jerald Sulky Co., of Waterloo, Iowa.

An early style of SURREY, showing the lines of the *Whitechapel.* Wheels, 44-inch and 48-inch; body and gearing, black striped with carmine; trimming, green cloth. (*The Carriage Monthly,* October 1880.)

SURREY, still showing the *Whitechapel* lines, but now four-passenger. Wheels, 43-inch and 46-inch; body and gearing, black striped with red; trimming, brown cloth. (*The Carriage Monthly,* January 1883.)

Straight sill SURREY. Wheels, 42-inch and 46-inch; body, black with pillars and seat panels of deep green; moldings, black, fine-lined with carmine; gearing, carmine striped black; trimming, blue cloth. (*The Hub*, August 1896.)

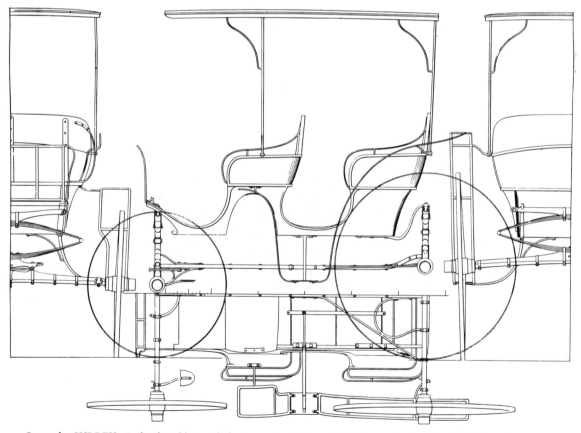

Cut-under SURREY. Body, deep blue, with dusters and moldings of light blue, striped with a single line of cream; gearing, deep blue striped cream; trimming, dark blue cloth. (*The Hub*, January 1895.)

ACME STREET SWEEPER, with water tank and sprinkling mechanism added to keep down dust. (Studebaker Bros. Mfg. Co., South Bend, Indiana, 1909.)

Bottom view of ACME STREET SWEEPER, showing how bevel gears drive the jack shaft from the axle, and the jack shaft sprocket from which a chain ran to the broom sprocket (not shown). The broom shaft was supported in bearings A and B.

T

TABLEAU WAGON — A type of parade CIRCUS WAGON. These were elaborately painted and often had carvings on the sides. The principal type had elaborate spectacles on top which employed both large carved figures, or costumed persons. Baggage was carried inside these WAGONS. (O)

TALIKA — The public CAB of Constantinople. (H)

TALLY-HO — A popular American term for a four-in-hand COACH, derived from the individual name of the ROAD COACH introduced in 1876 by Colonel Delancey Kane for service between the Brunswick Hotel in New York City, and Pelham. "Some newspapers," wrote Fairman Rogers, "in writing about the Pelham coach, called it *the* 'Tally-ho,' and others, less well informed, called all four-horse coaches 'Tally-ho's.' " The error was perpetuated by an American dictionary, the *Century*, which in 1891 gave as one of the meanings of the term, "a four-in-hand pleasure-coach." The application of the term to any ROAD COACH is entirely incorrect, and should be avoided. (H & O)

TALLY-HO BREAK — See BREAK.

TANDEM-CART — A DOG CART intended to be driven with a tandem team. It is usually hung quite high, and is elaborate in its appointments. Sometimes this vehicle is equipped with brakes, unusual for a two-wheeler, and it may also have a shifting device to move the body backward or forward, so that it will be evenly balanced with either two or four passengers. (H)

TAPCU — (From the French, meaning, literally, seat thumper.) A term applied in some parts of France to a variety of CART . (H)

TARANTASS — A four-wheel Russian traveling carriage. The body resembles a flat-bottom punt, and is placed on a series of long poles which connect the axles. A hood and apron protect the passengers from the weather. (H & O)

TAR-WAGON — A WAGON, provided with a furnace, used in heating tar or asphalt for application to pavements. (H)

TAX-CART — A term which is said to have come into use in England in 1843, when a law was introduced exempting from tax any two-wheel vehicle costing no more than twenty-one pounds. The law also required that the name of the owner be painted on the vehicle. Any CART answering the above-named conditions was originally called a TAX-CART; later the term was applied to any inexpensive CART. See DOG CART. (H)

T-CART — A variety of PHAETON, being a modification of the STANHOPE PHAETON. According to Sidney, in his *Book of the Horse*, the T-cart seated only one groom behind, and was much used by sporting men among the military. Why it was so-called is not understood, but we are led to believe by Mr. Sidney that it may have come from the resemblance of the ground plan of the vehicle to the letter T. This is also another example of the improper application of the word CART to a four-wheel vehicle. The modern T-cart is a four-wheel, square-body, open PHAETON, accommodating four passengers. It is hung on four elliptic springs, and has a cut-under body. The lighter build of the T-cart distinguishes it from the STANHOPE PHAETON. (H)

TEAM — This term is strictly applicable to a pair of horses, but in America it is frequently applied to a vehicle with horses, either for industrial, agricultural, or pleasure purposes. (H)

TELEGA — A crude, springless, Russian passenger vehicle. (O)

TELEGRAPH — A name applied in England to a vehicle of the CHAISE family, in the early part of the nineteenth century. Felton gives a plate and description of one in the second edition of his *Treatise on Carriages* (1805), but it is not mentioned in the edition of 1796. The plate represents a two-wheel CHAISE or

270

TANDEM CART with 60-inch wheels. The body is black with carmine slats, while the gearing is carmine striped with black; trimming, dark green corduroy. (*The Carriage Monthly,* June 1894.)

T-CART. Wheels, 33-inch and 41-inch; body and gearing, black, with gearing striped with a broad line of blue, edged with fine lines of canary yellow; trimming, blue cloth. (*The Hub,* May 1887.)

WHISKEY, with a lower body resembling a Salisbury boot (said to be used for the conveyance of dogs for sporting purposes), surmounted by a seat known as the Barouche seat. The term TELEGRAPH was later applied to one of the early English MAIL COACHES, probably because of its superior speed. The name probably derives from the French *telegraph*, a 1792 invention of semaphore type used for the rapid conveyance of messages. (H & O)

THENSA — The sacred car of the Romans, used to carry the images and sacred vessels of the gods to the races (illustrated in *Scribner's Monthly*, May 1879). The heavy body was similar to that of the Roman CHARIOT, attached rigidly by three supports to the axles, and borne by four, four-spoke wheels. (h)

THESPIS, CAR OF — The vehicle in which Thespis, an early Greek dramatic poet, generally regarded as the inventor of tragedy, is said to have traveled from place to place, and to have used as a stage for his presentations. Dryden's prologue to Nathaniel Lee's *Sophonisba*, reports, "Thespis, the first professor of our art, at country wakes sung ballads from a cart." (H)

TILBURY — A two-wheel vehicle that originated about the same time as the STANHOPE, said to have been designed by Fitzroy Stanhope, but named after the London carriage builder who first constructed it. The Tilbury resembles the STANHOPE, except in the manner of suspension, and the absence of the locker under the seat that characterizes the STANHOPE, although some Tilburys were built with a drop-box under the seat. This carriage became popular soon after its introduction, then lost popularity, but regained it toward the end of the nineteenth century. Deceptively light in appearance, the Tilbury is said to have been one of the heaviest two-wheelers because of the suspension and ironwork. Suspension of the Tilbury was on seven springs and a pair of braces. The rear ends of the shafts rested on a pair of half-elliptic springs set transversely on the axle, and the body hung on four single elbow-springs (those in the rear might more appropriately be called *loop-springs*), the rear pair hanging from braces suspended from the ends of a high cross-spring. Originally the axle springs were not used, carved wooden blocks being used in place of them, but Adams implies in 1837 that the half-elliptics had already been in use for some time. In later years the term was applied to similar-appearing vehicles that used platform springs or some other method of suspension. As previously stated, the no-top body of the Tilbury was much like that of the STANHOPE, except for the absence of the large locker that was situated under the seat of the STANHOPE. The seat was often of the spindle type. Since the Tilbury was owner-driven, the groom sat to the driver's left. Adams was critical of the Tilbury's appearance, and also complained that the suspension caused an unpleasant motion to the body, and lessened the driver's control over the horse. On the other hand, Underhill, in 1896, wrote that it was one of the handsomest two-wheelers, and felt that there was "no carriage more smart for use with a brilliant goer." The popularity of the carriage supports Underhill's comments rather than those of the overly critical Adams. Also see STANHOPE. (H & O)

TILT-CART — Same as TIP-CART, or DUMP-CART.

TIMBER-BOB — See NIB. (O)

TIP CART — Same as DUMP CART

TOBOGGAN — (Derivation: from a number of similar sounding Algonquian words (such as *odabanak*), describing a light sledge used by certain Canadian Indians. The term PUNG comes from the same source.) Of simple construction, the toboggan consisted of a single board curved upward in front to form a dasher, and was provided with hand rails or ropes. This same sled is used much today for sporting purposes. The term was also occasionally applied to certain vehicles with runners. (H & O)

TO-CART — The term by which the DOG CART is known in France. (H)

TOMBEREL — An old French term equivalent to, and corrupted into, TUMBREL or TUMBRIL. (H)

TONGA — A two-wheel passenger vehicle of India, carrying three passengers and a driver. Drawn by two horses, it has a broad seat and is hung quite low, with the sides and slatted top covered by heavy canvas. (H, Aug. 1897, p. 285)

TORTOISES, FOUR-WHEELED — A slang term of reproach, like HELCART, applied to public COACHES by John Taylor, the Water Poet, in 1623. See HELL-CART. (H)

TOWER WAGON — A specially equipped vehicle, often an EXPRESS WAGON, having an elevating platform that could raise workmen to a convenient height for working on overhead wires. They were frequently used by street railway companies to repair trolley lines. (O)

TRAINEAU — A French term, meaning a sleigh or sledge. (O)

TRAM — This term, as early as the sixteenth century, was applied in England to a sledge-like frame used to carry baskets of coal out to the mines. Eventually these were provided with low wheels. The word is derived from the Scandinavian *traam*, the Low German *traam*, and the High German *dram*, meaning a beam of wood.

A RUSSIAN TELEGA. (M. M. Kirkman: *Classical Portfolio of Primitive Carriers*, 1895.)

TELEGRAPH of 1805. (Ezra M. Stratton: *The World on Wheels*, New York, 1878.)

A bronze THENSA, or ROMAN SACRED CAR, illustrated in *Scribner's Monthly,* May 1879.

Thus the original trams were so-called because they were constructed of several light beams. The supposed derivation from the name of Benjamin Outram, an early railway engineer and contractor, is erroneous. Late in the eighteenth century the term became associated with the words *road* and *way* (*tram-road* and *tram-way*), but it is uncertain if this was because a tram-way was built of wooden beams, or rails, or because trams traveled over them. It does appear certain that the term TRAM was applied to vehicles before these vehicles ran on rails. Finally, in the nineteenth century, the terms TRAM, or TRAM-CAR were applied to street railway cars in England, just as the terms HORSE-CAR and STREET-CAR are used in America. (H & O)

TRANSFER WAGON — See DRAY. (S226)

TRAP — A colloquial term used in both England and America as early as the late eighteenth century, to designate any light pleasure vehicle. At about the same time, the name was also applied in England to a shooting GIG, which later came to be known as a Dog Cart. According to the October 1897 issue of *The Hub*, this GIG had a box behind the seat in which a hunter's dogs could be carried. The back end of this box had a lid, or trap door, that opened downward, and the vehicle thus came to be known as a *trap*, by the omission of the word *door.* While the name TRAP continued to be applied to an enlarged body of this type on four wheels, the shooting GIG came to be known as a DOG CART.

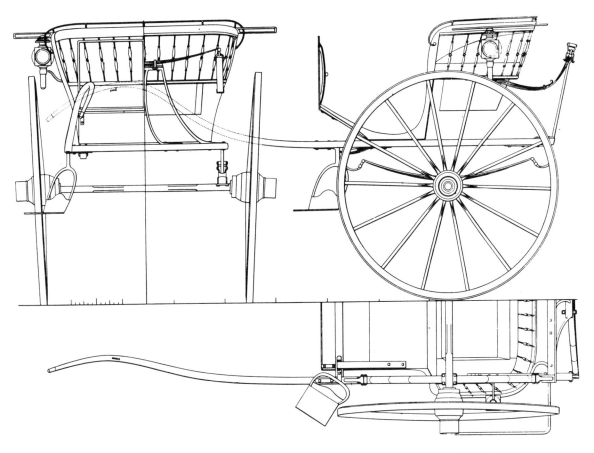

TILBURY. Wheels, 52-inch; body, black, with carmine seat spindles; gearing, carmine, striped black; buff corduroy trimming and silver mountings. (*The Hub,* June 1894.)

India's TONGA, as illustrated in *The Hub,* August 1897.

In America, a four-wheel sporting carriage had become popular by the mid-nineteenth century, but it was not yet known as a TRAP. It was known by various names such as DOS-A-DOS, SLIDING-SEAT WAGON or CARRIAGE, DOG CART PHAETON, FOUR-WHEEL DOG CART, or simply, DOG CART, though this last usage confused the vehicle with the true DOG CART. This carriage was characterized by some type of shifting seat, though the mechanical features of this arrangement were by no means standardized. *Trap* appears not to have been applied in America to this specific type of vehicle until 1892, when plates bearing the name appeared in both *The Hub* and *The Carriage Monthly*. In 1895 *The Hub* advised, "the trap family is large and the Christian names many, but as a rule they indicate a whim of the builder, rather than give any indication as to the specific character." The trap generally had specific characteristics, however, the principal feature being the sliding, swinging, or pivoting seats which permitted conversion of the carriage for the accommodation of either two or four passengers. The rear seat was often arranged so that it could be made to face either the front or back, or slide entirely out of sight, and the front seat could sometimes be moved backward or forward, or half of it could be swung or tipped aside to permit access to the rear seat. If arranged dos-a-dos, the hinged rear end-gate opened downward to serve as a foot-rest for the rear passengers. Hundreds of patents were issued on the various mechanical arrangements for these movable seats. Other features characteristic of the trap were a short wheelbase, with the body and seats having an overall appearance of height, this being accentuated by the short wheelbase. Suspension was frequently on elliptic springs and for a time in the early 1890s traps with a natural-wood finish were in style. No tops were provided. Vehicles capable of carrying only two persons were also called TRAPS, because they exhibited the general outline of the convertible trap, but lacked the additional movable seat. As a writer in *The Hub* pointed out, this usage might best be avoided, yet it cannot be ignored since many builders did assign the name TRAP to these two-passenger vehicles. (H, M, & O)

TRAVELING-FORGE — A complete blacksmith's establishment, which accompanies an artillery battery for the purpose of making repairs and shoeing horses. It consists of a body, upon which is constructed the bellows-house, etc., and the limber, which supports the stock while the forge is being transported. The bellows-house is divided into the bellows-room and the iron-room. Attached to the back of the house is the coal-box, and in front of it is the fireplace. From the upper and front part of the bellows an air-pipe proceeds in a downward direction to the air-box, which is placed behind the fireplace. The vise is permanently attached to the stock, and the anvil, when in use, is supported on a stone or wooden base, and when transported is carried on the hearth of the fireplace. The remaining tools are carried in the limber-chest. When in working order the end of the stock is supported by a prop. (W)

TRAVELING-KITCHEN — A WAGON or CART equipped with an oven and cooking range, which accompanies troops on the march. (W)

TRAVOIS (or TRAVOISE) — (Derivation: a corruption of *travail*, a frame in which unruly horses are confined while being shod. Also, in Canada, the word *travail* was applied to the space between the shafts of a vehicle, in which the horse was positioned for draught purposes.) Because of these similarities to the primitive means of transportation employed by certain North American Indians, the word *travois* came to be applied to the two long, slender poles that were attached to the sides of an Indian pony or dog, and dragged behind the animal to carry either goods, or aged and infirm persons. Between the poles, at the rear, was arranged a crude platform, or perhaps a net of rawhide, vines, or rope, on which the burden was placed.

Travois was also used to designate a type of SLEDGE used in the lumbering industry. One end of a log was supported by this SLEDGE, while the rear end trailed on the ground. (O)

TRAVOY — This spelling is sometimes used for the variety of TRAVOIS used by the lumbering industry, described above. (O)

TREE TRANSPLANTING WAGON — A special purpose WAGON, the use of which is self-descriptive. (O)

TRESTLE WAGON — Part of the equipage of a canvas pontoon train. The bridge trestles were carried on this wagon. (O)

TRIBULUM — An ancient Roman threshing machine, consisting of loaded boards shod with iron bars, and drawn by oxen. (H)

TRIBUS — A two-wheel, three-passenger public CAB, introduced by a Mr. Harvey, of London, in the spring of 1844. It was similar to a HANSOM CAB, with the driver sitting at the rear, and on a level with the roof. His seat was to the right, and the door was at the rear, to the left of the driver, so that he could open it without leaving his seat. It had five windows, two in front, one on each side, and one in back, underneath the driver's seat. It was drawn by one horse. Mr. Harvey also built a CURRICLE TRIBUS, for two horses, across

A common variety of TOWER WAGON used throughout the United States during the early years of the twentieth century.

PARK TRAP made for two passengers only. Wheels, 40-inch and 46-inch; panels, blue with red stripe, balance of body, black; gear, red with black stripe; trimming, drab cloth or corduroy. (*The Hub,* April 1895.)

PARK TRAP for four passengers, with a rear seat that will face in either direction. Wheels, 43-inch and 47-inch; panels, blue; moldings, black; slat work, seat risers, and toe-board, vermilion; gear, vermilion striped with black; trimming, drab Bedford cord. (*The Hub*, July 1895.)

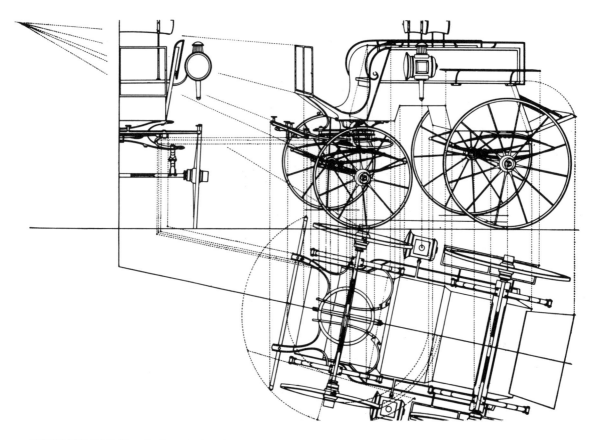

KENSINGTON TRAP. Wheels, 36-inch and 46-inch. (*The Hub*, February 1898.)

A group of Southern Cheyenne using a Travois. (Smithsonian photo.)

Tree transplanting wagons of this type could move a tree with a ball of earth weighing up to 14 tons. (*Manufacturer & Builder*, May 1870.)

Side view of Mr. Harvey's TRIBUS. (H. C. Moore: *Omnibuses and Cabs*, 1902.)

whose backs was a horizontal steel bar, similar to that of the earlier CURRICLE. Neither type was very successful. (H & O)

TRICYCLE — A three-wheel VELOCIPEDE. (H)

TRIOLET — A two-wheel French carriage patented in 1826. It carried three passengers (from whence the name apparently comes), and a driver, and was suspended by a single half elliptic spring. (H)

TROIKA — A Russian term meaning a team of three horses harnessed abreast; specifically it came to be applied to the SLEIGH drawn in this manner. (H)

TROLL — (Derivation: from Middle English *trollen*, to roll.) A hawker's CART. Also sometimes synonymous with TROLLEY. (O)

TROLLEY (or TROLLY) — An English term applied to various types of CARTS and TRUCKS. It was sometimes applied to hand CARTS and BARROWS, and occasionally to certain SLEDGES that were used for farming. (O)

TROTTLE-CAR — A two-wheel vehicle used in Ireland from the late eighteenth century through the twentieth century. It was somewhat like the WHEEL-CAR, but had spoked wheels outside the shafts, the wheels turning freely on a fixed axle. Sometimes called a CART-CAR, it was the forerunner of the Irish JAUNTING-CAR. (O)

TROY COACH — A popular type of public traveling COACH built by several different firms in Troy and Albany, New York. This type was apparently nearly identical to the CONCORD COACH, and in fact seems to pre-date the CONCORD, which was first built in July 1827, while the Troy is known to have existed at least in May of that year. Charles Veazie of Troy seems to have made the first improvements, including a roof railing for luggage, while Orsamus Eaton added a roof seat a short time later. From available figures, it would appear that the Eaton & Gilbert firm built more COACHES than Abbot, Downing & Company, although the former gave up COACH construction much earlier than the latter, in order to concentrate their efforts on railway car construction. (O)

TRUCK — (Derivation: believed to have come from the Latin *trochus*, a hoop). (1.) As early as the seventeenth century, the term came to be applied to a small strong wheel of wood or iron, particularly one supporting a ship's gun carriage.   (2.) In the eighteenth century, *truck* was also applied to a strong WAGON or CART used for heavy hauling, and has been rather generally applied ever since to both two- and four-wheel vehicles used for this purpose. In America the two-wheel variety of truck, synonymously called a DRAY, was predominant over the four-wheel type until well after the mid-nineteenth century. Gradually, then, the four-wheel truck supplanted the two-wheeler, but never to the extinction of the latter, which continued in use until the end of the horse-drawn era. The term is frequently used to designate any of that family of vehicles intended for delivery work, but might more specifically be applied to that class of medium to medium-heavy types used for bulk deliveries from wholesaler to retailer, or from the latter to the consumer. While styles vary to some extent, the later truck is generally characterized by having a full-circle fifth-wheel, which provides great maneuverability in crowded city areas. The seat is high above the body, and usually turns over forward to be out of the way while loading. The floor is often sloped downward toward the rear, to facilitate the loading and unloading of goods. Suspension is most frequently on springs, sometimes of the platform variety, but again on a special truck spring resembling a half elliptic that is shackled only at one end, while the opposite end slides freely in a box. On some light trucks, ordinary elliptics are sometimes used on the front. The sides are often formed of racks, or of standards with chains. Painting is likely to be bright, with rather heavy striping. (3.) A two-wheel hand-barrow, or hand truck, such as those used in warehouses. (4.) Various types of small, sturdy, four-wheel vehicles used in warehouses, and around railway depots. (5.) The swiveling carriage of a railway car, having two or more pairs of wheels, with axles and springs. (H, C, M, & S226)

TRUCK, HOOK-AND-LADDER — A long WAGON used for conveying fire ladders. (H)

TRUCK, HOSE — A carriage used to carry fire hose. (H)

Mr. Harvey's TRIBUS [rear view]. (H. C. Moore: *Omnibuses and Cabs*, 1902.)

Mr. Harvey's CURRICLE TRIBUS. (*The American Coach-Makers' Illustrated Monthly Magazine,* November 1855.)

TRIOLET, invented in 1826 by M. Avril, of Paris. The spring rests on the axle, and the ends protrude through the sides of the body, which was suspended from a bar running across the roof. (Ezra M. Stratton: *The World on Wheels,* New York, 1878.)

TROTTLE-CAR. (G. B. Thompson: *Primitive Land Transport of Ulster*, 1958.)

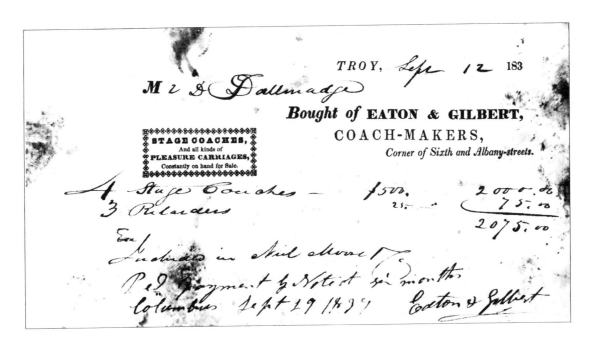

While no authenticated illustrations of *Troy Coaches,* enumerated on this Eaton & Gilbert invoice, have been discovered, it is likely that they were nearly identical to *Concord Coaches.* It seems a good possibility that some unmarked *Coaches* that are attributed to *Concord,* may in reality be *Troy Coaches.* (Invoice in Smithsonian collection.)

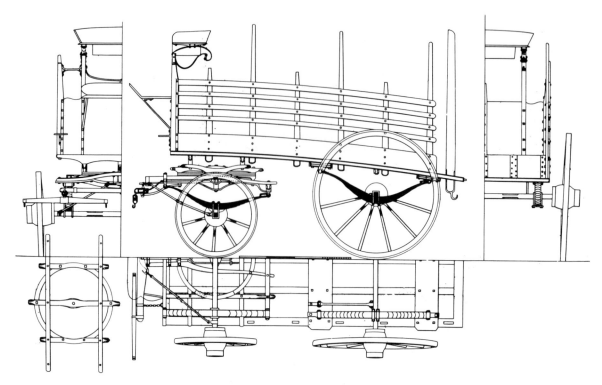

A medium-size TRUCK built by J. Sebastian of New York City. Wheels, 34-inch and 52-inch; body, Milori green striped black and edged with white; gearing, light vermilion striped with black and fine lined with white. (*The Hub*, October 1883.)

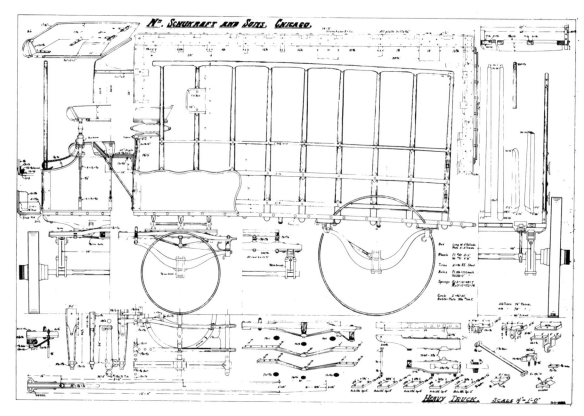

Heavy TRUCK built in 1898 by William Schukraft and Sons, Chicago. Wheels, 38-inch and 56-inch; dark green body on red or yellow gearing. (Drawing from Smithsonian collection.)

Heavy TRUCK on truck springs, equipped with winch for drawing on heavy loads. Wheels, 40-inch and 48-inch; body and gearing, carmine striped white and black. (*The Carriage Monthly,* December 1899.)

Standard two-horse TRUCK, built by Wm. Schukraft & Sons, of Chicago, about 1900. (Smithsonian photo.)

TRUCKLE-CAR — A primitive type of WHEEL-CAR used in Ireland and Wales during the eighteenth and nineteenth centuries. (O)

TRUCK, WAREHOUSE — A hand-truck intended for indoor use, or for use around warehouses. (H)

TRULL — A variant spelling of TROLL. (O)

TUB-CART — A term sometimes applied to a GOVERNESS CART, particularly the heavier variety. (O)

TUB-PHAETON — A PHAETON having a tub-body, characterized by a short, curved bottom line, and a deep, heavy quarter. (H)

TUMBLER — A colloquial expression, meaning TUMBREL. (H)

TUMBREL (or TUMBRIL) — (From the Old French, *tomberel*.) An agricultural cart, commonly used for hauling dung. During the French Revolution, this vehicle acquired a melancholy celebrity from the fact that it was used to carry the unfortunate victims to the guillotine. The term was also applied to the Carts used by the artillery for the conveyance of tools and ammunition. (H, C & W)

TURNOUT — Usually spoken of as an American slang expression, applied in a general way to a horse and carriage. However, we also find the following quotation in Adams's *English Pleasure Carriages*, London, 1837 (page 55): "Within our own time may be remembered the uncouth 'turns-out,' [sic] as a coachman would emphatically call them, both in private and public conveyances: — heavy coaches, laden with ill-packed luggage; miserable horses, bound in worse harness, . . ." (H)

TWO-WHEELER — American synonym for CART. (H)

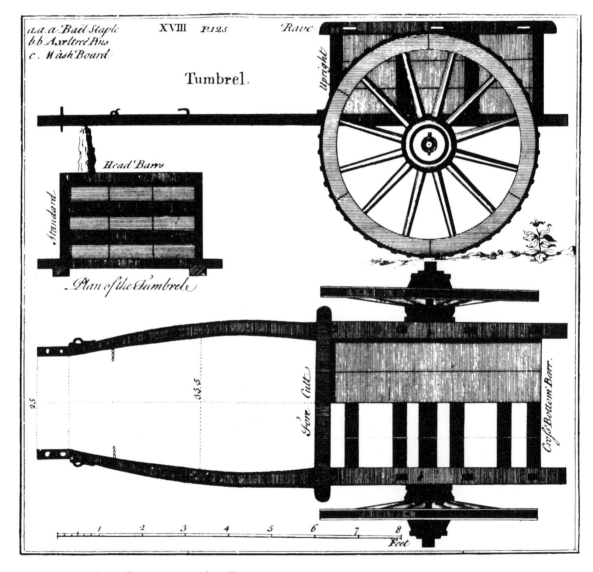

TUMBREL. (John Muller: *A Treatise of Artillery*, London, 1780.)

U

UNDERDRAW-WAGON — A freight WAGON wherein the draught is from the bottom of the axle, below the tongue. (H)

UNDERTAKER'S WAGON — A somewhat general term, applied to a variety of styles of WAGONS used by undertakers. The principal types were also called CASKET WAGONS, these being similar to delivery WAGONS, and were used for transporting caskets and other funeral accessories at times other than in funeral processions. They could also carry flowers in a procession. For several decades after the mid-nineteenth century, these WAGONS were open, with a seat on risers, tastefully designed and finished, and suspended on either elliptic or platform springs. A plated rail often surmounted the rear portion of the body. By 1890 there had begun a trend toward closed UNDERTAKER'S WAGONS, very much like a panel-side DELIVERY WAGON, with rear doors, covered driver's seat, beveled glass windows, and mourning draperies in the windows. Some of this latter type, in fact, approached the HEARSE in trimmings and finish. A lighter type of vehicle, frequently called an EMBALMING WAGON, was likewise called an UNDERTAKER'S WAGON. (O)

UNICYCLE — A one-wheel VELOCIPEDE. (H)

UTILITY CART — See BREAKING CART. (S226)

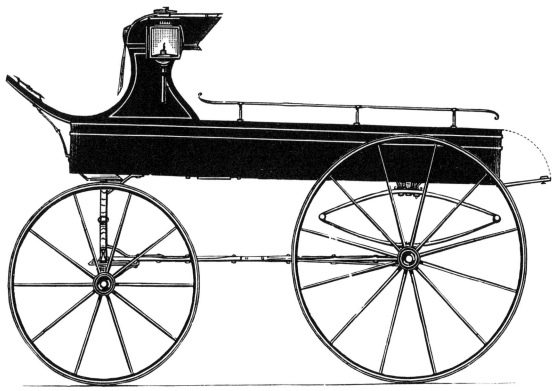

UNDERTAKER'S WAGON. Wheels, 42-inch and 50-inch; body and gearing, black; moldings may be silver-leaf or bronze. (*The Hub,* February 1881.)

UNDERTAKER'S WAGON. Wheels, 33-inch and 48-inch; body, black; bracket front, and pillar, deep green striped pale green; gearing, deep green striped black, or pale green. (*The Carriage Monthly,* December 1895.)

V

VAN — A modern abbreviation of *caravan*. A general term, used at least as early as 1829, for any large, covered, boxlike vehicle, either road or railway, intended for the conveyance of freight. The term is used in numerous self-descriptive compounds, such as CIRCUS-VAN, BRAKE-VAN, GOODS-VAN, GUARDS-VAN, etc. (H)

VARDO — See GYPSY-WAGON.

VASOK — A Russian carriage mounted on runners. (H)

VEHICLE — (From the Latin *vehiculum*, a carriage.) A general term for any form of carriage, cart, wagon, or sled. (H)

VELOCIPEDE — (From the Latin *veloci*, crude form of *velox*, meaning swift; and *ped*, stem of *pes*, the foot, thus meaning swift-footed.) A light vehicle intended to be propelled by the feet, this class of vehicles includes the UNICYCLE, BICYCLE, TRICYCLE, etc. Specifically the term is commonly applied to the BONE-SHAKER, which see. (H)

VETERINARY AMBULANCE — A vehicle for transporting sick or injured horses, or other large animals. The body, often a huge, padded box with high sides, was hung very low by means of a cranked rear axle, so that the floor came within a few inches of the ground. The equipment included some sort of windlass which drew up a sling to help support the animal. Some of these ambulances were constructed with only two wheels. (H & O)

VETERINARY CART — A two-wheel variety of VETERINARY AMBULANCE. (O)

VETTURA — A four-wheel carriage used in Italy as a public conveyance. Used as early as the late eighteenth century, its driver was known as a *vetturino*, a term also applied in the seventeenth century to one who rented out carriages and horses. (C & O)

VICEROY — A modern, four-wheel show carriage, highly finished and with numerous chrome-plated parts. This light-weight vehicle is mounted on wire wheels fitted with ball bearings and pneumatic tires. The running gear is generally similar to that of the BIKE WAGON. The skeleton body carries a spindle seat, and is provided with patent leather dash and fenders, the latter being attached to the seat rail. (O)

VICTORIA — This carriage is believed to have had its beginnings in England, in both the GEORGE IV PHAETON, a variation of which was built for Queen Victoria in 1850, and in the CAB-PHAETON. Some of the designs of these carriages migrated to the Continent, where, being more favorably received than in England, they resulted in the vehicle known as the MILORD. First used as an aristocratic pleasure carriage, the MILORD soon degenerated into a public hack, and after 1850 lost favor among the gentry. The term VICTORIA was applied by the French to some of these carriages at least as early as 1844, in honor of the English Queen. In 1869 the Victoria returned to England when the Prince of Wales imported one from Paris, and Baron Rothschild imported one from Vienna, following which it gained an immense popularity among both the English and American aristocracy.

Adaptable for use with either one or two horses, the Victoria has but one seat-board, and a curving dash that is reminiscent of the GEORGE IV PHAETON, with the driver's seat being supported by an iron framework over the dash. Occasionally a rumble was added. A variation of the Victoria is the panel-boot Victoria, sometimes called a CABRIOLET. The forward portion of the body differs from the true Victoria in having a paneled driver's seat, framed to the body proper, with a straight, conventional dash standing in front of the seat. Many have a child's seat that folds out of the rear of the panel-boot. Suspension of both these vehicles may be on four elliptics, elliptics and platform springs, C-springs, or double suspension.

The Victoria was mainly a park carriage in England and the United States, and was more stately in form

An American VELOCIPEDE of 1869, popularly known as a BONESHAKER. (Smithsonian collection.)

The Jerald Devon VICEROY is built by the Jerald Sulky Co., of Waterloo, Iowa. Priced at $1550 in 1977, the standard color is a deep maroon with vermilion striping. This one has chrome wheels.

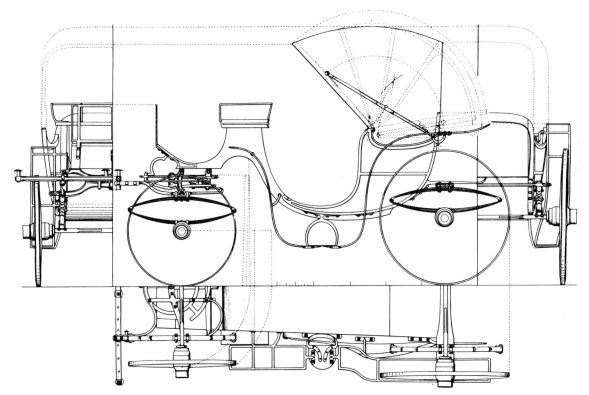

Paneled-boot VICTORIA. Wheels, 36-inch and 44-inch. (*The Hub,* February 1908.)

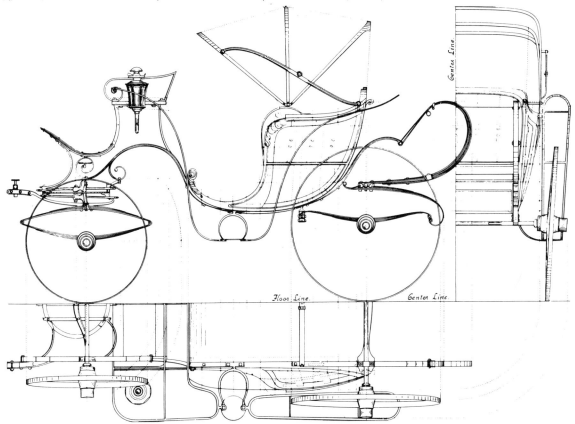

VICTORIA, with double suspension in the rear. Wheels, 35-inch and 44-inch; body, ultramarine blue with black moldings; spindles, pale blue shaded with deep blue and yellow; moldings striped with a fine line of yellow; gearing, ultramarine blue striped with two 1/8-inch lines of pale blue with a fine line of yellow in between; trimming, blue cloth. (*The Carriage Monthly,* April 1893.)

Double-suspension GRAND VICTORIA, employing C-springs on elliptics, exhibited at the World's Columbian Exposition in Chicago by Studebaker Brothers. Wheels, 34-inch and 45-inch; body colors not given; gearing, dark maroon striped with three lines of silver bronze blue; trimming, drab cloth with maroon velvet broad lace. (*The Hub,* March 1894.)

A VICTORIA-HANSOM of 1887. (*The Hub,* April 1890.)

than the CABRIOLET. In Europe the driver's seat was sometimes removed from the Victoria, and the carriage was driven with postilions. (H & O)

VICTORIA-HANSOM — An improved form of HANSOM CAB, mainly distinguished by an adjustable top, invented by Jas. C. Robinson of London, and patented in 1887. (H)

VILLAGE CART — A variety of ROAD CART having a deep body. It is usually suspended on platform springs, and is capable of carrying one or two passengers, and a small trunk or other luggage. (O)

VILLAGE WAGON — A name sometimes applied to a SPRING WAGON. (S226)

VINAIGRETTE — A ROLLING-CHAIR used in Paris late in the seventeenth century and throughout the eighteenth, called also ROULETTE and BROUETTE (see the latter), having a body like that of the SEDAN CHAIR, but resting on springs and two wheels, and drawn by a man. (H)

VIS-A-VIS — (From the French, *vis-a-vis*, meaning face to face.) The French, in the course of cutting down the COACH into various forms (as fully described under COUPE), halved the COACH longitudinally about the middle of the eighteenth century. The resulting vehicle, accommodating only two persons who sat face to face, was called a VIS-A-VIS. The term later came to be applied in a general way to seating arrangements wherein the occupants of a vehicle sat face to face. In the United States, by the late nineteenth century, the vehicle formerly known as the SOCIABLE had come to be known as a vis-a-vis. These later carriages were made either with or without doors, with or without driving seats (generally the former), without tops or with folding or umbrella tops, with paneled, caned, or basketwork bodies, and sometimes with a rumble. Suspension was most often on four elliptic springs, but platform springs were occasionally used in the rear. (H, C, & O)

VOITURE — A general French term for any carriage, conveyance, coach, wagon, etc.

VOLANTE — A two-wheel vehicle peculiar to some Spanish countries, particularly Cuba, used from the late eighteenth to the early twentieth century. It had a CHAISE-like body, hung forward of the axle, and long, flexible shafts. Sometimes one or two extra horses were harnessed outside the shafts.

From the general appearance of the Volante, one might be inclined to think that it was an awkward construction, putting too much weight on the horse, and possibly being unpleasant to ride due to the movements of the horse. On the contrary, users of the vehicle claimed that it was a light and easy carriage, most pleasant to ride — in fact far superior to the CHAISE — and easy on the horse. The large wheels enabled it to readily overcome obstacles. A comparatively expensive vehicle, it was not built in Cuba, but in England and the United States during the latter part of the nineteenth century, and exported to Cuba.

The volante was driven by a postilion, who was frequently a gaily dressed *calisero* wearing a scarlet jacket, high jack boots with silver buckles at the knees, and a gaudy hat with a cockade. (H & O)

VILLAGE CART. (The George C. Miller Sons Carriage Co., Cincinnati, Ohio, 1883.)

VIS-A-VIS, with ogee quarters. Wheels, 33-inch and 42-inch; panels blue; moldings, black, with fine straw-color striping; carriage part, dark blue with straw-color striping; trimming, blue goatskin. (*The Hub,* January 1887.)

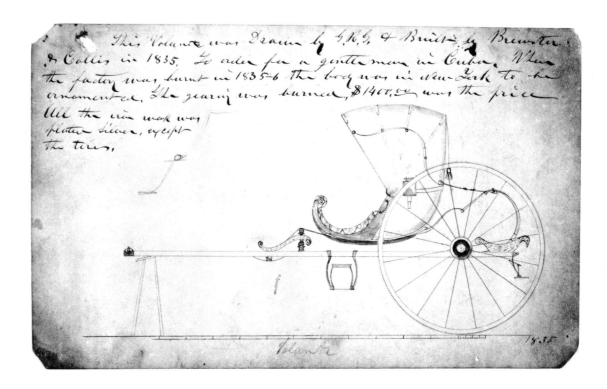

A Cuban VOLANTE built in 1835 by Brewster & Collis, New Haven, Connecticut. (From an original drawing in the Smithsonian collection.)

Cuban VOLANTE, showing the outrigger hitch for two animals outside the shafts. (Smithsonian photo.)

WAGON — (From the Dutch and German *wagen*, a general term for wheeled vehicles. The word was adopted into English in the sixteenth century, in this same general sense. The spelling, *waggon*, known in seventeenth-century England, was commonly used there during the eighteenth century, but this usage became rare in America after the early part of the nineteenth century.) The first vehicle of this type (though the term itself did not come into use for many centuries) is believed to have resulted from the coupling together of a pair of two-wheel carts, thus producing a four-wheel vehicle that was comparatively rigid, necessitating that it be dragged sideways in rounding a corner. It is uncertain when the pivoted front axle was developed, though several theories date the invention from about 1000 B.C. to A.D. 1. The wagon consists of a box placed on a running gear, and it is a relatively heavy vehicle used for carrying freight, agricultural products, etc. Many of these were built without springs, but many late wagons were provided with elliptic, platform, truck, bolster, or some other form of spring. The term WAGON eventually came to be applied not only to various types of lighter delivery vehicles, but also to numerous PASSENGER WAGONS, from the very light ROAD WAGON to the heavier DOUGHERTY WAGON, and the many similar types of STAGE WAGONS built by such firms as the Abbot, Downing Company, and M. P. Henderson & Son. Nearly all of the passenger-carrying wagons had some sort of spring-suspension, if not under the body, then under the seat. See specific types, such as BEACH WAGON, BEER WAGON, BOLSTER WAGON, BRACKET-FRONT WAGON, BRITTON WAGON, BUCKBOARD WAGON, DOUGHERTY WAGON, DELIVERY WAGON, ESCORT WAGON, EXPRESS WAGON, LUNCH WAGON, FARM WAGON, MOUNTAIN WAGON, ROAD WAGON, SURREY WAGON, WHITECHAPEL WAGON, etc. (C, H, & O)

WAGON, BARGE — A plank-sided FARM WAGON which came into use in England late in the nineteenth century. The sides were vertical, or nearly so, and it was equipped with rave-boards. The wheels were smaller than those on conventional WAGONS, and sometimes the front wheels were small enought to turn under the body. Barge wagons succeeded the earlier types of BOX and BOW WAGONS. (J)

WAGON, BOAT — (1.) A type of FARM WAGON which came into use in England in the last decade of the nineteenth century. The body was shallow and the sides sloped outward from the floor. Frequently they were mounted on leaf springs, and had no coupling pole. (J) (2.) A running-gear supporting a cradle which is designed for carrying a small boat, such as boats used by life-saving stations. (O)

WAGON, BOW — A type of English FARM WAGON, commonly found in southwestern England during the late eighteenth and nineteenth centuries. It is characterized by a shallow body, either panel-sided or spindle-sided, with large sideboards arching over the rear wheels. (J)

WAGON, BOX — A type of English FARM WAGON, commonly found in central and southeastern England during the late eighteenth and nineteenth centuries. It is characterized by a deep rectangular body, with paneled or planked sides. (J)

WAGONETTE — The first vehicles of this type were built in England about 1842 or 1843, following which, in 1845, one was built under the supervision of Prince Albert, Consort of Queen Victoria, for the use of the royal family. In a short time wagonettes became popular as family carriages, particularly in the country. As family carriages, the first wagonettes were comparatively large, but in time a great variety of styles developed, both large and small, open and closed, and for public or private use. The principal features of a wagonette are the longitudinal seats behind the driving seat, facing each other as in an OMNIBUS, and a door in the rear. The driving-seat is generally on the same level as the rear seats, but at times is slightly elevated. Some of those intended for private use were enclosed with glass windows like an OMNIBUS, while others had standing tops, and open sides that could be closed by curtains in foul weather. Others, of both the large

BARGE WAGON. (J. G. Jenkins: *The English Farm Wagon*, 1961.)

BOAT WAGON. (Photo from the University of Reading, Museum of English Rural Life.)

A BOAT WAGON built for the United States Government in 1898, by the Studebaker Bros. Mfg. Co.

BREAK variety and the smaller TRAP-type, frequently had no top, but in all instances the rear entrance and the side seating arrangement prevailed. One of the larger types was the public wagonette — a sort of summer Omnibus — which was most popular in the United States, and continued in use until the end of the horse-drawn era. Wagonette suspension was on elliptics or platform springs. Ezra Stratton said of Wagonettes, "They possess the advantage of carrying a greater number of persons in a carriage of given weight than any other on four wheels." (H, C, & O)

WAGONETTE-BREAK — See BREAK.

WAGONETTE-OMNIBUS — A vehicle with COACH driving-seat and WAGONETTE body, either with or without a top. (H)

WAGONETTE-TRAP — A TRAP having rear seats which can be shifted so as to face each other in the manner common to the WAGONETTE. (O)

WAGON LOADER — See EXCAVATOR.

WAGON, SELF-LOADING — See EXCAVATOR.

WAGON, TURN-OVER SEAT — Synonymous with JUMP-SEAT WAGON. (H)

WAIN — (From the Anglo-Saxon *waegn*, by the omission of the letter *g*.) This term was commonly applied in England to WAGONS, being vehicles used to carry heavy loads. While WAGON and WAIN both refer to a four-wheel vehicle, WAIN was also frequently applied to a heavy two-wheeler. (H, C, & O)

WARDO — Varient spelling of VARDO.

WATER CART — A vehicle carrying a large tank for the purpose of sprinkling the streets. A similar CART, without the sprinkling apparatus, was also used by the army for carrying drinking water to troops in the field. (C & O)

WHEELBARROW — (1.) (Spelled *whelbarewe* in fourteenth-century England.) A small, man-powered conveyance, consisting of a platform or open box mounted on a pair of shafts that protrude both front and rear, carrying a small wheel between them in front, and serving as handles in the rear. The word is also applied to a number of similar vehicles having one or more wheels. (2.) In seventeenth-century England the term came to be applied to a horse-drawn carriage of unknown description. It is believed to have been a light and inexpensive type. (H & O)

WHEEL-BOAT — A boat provided with wheels, and intended to be used on both the water and inclined planes or railways. (H)

WHEEL-CAR — A primitive vehicle used in Ireland in the nineteenth and twentieth centuries. It is essentially a TRAVOIS, with wheels added. During the earlier period it was often an altered SLIDE-CAR, with a wheel unit added in such a manner that the back end of the car slid along the ground, while the front only was supported by wheels. Later types put all the weight of the car on two wheels. The wheels were of the solid type, generally were located inside the shafts, and were fixed to a free-turning axle. (O)

WHIRLICOTE — (Derivation unknown, but Bridges Adams, in his *English Pleasure Carriages*, suggests, "a whirling cot, or moving house.") It was an old English vehicle, frequently referred to by writers of the fourteenth

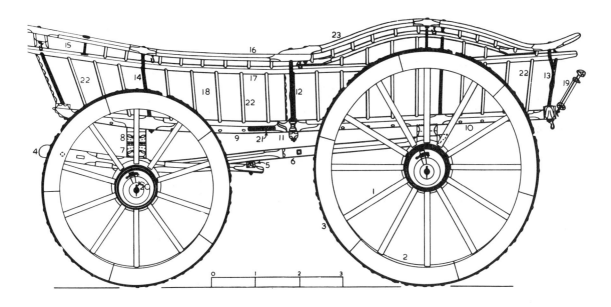

South Midlands BOW WAGON. 1, Spoke; 2, Felloe; 3, Strake; 4, Hound; 5, Slider bar; 6, Coupling pole; 7, Bolster; 8, Pillow; 9, Forward side frame; 10, Rear side frame; 11, Waist; 12, Main side support; 13, Rear side support; 14, Intermediate side support; 15, Sideboard; 16, Outer top rail; 17, Inner top rail; 18, Side planks; 19, Tailboard; 20, Linch pin; 21, Wearing plates; 22, Body spindles; 23, Wheel arch. (J. G. Jenkins: *The English Farm Wagon*, 1961.)

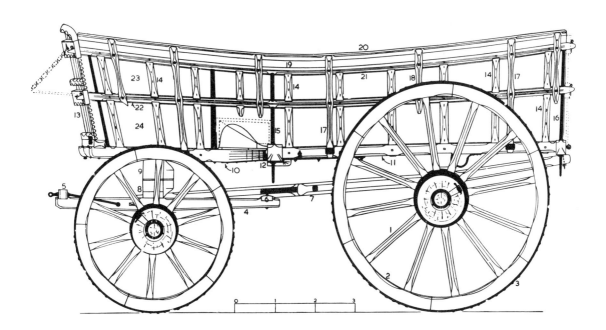

East Anglian BOX WAGON. 1, Spoke; 2, Felloe; 3, Strake; 4, Hound; 5, Splinter bar; 6, Slider bar; 7, Coupling pole; 8, Bolster; 9, Pillow; 10, Forward side frame; 11, Rear side frame; 12, Waist; 13, Frontboard; 14, Body standards; 15, Main side support; 16, Rear side support; 17, Sideboard supports; 18, Intermediate side supports; 19, Sideboard; 20, Outer top rail; 21, Inner top rail; 22, Midrail; 23, Upper side plank; 24, Lower side plank. (J. G. Jenkins: *The English Farm Wagon*, 1961.)

century. Stowe, in his *Chronicles*, relating the history of Wat Tyler's Rebellion (1380), tells us "that Richard II, being threatened by the rebels of Kent, rode from the Tower of London to the Mile's End, and with him his mother, because she was sick and weak, in a whirlicote of old time" (showing that they had then been long in use); and he adds, "Coaches were not known in this island (England), but chariots, or whirlicotes, then so called, and they only used of princes, or men of great estate, such as had their footmen about them. . . . But in the next year after Richard had married Anne of Bohemia, she introduced the fashion of riding on horseback; and so was the riding of these whirlicotes and chariots forsaken, except at coronations and such like spectacles."

The mechanical characteristics of the whirlicote are unknown, but it is believed to have been identical to the LONG-WAGON. (H & O)

WHISKEY (or WHISKY) — William Felton, in his *Treatise on Carriages*, wrote in 1796, "whiskies are one-horse chaises of the lightest construction, with which the horses may travel with ease and expedition, and quickly pass other carriages on the road, for which reason they are called *whiskies*," the name implying a light, rapid movement. Suspended on some light form of spring, such as the grasshopper spring, the whiskey seldom was equipped with a top, and only occasionally had a dash or apron, thus keeping the weight to a minimum. (H & F)

WHISKEY CURRICLE — A WHISKEY with a carriage similar to that of a GIG CURRICLE, constructed in such a way that the shafts can be moved together to form a pole, thus allowing the use of either one or two horses. (F)

WHITECHAPEL (or WHITECHAPEL CART) — An English form of DOG CART, for gentlemen's use, characterized by a deep, square-box body and bracket front. The name is derived from its having originated as a butcher's CART known as the WHITECHAPEL, which was afterwards modified for use as a gentleman's vehicle. The first Whitechapel cart was introduced in the United States about 1867, by Burton Mansfield of New York, it having been built by Peters & Sons of London. The mechanical characteristics of the Whitechapel correspond with those of the DOG-CART. The lines of the body were later followed in the construction of the SURREY, which was, in fact, sometimes called a Whitechapel, thus confusing the two-wheeler with the four-wheeler. (H & O)

WHITECHAPEL WAGON — An American modification of the English WHITECHAPEL CART, by the adaptation of the body to the carriage of a four-wheel ROAD WAGON. First introduced by James B. Brewster & Company of New York in about 1872, the vehicle gained great popularity, and was best known by its other name — SURREY. It was sometimes called, simply, a WHITECHAPEL, which confuses it with the two-wheel WHITECHAPEL. See SURREY. (H)

WIND-CARRIAGE — Attempts have been made for centuries to propel carriages by the force of the wind. W. B. Adams, in his *English Pleasure Carriages*, wrote that the Chinese reportedly used or experimented with sail-carriages at a very early date. Numerous others have likewise built wind-carriages that were either equipped with sails, attached to a kite, or fitted with a windmill. None of these efforts were attended with any measure of success, however, due to the narrowness of the roads, overhanging trees, etc., which prevented the necessary manipulations of the sails, kites, or carriages. Also see CHARVOLANT. (H & O)

WINDSOR WAGON — A term applied to a particular form of American square-box BUGGY, hung on cross-springs and side-bars. (H & M)

WIRE WAGON — A WAGON, similar in form to a BATTERY WAGON, which carried a reel of wire for a United States Army field telegraph train. See BATTERY WAGON. (O)

WOOSTEREE — A local New England term applied to an old pattern of two-wheel CHAISE, having a TILBURY body hung on Boston CHAISE-springs. (H & M)

WORCESTER CART — A variety of DOG CART.

WORTHING CART — A variety of DOG CART.

WURST-WAGON (also called a WURST-CAISSON) — (From the German, *wurst*, a sausage; supposedly applied to the vehicle because of the resemblance of the long, narrow body to a sausage.) A four-wheel vehicle used in the horse artillery to convey the cannoneers with ease and rapidity. The body of this carriage, which carried ammunition, was hung on springs, and the rounded top was padded and covered to provide a seat for the men; small boards were fixed alongside to serve as footrests, and a pommel was affixed to each end, similar to that on a saddle. The gun crew rode astride the body in the manner of men on horseback. The term WURST-WAGON was also applied to a long hunting WAGON, similarly constructed, on which the hunters likewise sat astride the body of the vehicle. The military type is described in *Practical Instructions for Military Officers*, by E. Hoyt, Greenfield, Massachusetts, 1811. (H & O)

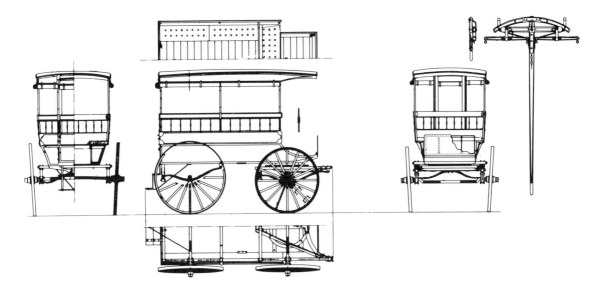

WAGONETTE built about 1906 by George E. Gould, of Lake Grove, Long Island. Wheels, 41-inch and 46-inch; body, dark green striped in gold or red; gearing, red or yellow, striped black; trimming, black imitation leather; roof, black. (Smithsonian drawing from original vehicle at Suffolk Museum, Stony Brook, Long Island.)

WAGONETTE OMNIBUS. Wheels, 33-inch and 42-inch; painting generally in light and bright colors, with fine scrolls and much striping; trimming, plush, with driver's seat of black leather. (*The Carriage Monthly*, February 1899.)

WAGONETTE-TRAP. Wheels, 28-inch and 34-inch; panels, black or dark green; basket seats, natural finish; gearing, carmine, yellow, or green, striped with black; trimming, Bedford, whipcord, or Broadcloth. (J. A. Lancaster & Co., Merrimac, Massachusetts, ca. 1910.)

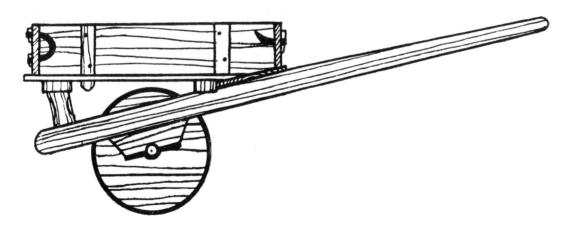

WHEEL-CAR. (G. B. Thompson: *Primitive Land Transport of Ulster*, 1958.)

CANE WHISKEY. Felton said of this type, "the lightest and cheapest of all others, and have, for summer use, a light, airy appearance; they are not so strong as pannel bodies, but are less in the expense for painting and lining, and are principally intended for country use in fair weather..." (W. Felton: *Treatise on Carriages,* 1796.)

This half-panel WHISKEY is essentially the same as the cane variety, except that panels are substituted for cane, giving the body the strength to carry a falling top. The body is also trimmed inside, and the exterior is ornamented with plated moldings and a sword case. (W. Felton: *Treatise on Carriages,* 1796.)

WHITECHAPEL CART, with tandem hitch. Wheels, 54-inch; body, dark green, with black moldings; gearing, light vermilion striped with half-inch line of black. (*The Hub,* September 1881.)

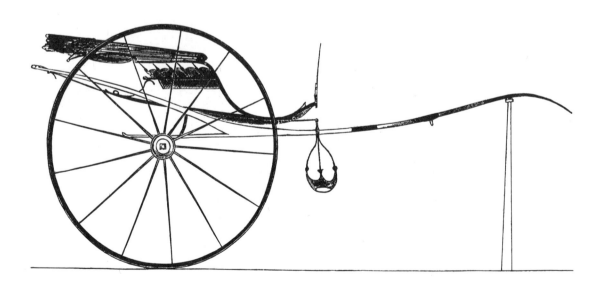

WOOSTEREE with 60-inch wheels. (*The New York Coach-Maker's Magazine,* September 1858.)

Y

YACHT, PRAIRIE — (Not to be confused with PRAIRIE SCHOONER) A vehicle invented by Dr. H. M. Wheeler, of Grand Forks, North Dakota, for skimming over snow-covered plains by the aid of the wind, and patterned after the ICE-YACHT. The chief difference between the ICE-YACHT and the prairie-yacht is that the runners of the former are narrow, and shod with metal, while the runners of the latter are similar to skis. The prairie-yacht was illustrated and described in the *Herald*, Watertown, New York, March 19, 1887. (H)

YANDELL DOCTOR'S BUGGY — A variety of physician's PHAETON characterized by its peculiar top, called the Yandell top, after the Louisville, Kentucky, physician to whose order it was first built by a carriage builder of that city. (H)

YELLOWSTONE WAGON — A thoroughbrace PASSENGER WAGON used as an excursion or sightseeing WAGON. Originated by the Abbot, Downing Company, of Concord, New Hampshire, it had an open body with three or four seats, and a canopy top, and was hung on a running gear like that of a regular Abbot, Downing PASSENGER WAGON. The driver's seat was elevated like that of a MAIL COACH. They were equipped with a baggage rack on back, and were built to carry either nine or twelve persons, including the driver. (O)

YELLOWSTONE WAGON. (Abbot, Downing Co., Concord, New Hampshire, ca. 1886.)

CARRIAGE NOMENCLATURE

A

ABELE-TREE — The white poplar. (C)

ABOUT-SLEDGE — The largest hammer used by smiths. (C)

ACCIDENTAL LIGHTS — (Painting) Secondary lights; effects of light other than ordinary daylight. (C)

ACKERMANN STEERING — A system of steering a vehicle by means of two stub axles that pivot separately, with arms and connecting linkage so designed that the inside wheel turns more sharply than the outside wheel, in order that the angle of each wheel will agree with the radius of its curve. While this arrangement honors the name of Rudolf Ackermann, the well-known English print seller, it was not really his invention. Ackermann held the English patent for his friend Georg Lenkensperger, a German coach builder who worked out the mechanical proportions of this steering in 1816. Although the invention was not extensively used in carriage construction, it later came to be the steering method commonly employed in automotive design. (O)

ADORN — To decorate with ornaments or gilding. (C)

ADZ — A tool for hewing wood, formed with a thin arched blade having its edge at right angles to the handle. The edge is only beveled on the inside. Adzes are made with long handles, as an ax, so that the work may rest on the ground; or with short handles and lighter blades, so that they may be used with the work in other positions. (C)

AFGHAN — A lap robe made of worsted in a variety of bright colors. The finest are ornamented with flowers or other devices needleworked upon a plain ground. (C)

AFTER-BED — The name sometimes applied to the bolster of a heavy wagon.

ALBURNUM — The white and softer part of the wood between the inner bark and the heartwood; sapwood. (C)

ALLIGATOR — A term used to describe varnish that has crawled together on a surface, leaving a series of cracks, ridges, and bare spaces. In England, the term *syssing* is sometimes used. (O)

AMEL — The same as enamel. (C)

ANCHOR-HEAD BOLT — A bolt with a projecting head, in which there is a slit that will receive a small strap. (O)

ANGLE-BAR — See *angle-iron*.

ANGLE-IRON — A rolled bar of iron having an L-shaped section. It is used for forming the edges of iron chests, corners of boilers and tanks, connecting side plates, etc., where great strength is required with light weight. (C)

ANGLET — See *sham-door*.

ANTIRATTLER — See *shaft-coupling antirattler*.

ANTISPREAD CHAIN — Synonymous with *spread-chain* and *tie-chain*. (O)

ANVIL — An iron block, usually with a steel face, upon which metals are hammered and worked. (C)

APRON — A piece of leather, enameled cloth or rubber cloth, attached to the dash or front of a carriage, used as a lap cover to protect the occupants from rain or snow. Occasionally the leather or cloth cover to a baggage rack on a stage is called an apron, and in some localities the seat fall is known as an apron. (T)

APRON FALL — A piece of leather attached to the top of the dash, used to cover the folded apron. (T)

APRON HOOK — A small metal hook, attached to a strap, for holding the apron in place when it is folded. (T)

APRON RING — A metal ring, used in connection with the *apron hook*, to secure the folded apron. (T)

APRON STRAP — A narrow strip of leather, used in connection with the *apron hook* and *ring* to hold the folded apron in place. (T)

ARABESQUE — A species of ornament used for enriching flat surfaces, either painted, inlaid, or carved in low relief. It consists of an ideal mixture of vines, foliage, fruit, figures of men, animals, birds, etc. (C)

ARCAGRAPH — An instrument for drawing a circular arc without the use of a central point. (C)

A PHAETON of 1819 equipped with Ackermann steering. (Smithsonian photo.)

ARCHIBALD WHEEL — An iron-hub wheel invented by Edward A. Archibald of Methuen, Massachusetts, who was granted patents on it in 1871 and 1872. In some respects it appears to resemble the Sarven wheel, but differs in several ways. First, it has no wood interior to the hub, which is entirely of iron. The spokes meet one another with tapered surfaces, as in the Sarven, yet no tenon is necessary, as the spokes bear directly against the exterior of the hub box. The inner flange, against which the backs of the spokes bear, is integral with the box The outer flange is pushed over the box, against the fronts of the spokes, and the two flanges are then bolted together with carriage bolts, instead of being riveted as is the Sarven. A front hub-band then screws onto the outer end of the box, and against the outer flange.

In addition to being one of the strongest patent wheels, little affected by climatic changes, the simple construction of the Archibald wheel enables it to be readily repaired. Replacement spokes and hubs can be installed with relative ease, and with a minimum of disassembly.

The United States Army began tests with various iron-hub wheels about 1873, and in 1880 authorized the use of Archibald wheels on Dougherty wagons. Soon the use of these wheels became common on other army vehicles. In civilian use the Archibald wheel was used on the heavier types of commercial vehicles. (O)

ARCH PANEL — The bent panel-board fitted to the rounded underside of a boot. (O)

ARM-RAIL — The piece of timber forming the top or shape piece to a lower quarter of a carriage body. Also called *arm-piece*, *waist-rail* and *belt-rail*. Sometimes synonymous with *seat-rail* on a light carriage. (C, O, & h)

ARMREST — A small projecting cushion at the side of a carriage body, on which to rest the arm. Also a cloth or broad lace loop, attached to the standing pillar, used for the same purpose. (h)

ASPHALTUM — A native bitumen of a dark brown or black color. It is very brittle and when broken has a high luster; it melts or burns without leaving a residue, and is used as one of the components of black japan. (C)

AWL — A pointed metal instrument for piercing small holes, used by saddlers and other leather workers; also used by woodworkers. (C)

AXLE — A transverse bar of wood or metal, forming the shaft on which a pair of vehicle wheels revolve. (h)

AXLE-ARM — That portion of the end of an axle which slips into the hub-box, and on which the wheel revolves. (C & h)

AXLE-BED – (1.) The wooden support to which an iron axletree is secured, generally by clips. The axle was formerly bedded in the wood, but at the present time it is fastened underneath. (2.) That part of an axletree inside of the arms. Iron axles frequently have square beds, though variations are found. A coach or platform bed is square for about nine inches behind the collar, then octagonal for about six inches, and the center is round. Fantail beds are square behind the collar, but then become flatter and wider. (C, Y, & h)

AXLE-BOX – A cylindrical tube of iron fastened in the hub and fitted to the axle-arm, to protect the hub from wearing on the axle-arm. (C)

AXLE-CLIP – A strap of iron with round ends upon which threads are cut, which is put around the axle and bed to draw them firmly together, and also for fastening the ends of the stays. (C)

AXLE-COLLAR – The projection on the end of an axle-bed, forming the bearing against which the axle-box wears. (h)

AXLE, COLLINGE – The most complicated and complete of carriage axles, invented in England about 1787, well adapted to the heavier classes of vehicles. It is also called the *patent axle*. Its features are: 1. A large collar, against and in which the box revolves; 2. A parallel arm (generally); 3. A loose collar, known as a collet, on the outer end of the arm, against which the recess in the front of the box runs, the collet being prevented from turning by a flat side on that part of the arm; 4. Right- and left-hand nuts, with the collet bearing against the inner nut; 5. A linch pin in front of the outer nut; 6. A cap, containing oil, screwed into the box, and turning with the wheel. (h, FR, & A)

AXLE, CRANKED – An axle which has the bed made after the fashion of a crank, so that the body of the vehicle may be hung lower. (Y)

AXLE, DROP – Same as *cranked axle.* (Y)

AXLE GATHER – The forward inclination given an axle-arm in order to make the wheel hug the shoulder. (C)

AXLE, HALF-PATENT – A term believed to have been first introduced into the United States about 1835, to distinguish from the old patent screw axle. The characteristics of the half-patent axle are: a single nut on the end, and a box which covers the collar. (h)

AXLE, MAIL – An axle in which the box is secured to the hub by three bolts, running through the hub, and into a flange at the back of the collar. It is so named because of its frequent use on English mail vehicles. (h)

AXLE, MATHER THOUSAND MILE – A well-known type of axle patented in 1898, guaranteed to run 1000 miles on one oiling. It features a system of oil grooves, whereby the oil lies in a groove running almost the entire length of the axle-arm, and is fed into the axle-arm at a point about midway in its length, and is returned to the ends of the long groove by spiral grooves at both ends. (O)

AXLE-NUT – A square or hexagonal piece of threaded iron, used in place of a linch pin to hold the wheel on the axle-arm. It differs from an ordinary nut by having a flange on one end. It sometimes has a blind hole, so that the end of the axle does not show. (C)

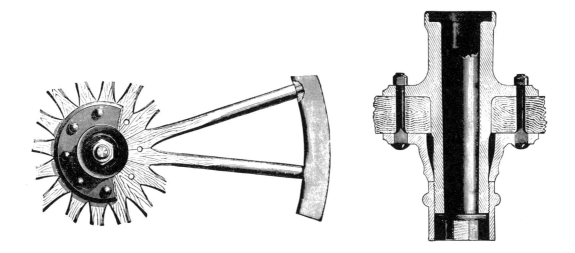

An Archibald hub, with a sectionalized view showing how the iron parts are drawn together on the wooden spokes. (From Edward S. Farrow: *Military Encyclopedia,* published in New York by Military-Naval Publishing Co., 1895.)

AXLE, NUT — An axle in which the box is secured to the axle-arm by a nut instead of a linch pin. (h)

AXLE, PATENT — See *axle, collinge*.

AXLE-SHOULDER — The projection or swell on the arms of certain axles, at the point adjoining the collar. (h)

AXLE-SKEIN — A strip of iron sunk into and on a line with the top or bottom of an axle-arm of a wooden axle, to protect the arm from wear. Also called a *clout*. (C)

AXLE, TAR-SKEIN — A term frequently applied to that variety of axle made with axle-skeins. So called because the lubricant used with this type of axle was often pine tar. (O)

AXLE, THIMBLE-SKEIN — A variety of axle, for either nut or linch pin, the arm of which is cast hollow to admit a wooden axletree. (h)

AXLETREE — A piece of timber or an iron bar, with ends rounded or turned for insertion into the hub, on which the wheels revolve. Synonymous with *axle*. (C & h)

AXLE WASHER — A ring of metal or leather, placed at either end of a hub, to prevent it from wearing against the axle-collar at the rear (when it is known as a *collar washer*), or against the axle-nut or linch pin at the front (when it is known as the *nut* or *front washer*). In some cases the term is applied to the metal band, or collar, placed at the back end of a wooden axle-arm, to act as a shoulder to the axle-box. (C, T, & h)

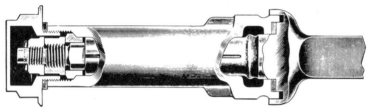

Full Collinge axle. (Dalzell Axle Co., So. Egremont, Massachusetts, ca. 1900.)

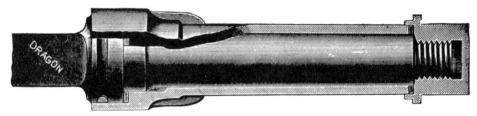

Half Patent double collar steel axle. (J. E. Sawyer & Co., Glen Falls, New York, ca. 1898.)

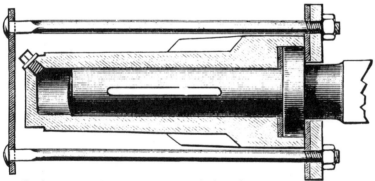

Mail axle. (From Fairman Rogers: *Manual of Coaching*, 1901.)

Mather's Thousand Mile Axle, supposedly required lubrication only that often. (From an advertisement in *Carriage Monthly*, November 1898.)

B

BACK-BAR — The carved or ornamented bar that ties the hind springs of a platform carriage, and on which the body loops are bolted. Also called a *saddle-bar*. (C)

BACK CROSS STRAP — One of a pair of ornamental straps, placed crosswise of the back end of a carriage body. The upper ends are secured to the body at points near the back ends of the arm-rails, and the lower ends either to the pump handles at points near the scrolls, or to corresponding points on C-spring and other types of carriages. (T & C)

BACK CROSS STRAP BUCKLE — A buckle, having a screw or other device attached to the bottom bar, whereby it is secured to the body. It is used for attaching the cross strap to the body. (T)

BACK CROSS STRAP CENTER — A metal ornament, attached to the cross straps at the central point, where they cross each other. They are made in a great variety of styles and patterns. (T)

BACK CROSS STRAP LOOP — A metal loop, to which the lower end of the cross strap is attached. It is sometimes used in place of a buckle. (T)

BACK CURTAIN — A curtain used on the back end of a carriage top. (T)

BACK LIGHT — The small window in the back panel or back curtain of a carriage. (T)

BACK QUARTER — A strip of leather or cloth at the back corner of a top, two of which, together with the back curtain, form the back of a carriage top. (T)

BACK STAY — (1.) A web or metal strip, one end of which is attached to the back bow, the other to the seat or rail, to hold the back bow in position and prevent the top from being thrown forward. (T)   (2.) Also a metal brace used as a support to a single perch carriage; a brace extending from the back end of a body to the top of the spring. (C)

BAND — A metal rim fastened on the end of a hub to prevent the hub from splitting. Those made for the front ends of hubs, known as *point bands*, are also intended to be ornamental, and are either silver or gilt plated, or are painted to correspond with the other portions of the hubs. (C)

BANQUETTE — The rear top seat of a continental diligence, protected by a hood. (O)

BAR — Technically all cross or tie pieces to carriage framework. See *side-bar, horn-bar*, etc. (C)

BASKET, UMBRELLA — A cylindrical basket for carrying umbrellas. It is attached to the iron of the rear roof seat, on the near side of a public COACH or a DRAG. (FR 93)

BASKET WOOD — A machine-carved wood, originally in imitation of basketwork, at one time very popular as ornamental moldings on belts to carriage bodies. (C)

BATTENS — Narrow strips of wood, secured to the framing of a carriage body, used as cross pieces inside of panels, to keep them to their shape; technically used to designate strips fastened on the outside of panels on which a molding was worked. (C)

BEAD — A small half-round molding worked or nailed upon bodies and carriage parts to ornament them. (C)

BEAD PLANE — A thin plane with a cutter shaped like that of a bead tool. (C)

BEAD TOOL — A tool having a cutter shaped to the style and size of the bead required. (C)

BEAM — See *bob-sleigh* (gearing nomenclature).

BED — This term is synonymous with *transom*, but it can also mean *axle-bed*.

BED PLATE — A plate secured to the top of an axle-bed. Together with the head block plate it serves as a bearing surface around the king-bolt. These plates are also called *transom plates*. In England they are often called *locking plates*, which tends to confuse them with *wear irons*. (K)

BEGILD (or GILD) — To cover or overlay with gold. (C)

BELL or CALL BELL — A gong, usually placed under the driver's seat, to which the bell cord is attached, by which the occupant of a closed carriage can signal the driver. (h)

BELL CORD — The cord attached to a bell or gong, by which an occupant of a closed carriage may ring the bell to attract the driver's attention. (O)
BELL PULL — A handle attached to the cord of a gong or bell; made of metal, wood, ivory, or other material. (T)
BELT — A sunken or raised band or strap passing around the outside of a body as an ornament. (C)
BELT-RAIL — See *arm-rail*.
BENCH — A craftsman's work table. Also see *bob-sleigh* (gearing nomenclature).
BENCH-DOG — An iron bolt with a flat top made at right angles with the body of the bolt, having square or saw teeth on the edge. It is used to hold the wood in place on the bench when being planed. (C)
BENT TIMBER — The general term given to all carriage timber that has been steamed and bent, such as bows, shafts, felloes, etc. (C)
BENZINE — An oily substance obtained from bituminous coal, consisting of twelve parts of carbon and six of hydrogen; at one time it was used extensively in place of turpentine in making varnishes and mixing paints. (C)
BERLIN BLUE — Same as Prussian blue. (C)
BESSEMER STEEL — Steel made direct from the iron ore by forcing a blast of air through the molten iron to remove carbon and other impurities. It takes its name from the inventor, Henry Bessemer. (O)
BEVEL — An instrument with two arms so joined as to allow the arms being set to any required angle. (C)
BEVEL-GAUGE — A gauge with a long head and spur, used to gauge thicknesses on beveled pieces of timber. (C)
BILLET — A short strap which enters into a buckle. (O)
BINDING HOOKS (or KNOBS) — Fittings fastened to the sides of freight vehicles, for securing ropes when binding on the load. (I)
BLANCHARD LATHE — A lathe for turning irregular forms, such as spokes, etc.; named for the inventor. (C)
BLIND — A movable frame, with or without slats, used as a substitute for a carriage window. Blinds are intended to exclude the passengers from view, but are generally constructed so as to admit air. Blinds may be slatted, covered with wire gauze, or paneled. The paneled variety is also known as a *stable shutter*. Blinds are but little used. (T, h, & C)
BLOCKS — Wooden or iron risers placed under or above the springs and other parts of the carriage to level it. They are generally carved or beaded. (C)
BLOWER — A fan enclosed in a housing and propelled by power, used to force a blast of air into a furnace or forge. (C)
BOBS — A term, apparently originated in New England, referring to short sleds which are joined together and used as one, but are actually separated. The front unit turns around a king-bolt, and often has a fifth-wheel. Bobs are used under sleighs, farm sleds, or carriage bodies. (C, h, & M)
BOB-SLEIGH or SLED (gearing nomenclature)
    Beams — The pieces of timber joining the tops of two corresponding knees, on opposite sides of the sleigh.
    Bench — A timber taking the place of a beam in bobs using camel-back runners, where no knees are employed.
    Camel-back runner — A runner sawed from a wide timber, being higher in the center than at the ends, and secured underneath the bench without the use of knees.
    Knees — The upright pieces that connect the beams and runners.
    Raves — Pieces running either across the ends of the beams, or the tops of the knees, and joining with the upturned end of the runners in front.
    Reach — Either of two parts joining the front and rear bobs. In some instances only a short reach is used, running from the rear roller to a beam or bench of the front bob. In other cases a long reach is used, running from a rear beam or bolster to the front bolster. In this case a short reach is also generally used, running from the rear roller to a point well forward on the long reach.
    Roller — A piece running between the front ends of the runners. In the case of the front bob the pole is attached to this, and on the rear bob, the short reach is attached to it.
    Runners — Pieces bent or sawed, with front ends upturned, on which the bob slides. They are generally, though not always, shod with iron or steel.
    Saddle — A piece running across the tops of the front beams, in center, through which the king-bolt passes.
BODY — (1.) The general term given to that part of the vehicle that contains the passengers or freight. This is the main part of the carriage, and gives name to the vehicle. (C) (2.) (Painting) Consistency, thickness; colors bear a body when they are capable of being ground so fine and being mixed so completely with oil as to seem only a very thick oil of the same color. (C)

BODY BRACE — An iron support which runs from the end of a cross-bar to the side of a wagon body, as a support to the side. (Y)

BODY CLOTH — The cloth used as the outside covering for cushions, squabs, falls, etc. (T)

BODY HANDLE — A handle attached to some part of the body of a vehicle to assist a passenger or driver in mounting or dismounting. (Y)

BODY LININGS — Materials of leather or cloth, etc., used for trimming. (C)

BODY LOOPS — (1.) Irons secured to the bottom of the body and projecting beyond the corners; used for securing the body to the carriage part. (C)   (2.) The term is also applied to the loops attached to the underside of a body, which receive the check-straps. These are also called *check-loops*, which term would best be reserved for them, so as to avoid confusion of the two different parts. (K)

BODY PLATES — The iron plates fitted to the inside of a rocker. (C)

BODY VARNISH — A superior quality of varnish made from the best gum copal, oil, and turpentine. (C)

BOLSTER — A support for the body of a lumber wagon, or other heavy wagon that is made without springs. It is sometimes called an *after-bed*. (C)

BOLSTER CHAINS — Chains running from the ends of a front bolster to a point further back on the coupling pole. They prevent excessive swivelling action of the bolster when the gear is used without a body. (O)

BOLSTER PLATES — Iron plates attached to the underside of the front bolster of a wagon, and to the topside of the sand-board, for the purpose of receiving the wear caused by the swivelling action around the king-bolt. In England these are often called *locking-plates*. (K & Y)

BOLSTER SPRING — A type of leaf spring, like a half-elliptic, used between the bolster and body of a wagon. Frequently arranged in pairs, one on either side of the bolster, their ends are supported by hangers which go across the top of the bolster. Coil springs are also often used as bolster springs. (Y)

BOLSTER SPRING HANGER — A yoke-shaped iron fitting, the center of which is attached to the top of a bolster, while the ends support the ends of the bolster springs. (Y)

BOLSTER STAKES — Short upright posts framed into and near the ends of bolsters, used to hold the wagon box in place on the gear. They are also called *wagon standards*. (C & K)

BOLT — (1.) A metal pin with a head on one end and a screw thread cut upon the other; used to fasten different parts of the carriage together. (C)   (2.) (British) One of the wooden bars (there are usually two) connecting the rear ends of a pair of shafts. (O)

BOLT PATENT AXLE — See *mail patent*.

BOND — A British term meaning *hub band*. (O)

BONE-BLACK — A black pigment made by burning bones in a closed vessel until they are thoroughly charred. The finest black used by painters. (C)

BONNET — Same as *hood*. (h)

BOOBY-BOOT — The name originally given in ridicule to the extra seat attached to the back part of a carriage for the use of the footman. It is now called a *rumble*. (C)

BOODGE — A term describing a variety of budget or boot (of which latter term it is probably a modification) that was used in the eighteenth century to carry swords, pistols, small parcels, etc. See *sword case*. (h)

BOOK-STEP — A folding step which is opened or closed by the action of the door; when they are closed up they have the appearance of a book. (C)

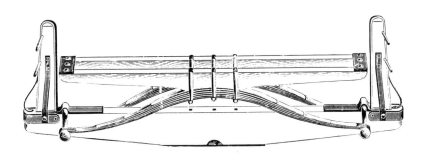

Bolster springs of this type could be applied to any unsprung farm or freight wagon. (Studebaker Bros. Mfg. Co., Catalog No. 222, 1903.)

BOOT — A term originally applied to almost any leather receptacle. When first applied to coaches, it described an extension of the body on either side of the coach, protected by movable curtains that formed the door; the two boots accommodated four persons, usually pages or court attendants (two on each side), who sat on a low seat near the floor of the coach, with their feet resting in a drop corresponding somewhat to an enlarged door-step. Old engravings frequently show these side projections, with the occupants facing the side of the road. This ugly contrivance must have been very dangerous in the case of an upset or collision. John Taylor, the 'Water Poet,' in his *World Ronnes on Wheels* (1630), in the course of his garrulous raillery against coaches, betrays himself into the following description, which fortunately has preserved a graphic likeness of this boot as it was wont to appear in Queen Elizabeth's time: "Besides, like a perpetuall Cheater, it weares two Bootes and no spurres, sometimes having two paire of legges in one Boote, and oftentimes (against nature) most preposterously it makes faire Ladies weare the Boote; and if you note, they are carried backe to backe, like people surpriz'd by pyrats, to be tyed in that miserable manner and throwne ouerboord into the sea. Moreouer, it makes people imitate sea Crabs, in being drawne sidewayes, as they are when they sit in the Boote of the coach." When coaches were improved, these boots were transferred to the fore and hind ends of the coach body. They were sometimes built of wicker, forming immense baskets, which afforded accommodation for cheap fares — though perhaps it could scarcely be called accommodation, since elderly or stout passengers were sometimes with difficulty fished up from their depths. In course of time, the coach being still further modified, the boot became simply a receptacle for luggage, and was a wooden box with a leather flap or cover.

    In the transformation of the coach into the modern pleasure carriage, the boot, as a receptacle for luggage, was naturally abandoned, and indeed disappeared from the hind end of the vehicle altogether, and in front was cut down into a shapely support for the driver's seat, being reduced to a mere relic of its former office. Thus it is that the boot of today is of wood and iron, and is often spoken of as a 'paneled boot' or a 'skeleton boot.' Also see *budget* and *boodge*. (h)

BOOT, BUGGY — A cover of wood, rubber, leather, canvas, etc., used on the open rear portion of a buggy body. (Y)

BOOT END WAGON BOX — A wagon box having a small box projecting from the rear end, and beyond the end gate. The boot may be either a separate attachment or it may be made integral with the wagon box. It serves for a variety of purposes, such as affording a place to stand while shoveling grain. (O)

BOOT SPRINGS — Coil springs which keep a buggy boot tightly closed, and prevent it from rattling. (Y)

BORAX — A salt formed by a combination of boracic acid with soda, it is of a white color and is used as a flux for the welding of metals. (C)

BOTTOM-BARS — The end bars that are framed into the bottomsides or rockers to hold the body together. (C)

BOTTOM-BOARDS — Boards that compose the floors of carriage bodies. (C)

BOTTOMSIDES — The main pieces of a body into which all the pillars are framed; they are the most important pieces in the body as they generally decide the shape. They are also called *sills*. (C & h)

BOTTOM-SLATS — Slats forming the floor of a vehicle, as in a slat-bottom buckboard. Also slats forming the bottom of a box, as in a seat box. (O)

BOW — A narrow strip of wood bent at two points, three or more of which form the frame or support of a carriage top. Also, of wider stock, the bent wooden supports for the cover of a wagon. (T & C)

BOW IRONS — Iron fittings on the sides of a wagon box for the purpose of holding the wooden bows which support the wagon cover. They serve the same purpose as *bow staples*, but differ in being riveted or bolted to the wagon box. (K)

BOWLS — Concave turned pieces inserted in the sides of buggy bodies as ornaments. By 1871 they were nearly obsolete. (C)

BOW REST — The portion of a seat rail which projects at the corners, acting as a point of attachment for the foot of the rear top joint, and a support to the top when folded down. (T)

BOW SEPARATORS — Devices of varying types, used to prevent the bows from crowding together when the top of a carriage is folded down. (Y)

BOW SOCKETS — Metal sockets into which the lower ends of top bows are inserted. They answer the same purpose as *slat irons*. (C)

BOW STAPLE — One of the iron staples driven into the side of a wagon box for the purpose of holding one of the wooden bows which supports the wagon cover. (O)

BOX — A term used in some areas to refer to the body of a wagon. Also see *axle-box, coach box*, and *boot*. (h)

BOXING-ON — (British) Fitting an axle-box into a hub. (O)

BOX LOCK — A thin door lock differing in its action from the spring lock by being moved only by the door handle. (C)

BOX ROD — An iron rod running through the sides of a wagon box, outside an end gate, to support the sides against the weight or the load. Special winged washers are often used with a box rod to provide a larger bearing surface for the head and nut. Sometimes box rods are permanently installed, and in other instances they are removable, having a *handle nut* instead of a square nut. (Y)

BOX STAPLE — An iron staple driven into the side of a wagon box, for the purpose of holding the top bows or the side board standards. Same as *bow staple*. (K & Y)

BRACE — (1.) A strap used to arrest motion, such as a *check brace*. (2.) A strap used to support a body, such as *thoroughbrace*. (3.) The term is sometimes used synonymously with *hound*; specifically, tongue brace, front brace, and hind brace. The hind brace is also called the reach brace. (4.) Generally, any part intended to lend support or rigidity to another part. (K & Y)

BRACKET — One of the bent pieces that supports the footboard of a heavy carriage body. In the case of the French boot, they constitute the risers to the seat. (C)

BRAD — A slender nail with a small head, used to attach moldings, etc. (C)

BRAD AWL — An awl used to pierce holes for the brad. (C)

BRAKE — (1.) A mechanism for retarding or arresting the motion of a vehicle by friction. The principal parts are the lever, by which the force is applied; the beam, which carries the shoes; the shoes, which come in contact with the rims of the wheels; and the rods, which transmit the motion of the lever to the beam. (2.) Also an iron shoe placed in front of and under the rim of the wheel, causing it to slide on the ground. See *shoe, drag-chain,* and *drag-staff.* (h, S226)

BRAKE, SELF-ACTING — A vehicle brake so constructed that it is operated by the action of the horses holding back. This type of mechanism applies just as much pressure as is needed in any given instance, and it may be put out of action so that the carriage may be backed. (O)

BRAKE BEAM — The wooden beam carrying the brake blocks. (O)

BRAKE BLOCK — The wooden or iron block which puts pressure on the tire when the brake is applied. These are known by a variety of names, such as brake friction block, brake shoe, rubber, etc. (I)

BRAKE BLOCK SHOE — An iron fitting attached to the brake beam, to which the brake block is attached. (I)

BRAKE LEVER — A bar operated by foot or hand pressure to apply the brake blocks to the wheels. (S226)

BRAKE RATCHET — A notched metal strip, straight or curved to suit any given brake mechanism, designed to hold the brake lever in any position. (K)

BRAKE RELEASE SPRING — A spring for withdrawing the brake blocks from the tires when pressure on the brake lever is released. (Y)

BRAKE ROD — One of the rods of wood or iron which transmit the motion of the brake lever to the brake beam or roller bar. The rod connections are known variously as connecting rod ends, rod yoke ends, or rod clevises. (O)

BRAKE ROLLER BAR — An intermediate part of the brake mechanism, transmitting the movement of the lever to the beam. With some types of brakes the roller bar and beam are one and the same, the bar being cranked and having the brake blocks on its outer ends. (I)

BRAKE SHOE — Same as *brake block*.

BRANCH — An iron stay used to connect two or more parts, as the branch to a step or draw-iron. (h)

BREAK — See *pump handle*.

BREAST-STROKE — (British) A fine line scribed into a hub while the latter is being turned, to mark the place for the front of the spokes. (O)

BREECHING HOOK — A metal hook attached to the upper side of a shaft, to which the breeching strap is attached. (T)

BREECHING LOOP — A metal loop attached to the underside of a shaft, used as a substitute for the *breeching hook*. (T)

Carriage door hardware: box lock, spring lock, and dovetail. (S. D. Kimbark, Chicago, 1876.)

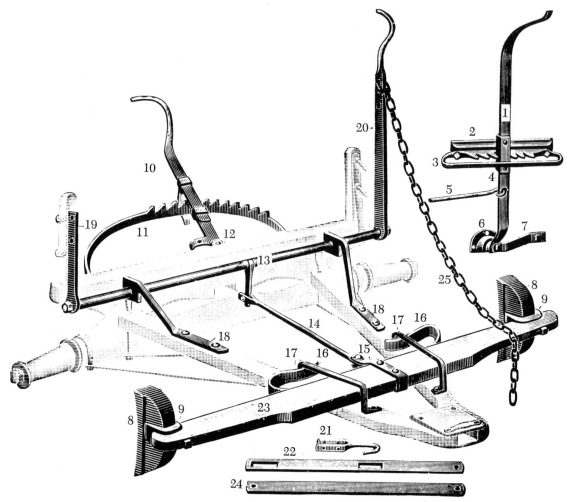

Detailed nomenclature of a brake mechanism, showing the names that Studebaker applied to the various parts: 1. Ratchet Lever; 2. Ratchet Block; 3. Wrought Iron Ratchet; 4. Lever Dog; 5. Ratchet Rod; 6. Foot Iron or Fulcrum; 7. Foot Iron Brace; 8. Brake Block; 9. Brake Clutch, right or left; 10. Rear Lever; 11. Rear Ratchet; 12. Rear Fulcrum; 13. Roller, no Lever; 14. Draw Rod; 15. Clip to attach Draw Rod to Bar; 16. Brake Bar Spring; 17. Hound Loop; 18. Roller Holder, right or left; 19. Short Roller Lever; 20. Long Roller Lever, no Chain; 21. Hook for Rear and Side Attachments; 22. Draw Rod for Rear and Side Attachments; 23. Brake Bar, no Irons; 24. Draw Rod for Rear Attachments only; and 25. Chain for Roller Lever. (Studebaker Bros. Mfg. Co., Catalog No. 222, 1903.)

BRETT-HINGE — A hinge of peculiar construction applied to a *Brett-lid*, and so made that when the lid is down, its top is perfectly level; when the lid is up, the hinge supports it in that position. (h)

BRETT-LID — The wooden lid used to cover and protect the front inside seat of a Brett, which when raised, forms the back to the seat. (h)

BREWER'S TOP — A duck-covered top for a wagon or truck seat. It is made similar to a buggy top, but with top quarters which come all the way down to the seat. (Y)

BREWSTER GEAR — A variety of side-bar gear, in which the body is mounted on the bars by means of two half-springs, in addition to the two half-springs which customarily unite the ends of the bars. (O)

BRIDLING SPRINGS — The practice of securing spiral springs by the use of wire, so that they will maintain a level surface when in cushions. (T)

BRITZKA PILLAR — (Derivation: from the traveling carriage of that name.) A graceful ogee-shaped pillar, finished with a scroll, which characterized the old BRITZKA body. (h)

BROADCLOTH — A twilled napped woolen or worsted fabric having a smooth lustrous face and a dense texture, used for trimming carriages. In the twentieth century the term came to be applied to materials made of cotton, silk, or rayon, but generally in the carriage trade, it was a woolen cloth. (O)

BROAD LACE — A woven close web, plain or ornamented, used for trimming carriages. Broad lace is used for cushion fronts, broad ornamental borders, armrests, lifters, etc. It is of uniform width, which is two and one-half inches. Other widths are made up to special order only. (T)

BROCATEL — A kind of coarse figured fabric, commonly made of silk or cotton, used for lining carriages. (C)

BROG — Same as *brad awl*.

BRONZE-POWDER — A metallic powder used for covering metal or wood to give it the appearance of bronze. (C)

BRUSSELS CARPET — A varicolored carpet made of worsted yarns that are fixed in a foundation web of strong linen thread. The worsted, which alone shows on the upper surface, is drawn up in loops to form the pattern. Much used on the floors of carriages. (O)

BUCKLE — A metal frame having a loose bolt or tongue, used to secure ends of straps to each other. (T)

BUCKRAM — A stiff and heavily sized coarse cloth, of linen or cotton. It is used as a lining to some parts of the carriage trimming. (C)

BUDGET — A term formerly used synonymously with *boot*, of which it is a development or modification. Budget described the boot at the period when it was used as a receptacle for luggage. (h)

BUFF — A color between a light pink and light yellow. (C)

BUFF LEATHER — A strong, supple oil-tanned leather, generally light in color and having a velvetlike surface. It is now made mainly from cattle hides. (O)

BUFF-STICK — A strip of wood covered with buff leather, used in polishing. (C)

BUMPER — See *horn and bumper*.

BUNK — The name applied to the exceptionally high bolsters of a lumber wagon or pair of logging wheels, or to the extra piece put on top of the regular bolsters. Bunks are sometimes made so that the top extends over the tops of the wheels, in order to protect the wheels when logs are being rolled off the side of the wagon. (O)

BURLAP — A coarse woven fabric of jute, hemp, or linen, used in the carriage trimming trade. It mainly serves as a foundation for squabs, etc., and is sometimes glued to the backs of panels to stiffen them. (C & T)

BURNISH — To make smooth and bright; to polish by rubbing with something smooth and hard. (C)

BURNISHER — A tool with a hard, smooth, rounded end or surface used in smoothing or polishing. (C)

BURR — A small iron washer placed on the tire bolt, between the felloe and the nut. Also a small washer put over the end of a rivet before peening it down. (h)

BUSH — A perforated piece of metal, usually a cylindrical lining, which receives the wear of pivots, journals, etc. *Axle-boxes* are sometimes called bushes. (C)

BUSHING — Same as *bush*.

BUSH METAL — An alloy of copper and tin used for making bushes. Babbitt metal. (C)

BUTT — The larger end of a piece of timber or of a fallen tree. (C)

BUTT BANDS — A hub band used on the rear of the hub. (O)

BUTTON — A small knob perforated or having an eye by which it can be attached to any part of the trimming. They may be japanned, or covered with cloth or leather. Japanned buttons are much used in place of tufts for trimming. Buttons are sized by lines, one-fortieth part of an inch.

BUTTON HANGERS — An ornamental attachment to footman holders; nearly obsolete by 1871. (C & T)

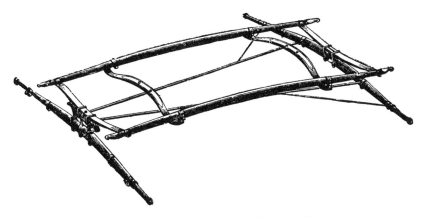

Brewster side-bar gear. (The Muncie Jobbing & Mfg. Co., Muncie, Indiana, 1898.)

C

CALASH-TOP — The falling top of a carriage, made of wooden bows covered with leather or fabric, and so constructed as to fold together, and fall upon a level with the top of the seat; the bow-top used on buggies and other vehicles. See *calash*. (C & T)

CALK — A piece of iron or steel attached to a horseshoe to give the animal better traction on slippery surfaces. (O)

CALL-BELL — A gong, usually placed under the driver's seat, to which a cord is attached, by which the occupant can attract the driver's attention. (T & h)

CAMBRIC — A fabric made of flax or cotton, sometimes used for slip linings to coaches. (C)

CAMEL BACK RUNNER — See *bob sleigh* (gearing nomenclature).

CAMEL BACK TOP — A fixed top covering the entire length of a delivery or business wagon, and rising in a hump over the driver's seat. (O)

CANE — Woven work of thin strips of rattan, or various bamboolike grasses, applied in place of certain panels in carriages, to lighten, ventilate, or create a pleasing effect; also used in place of seat-boards, to support a cushion. In some instances a decorative imitation is used on side panels in place of actual cane. (O)

CANOPY — (1.) An umbrella top, supported on a post attached to the back of the carriage body, used on light phaetons, and other light carriages. (2.) Also a standing top, generally fringed, supported by four or more iron rods on a surrey or spring wagon. The iron rods supporting the top are called standards, and the parts on the body into which they fit are called sockets. (T & Y)

CANOPY HOOKS — Specially shaped L-hooks that are screwed into a canopy top for the attachment of curtains. (O)

CANT — (1.) Technically, an inclination from a given line. (2.) The term is commonly used by the trade in place of the word *cantboard* and has approximately the same meaning as the word *projection*. (3.) The tire or rim of a wheel, though this usage is uncommon. (C & O)

CANTBOARD — A board used in connection with the outline drawing of a carriage body, upon which the view from the top is given, and from which the workman is enabled to obtain all bevels, turn-under, sweeps, etc.; the top view of a working drawing; a geometrical outline of the body, indispensable to bodymakers. Common trade usage meant the top view, rather than the board itself. (C & h)

CANT RAIL — Same as *top rail*. (O)

CANVAS — A coarse cloth made of flax, used in connection with glue, to cover the inside of the panels to give them greater strength. (C)

CAOUTCHOUC — A vegetable substance obtained from incisions made in several plants, affording a milky juice; it becomes dark on exposure. It is better known as India rubber, and is largely used on carriages in its various forms — in the soft state as springs, and in the hard state as whip-sockets, inside ornaments, covering to metal handles, etc.; buggy bodies have also been made of it. (C)

CAP — A piece of leather used to hide the lower ends of bows at the point where they are attached to the seat. Also a leather or metal covering for nuts or rivets. (T & C)

CAP BAND — A metal band used on the end of a hub, differing from the usual band in having a closed end, for use with mail axles. (K)

CARD CASE — A long, thin box open at one end, used as a receptacle for visiting cards. They are made of hard rubber, ivory, or metal which is generally covered with leather. (T)

CARMINE — A deep red pigment. It is derived from the coloring matter in cochineal. The cochineal is boiled with carbonate of potash and precipitated by means of an acid, generally alum. (C)

CARNATION – The natural or flesh color, made by mixing lake and white. (C)

CARPET – A heavy fabric, woven of wool or cotton, used for covering the floor of a carriage. (T)

CARPETING – The covering to floors of carriages and sleighs, whether ordinary carpet, oil cloth, or painted carpet. (C & T)

CARPET LACE – A coarse worsted fringe used for trimming the edges of pieces of carpet used for carriage floors. (T)

CARRIAGE – Same as *gearing*.

CARRIAGE BOLT – A bolt used in the carriage trade, having a square or fluted shank just under the head, which holds the head from turning while the nut is being tightened. The head itself is usually an oval head, but bevel heads were also used, particularly prior to 1900. Cone heads, steeple heads, and other types were also used. Common carriage bolts have the oval head with a short square shank. Philadelphia bolts have slightly higher oval heads and longer square shanks, and North carriage bolts have common oval heads with fluted shanks. Bastard-head Philadelphia bolts have oval heads that are nearly flat. Step bolts have very large, but thin, oval heads. Shaft bolts and whiffletree bolts have T-heads which curve around the work, and the former frequently has no square shank since the curving head is sufficient to keep the head from turning.

Tire bolts have short tapered heads and sleigh shoe bolts have long tapered heads. The tops of the heads are in both cases flush with the metal they secure. Dash bolts have countersunk heads and a square shank, and spring bolts have an ordinary oval carriage head, but without a square shank. (Y & B)

CARRIAGE BUILDER – Same as *carriagemaker*.

CARRIAGE HEATER – A metal box, often covered with carpet, having a drawer in which a block of charcoal burns to provide some heat for the feet. Also called a *foot warmer*. Some of the earlier types were made of wood and sheet metal, with a metal drawer inside. (Y)

CARRIAGE KNOB – A small metal knob resembling a capstan in shape. It may be made to be screwed, driven, or riveted in place, and is used to secure curtains or straps in place. (K & Y)

CARRIAGEMAKER – A mechanic who builds carriages; a manufacturer of carriages. (C)

CARRIAGE-PART – Any part of the vehicle except the body. (C)

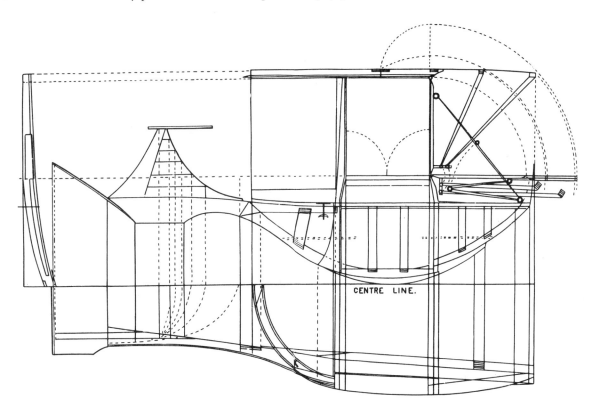

Drawing of a LANDAULET, showing the top view, or cantboard, below the side elevation. (From I. D. Ware: *Coach-Makers' Illustrated Hand-Book*, 1875.)

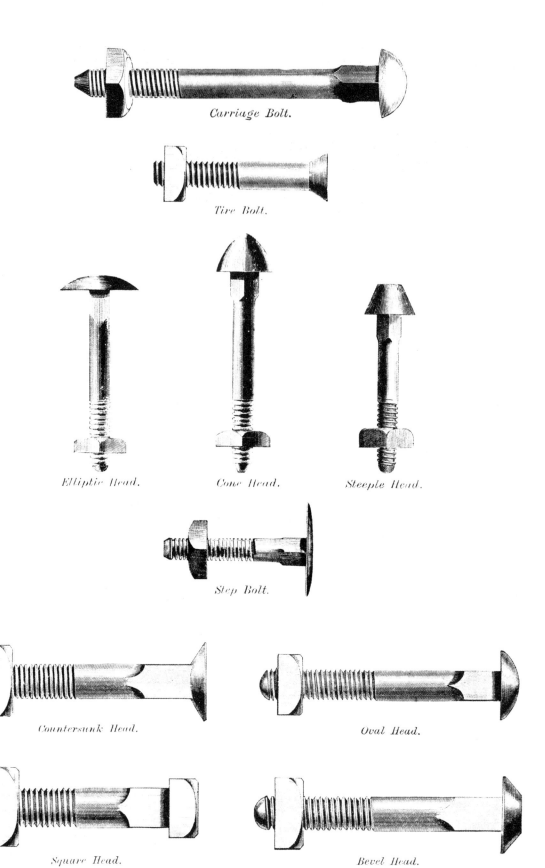

Assorted carriage bolts. (S. D. Kimbark, Chicago, 1876.)

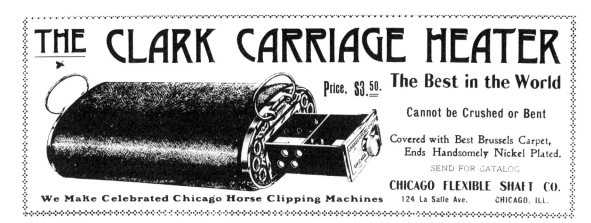

A charcoal-burning carriage heater offered in *The Hub,* 1901.

CART SHAFT EYE — A fitting used in place of the upper *cart shaft hook*, for the purpose of receiving the saddle chain. (I)

CART SHAFT HOOK — A hook attached to the shaft rod of a two-wheeled cart or dray. There are three hooks on each rod, one on each end for trace chains and hold back chains, and one on top for the saddle chain. (I)

CART SHAFT ROD — A rod on top of a cart shaft, running parallel to same, with its ends turned down at right angles to secure it to the shaft. The rod carries three hooks, intended for draught, hold-back, and support. (I)

CARVING — The art of cutting wood in an artistic or decorative manner. (C)

CASE-HARDENING — To harden the surface of iron or steel by carbonizing the surface to a small depth, while the interior retains its softness and toughness. By this method, the surface of iron is converted into steel, and the surface of mild steel is converted into hard steel. The process is accomplished by putting the iron or steel to be hardened into an iron box surrounded by animal or vegetable charcoal, and submitting it to a red heat; or, by heating the iron in a charcoal fire to a red heat and sprinkling the surface with prussiate of potash and cooling it in soft water. (C)

CASE LOOP — See *curtain loop*.

CASE RAIL — Same as *top rail*. (O)

CAST IRON — Iron run from the smelting furnace into pigs or ingots, or into molds. There are many varieties, differing by almost insensible shades, the two principal being gray and white. Gray iron is softer and less brittle than white iron. (C)

CAST STEEL — Steel made from blistered steel which is broken up into small pieces and melted in closed crucibles, from which it is poured into iron molds, afterward made into bars by hammering or rolling. (C)

CAUL — A strip or block of wood fitted to any part of a body, and used by the bodymaker to draw the panel down to its place by means of a thumb or hand screw. (C)

CEMENT — Any substance used for making bodies adhere to each other, such as glue. (C)

CENTER CLIP — A fitting around the center of a singletree. It consists of a thin, broad strap in front, changing into a loop in the rear, whereby it is attached to the doubletree. (K & Y)

CERUSE — White lead. (O)

CHAIN EYE — A metal eye mounted on some part of a vehicle, to receive a chain, such as an *end gate chain* or *feed box chain*. (Y)

CHAIR — An assembly attached to the front underbody of a platform spring vehicle, consisting of the upper half of the *fifth-wheel* and a number of cross pieces, known as *platform bars*. (O)

CHAMFER — To cut or grind anything to a bevel; to slope; to channel, to groove. (C)

CHAMOIS — A small goatlike antelope, the skin of which makes a soft and pliable leather, which is generally called *shammy*. This leather is highly water absorbent, and is much used by painters for wiping water from vehicles after washing. (O)

CHANNEL-TACK — The commercial name for a slim cut tack with a small flat head. (C)

CHANNEL-TIRE — Iron or steel tire for carriage wheels, made with a channel for holding rubber tire. (B)

CHARCOAL-BLACK — Black pigments made of burnt ivory, peach stones, white sugar, and other substances. (C)
CHASER — A short leather strap used to secure a buckle to a loop, or to a piece of other material.
CHECK-BRACE — Same as *check-strap*. (h)
CHECK-LOOP — A small metal loop by which the *check-strap* is attached to the body or other part of the vehicle. (T)
CHECK-PLATE — Generally a flat piece of iron, with one edge bent at a right angle, to assist in holding two wooden members in rigid connection with one another. It is bolted to one member while the bent portion bears against the side of the other. (O)
CHECK-STRAP — A leather strap extending from a carriage body to the perch or spring, to prevent the excessive movement of the body when the vehicle is going over uneven roads. In Felton's day the strap running from the perch to the underside of the body was called a *collar brace*, while the strap from the spring to the rear of the body was called a *check-brace*. Eventually the term *collar brace* became obsolete, and *check-brace* or *check-strap* came into general use for both types of strap. (C, T, & h)
CHECK-STRING — A cord or strap extending from the inside of a carriage to the driver, which the passenger pulls to attract the attention of the driver. (C & T)
CHILL — To produce, by sudden cooling, a change of crystallization at or near the surface of metal so as to increase its hardness. The interiors of axle-boxes are chilled by the use of iron cores when casting. (C)
CHILLED — (1.) Having undergone the process of hardening by chilling.   (2.) (Painting) Having a cloudiness or dimness. (C)
CHINCHILLA — A woolen fabric of a pearly gray color made in imitation of chinchilla fur. It is used to some extent for trimming sleighs, road wagons, and sporting wagons. (C)
CHINCHILLA BEAVER — The commercial name given to a woolen fabric with a long curly nap. It is made in a variety of colors and extensively used for lap robes and sometimes for carriage and sleigh trimmings. (C)
CHINTZ — Cotton cloth with a glazed surface which is printed with flowers and other devices. It is used by coachmakers for slip linings. (C)
CHISEL — A long flat tool made of, or faced with, steel, the cutting edge being at the end. Chisels are known by various names, such as firmers, framing, tanged firmers, socket firmers, socket framing, socket corner, etc. (C)
CHROME — (Chromium) A blue-white metallic element. (C)
CHROME GREEN — A brilliant green pigment consisting basically of some chromic salt. (C)
CHROME ORANGE — A pigment of dark orange color prepared from the subchromate of lead. (C)
CHROME RED — A basic lead chromate pigment. (C)
CHROME STEEL — The commercial name of a very fine quality of tool steel. (C)
CHROME YELLOW — The most valuable color produced from chromium. It is obtained by pouring a solution of chromate of potassa into a solution of the salts of lead. (C)
CIRCLE — The same as *fifth-wheel*. (h)
CIRCLE FRONT — The segment of a circle forming the front end of a standing-top carriage, known as full, half, and quarter circles. (C)
CIRCLE POST — A short support used between a full circle fifth-wheel and some other part of the gear. (B)
CLAMP — (1.) A pair of movable jaws, made of lead or other soft material, and used as caps to iron vise jaws to prevent injury to the object being held between them.   (2.) A long wooden bar with a movable catch at one end and a stationary screw at the other, used by carriage builders to draw different parts of the work together.   (3.) Any mechanical device for drawing or holding several parts together. (C)
CLAW — A small tool with a forked end, used for drawing nails. (C)
CLAW HAMMER — A hammer having a claw upon one end. (C)
CLEAT, LASHING — A double winged fitting attached to a wagon body, for securing ropes used to tie or cover the load. (O)
CLERESTORY — A wall raised above an adjoining roof, and pierced with windows so as to admit air and light, such as the clerestory on the roof of an omnibus or railway car.
CLEVIS — (1.) A U-shaped metal shackle with the ends drilled to receive a pin or bolt, used for attaching or suspending parts, such as the clevis attached to the end of an evener for the attachment of a singletree. (K)   (2.) Also a heavy, U-shaped iron part, with a few links of chain for attachment, used for the same purpose as a *rough lock*. (Searight)
CLIP — A piece of metal having a flat center, the ends being left round, and upon which a screw thread is cut; a U-bolt having a wide, flat center. The advantage is that the clip eliminates the necessity of boring a hole for

an ordinary bolt, which weakens the parts to be united. Clips have almost superseded bolts upon all parts of carriages where they can be applied. (C & h)

CLIP COUPLING — A coupling where the eye is clipped instead of being bolted to the bar or other point of attachment. (C)

CLIP, FRENCH — A broad clip, with lugs, used to secure a spring to the transom. (h)

CLIP KING-BOLT — A combined axle-clip and king-bolt, used for coupling the perch to the front axle. The clip king-bolt was invented about 1851 by Vel Reynolds, a New York carriage builder, who, not appreciating the great value of his invention, neglected to patent it. It was shortly afterward named and given to the world by the *New York Coach-makers' Magazine*. For an account of the legal complications that followed, see said magazine, and Stratton's *World on Wheels*, page 450. The clip tie used with this fitting frequently has a bolt on the underside, which secures a brace or braces running back to the reach or reaches. (Y)

CLIP PLATE — A specially formed plate, used as a bearing surface for regular U-bolts used in place of ordinary axle or spring clips. It is also called a *clip seat*.

CLIP SEAT — See *clip plate*.

CLIP TIE — See *clip yoke*.

CLIP YOKE — A straight bar of iron having a hole at each end, used to tie the ends of clips. It is also called a *clip tie*. (C)

CLOSE PLATING — A covering to iron or other metal, done with thin leaves of silver, sweated or soldered to the surface. Compare with *electroplating*. (O)

CLOSE TOP — A falling carriage top having permanently attached side-quarters, such as the top on a Victoria. (O)

CLOUT — An iron plate on an axletree, to guard it from wearing. Also called an *axle-skein*. (C & O)

CLOUT NAIL — (1.) A nail with a large flat head, used for securing clouts to axletrees. (2.) The commercial name for an annealed cut nail. (C)

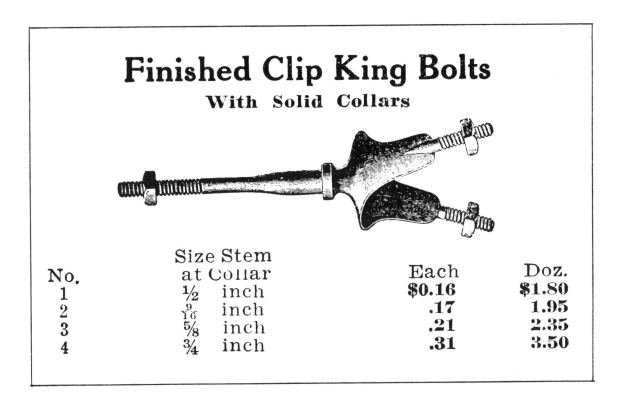

Clip king-bolt. (Cray Bros., Cleveland, Ohio, 1910.)

CLUB HANDLE — A T-shaped metal handle used on a carriage door. (K)
COACH BED — A heavy piece of timber, the center of which is bent so as to form a segment of a circle, used on platform carriage parts in order to throw the center or turning point forward of the axle. (C)
COACH BOX — The elevated seat in front of a coach, on which the driver sits. (h & C)
COACH HINGE — A metal hinge having a loose bolt, designed for use on doors of all kinds of vehicles. (C)
COACH-HORN — A long, slender, bugle-type horn used on a coach, both as a signal and to enliven the journey, In a booklet entitled *The Coach-Horn*, published by Kohler & Son, manufacturers of musical instruments in London, there is a considerable discussion of coach-horns and post-horns, the latter also being called tandem-horns. It is here stated that the true coach-horn should always be made of copper, always straight and having a conical or straight bell, varying in length from three to four and one-half feet. The post-horn should always be made of brass, twenty-eight to thirty-six inches long, having a curved bell, and either straight, or coiled so as to be a compact instrument. The coach-horn, in order to be compact, was sometimes made to telescope, so that the smaller top portion would slide down into the larger bottom portion. Fairman Rogers, in his *Manual of Coaching*, makes no such distinction, and in fact, even Kohler made horns of both types of either brass or copper. The post-horn was more likely to be used with a lighter vehicle, though customs changed from time to time. The tone of the post-horn is likely to be brilliant, or sharp, while that of the coach-horn is more mellow.

    Many horns used with stages in the United States were crudely fashioned, of tin, for any country tinsmith could make one, and they were far less expensive. They were often called a *yard-of-tin*. In mountainous areas the horn was sounded frequently as the stage ascended, for the ascending vehicle had the right of way over descending vehicles if the road was narrow, and the horn was the signal for the descending vehicle to pull aside at the first widening of the road, until the ascending coach had passed. (O)
COACH LACE — A broad lace with two selvages, woven of silk or worsted, the figure being raised upon the surface. (C)
COACH LAMP — A peculiar style of lamp made expressly for wheeled vehicles. They are, however, more for ornament than use. See *lamp*. (C)
COACH LOCK — A catch made expressly for doors to carriages. (C)
COACH-SCREW — (British) A large screw with a square head, turned into place with a wrench instead of a screwdriver. Also called a lag screw. (O)
COBALT GREEN — A preparation consisting of cobalt and iron, having a green color. (C)
COCHINEAL — A small insect found on several species of cactus. At a suitable season of the year the insects are gathered from the plants and plunged into hot water, and then dried by the sun. They have the appearance of small rough berries of a grayish purple color. Carmine, the most brilliant of all reds, is made from them. (C)
COCKADE — A badge worn on the hat of a coachman. In England it denotes that the owner of the vehicle is in some way connected with the service of the royal family. (h)
COCK-EYE — A type of iron fitting screwed into the end of a whiffletree, the head having an eyelet hole, to which the trace is attached. (K & h)
COD-PIECE — The flap or projection at the lower front corner of the side-quarter of a close leather top, covering that area where the bow sockets attach to the gooseneck iron. It is so named because of its similarity to the appendage attached to the tight-fitting breeches of the fifteenth to seventeenth centuries. (O)
COFFRE — See *Daumont, A la*.
COIL SPRING — A round spring made of heavy tempered wire, wound in a spiral form, used in seat cushions, and sometimes in the suspension of a body, such as bolster springs. (O)
COLLAR — (1.) The projecting ring at the rear end of an axle-arm. See *axle-collar*. (2.) A turned or swedged ornament on braces, etc. (3.) A sectional portion of improved top props. (C & h)
COLLAR BRACE — A leather strap extending from the perch to the underside of the body, to prevent excessive movement of the body when the vehicle is running over uneven roads. This term was used at an early period (see Felton), but was later replaced by the term, *check-brace*. In Felton's day, a *check-brace* ran from the spring to the end of a body. (O)
COLLAR BRACE RING — The eye welded on the end of a body loop to receive the end of a check-strap; but little used. (C)
COLLAR WASHER — See *axle washer*.
COLLET — Same as *axle washer*. Also see *axle, Collinge*. (O)

An ordinary post-horn, and a coiled one, the latter being handier for use with a DOG-CART or GIG. Post-horns were most popular for pair-horse driving. (From *The Coach-Horn: What to Blow and How to Blow It,* by an Old Guard, published by Kohler & Son, London.)

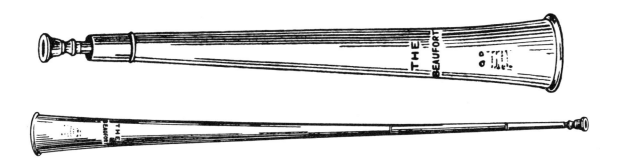

An ordinary coach-horn, and a telescoping one, the latter being easier to transport because of the telescoping feature. These were most popular for four-in-hand driving. (From *The Coach-Horn: What to Blow and How to Blow It,* by an Old Guard, published by Kohler & Son, London.)

An inexpensive tin coach-horn, 55 inches long, commonly used in the United States by many stage drivers. These longer horns were more practical if the driver wished to play the popular airs of the day, yet Kohler's "Old Guard" advises us that the instruments performed the task "very imperfectly." (Horn from the Smithsonian collections.)

On this and the facing page are useful calls for the coach-horn, and three airs adapted to that instrument. (From Fairman Rogers: *Manual of Coaching*, 1901.)

COLORS — The carriage painter was seldom required to produce a great variety of colors, since the basic carriage colors were generally of the proper shade without any addition. The colors commonly used in the latter part of the horse-drawn era were:

Brewster green	Ivory black	Raw umber
Burnt sienna	Lamp black	Ultramarine blue
Burnt umber	Lemon chrome	Venetian red
Carmine	Milori green	Vermilion
Chrome green	Orange chrome	White lead
Chrome yellow	Prussian blue	Yellow lake
Dutch pink	Raw sienna	Yellow ochre
Indian red		

On the other hand, the painter of commercial vehicles had need for a great variety of tints and shades, and required an ability for mixing these hues. A table for acquiring these varying shades is given in *Practical Carriage and Wagon Painting* by M. C. Hillick, published in Chicago in 1898. The formulae given correspond with those given in the 1883 work of Fritz Schriber, *The Complete Carriage and Wagon Painter*. The following table is presented not as an infallible guide, but more in the nature of a reliably helpful one.

REDS

Brick red: Yellow ochre, 2 parts; English vermilion, 1 part; white, 1 part
Carnation red: Red lake, 3 parts; white, 1 part
Cherry red: Carmine, 1 part; English vermilion, 2 parts
Claret: Carmine and ultramarine blue; or red and black
Deep rose: Victoria lake, 1 part; flake white, 6 parts
French red: Indian red and vermilion glazed with carmine
Imperial red: Yellow lake, 1 part; solferino lake, 5 parts
Maroon red: Lampblack, 1 part; Venetian red, 8 parts
Metropolitan red (also called New York red): Carmine and vermilion, glazed with carmine. A stunning and saucy panel color.
Solid crimson: English vermilion, 1 part; carmine, 2 parts
Superlative vermilion: English vermilion, 3 parts; orange mineral, 1 part
Transparent red: No. 40 carmine
Wine color: Carmine, 3 parts; ultramarine blue, 3 parts

YELLOWS

Acorn yellow: White and raw sienna, equal parts
Buff: White, 2 parts; yellow ochre, 1 part
Canary yellow: White, 6 parts; lemon chrome, 1 part
Cream color: White, 5 parts; red, 1 part; yellow, 2 parts
Cream tint: White, 150 parts; orange chrome, 1 part
Gold: White and medium chrome yellow; add a little vermilion and French yellow ochre
Jonquil yellow: Flake white and chrome yellow, with a bit of carmine
Lemon color: Lemon yellow, 2 parts; white, 5 parts
Maroon yellow: Carmine, 3 parts; yellow, 2 parts
Naples yellow: White, 150 parts; golden ochre, 9 parts; orange chrome, 1 part
Oak: Yellow ochre, 1 part; white, 8 parts
Pale orange: Orange chrome, 1 part; white, 5 parts
Primrose: Add a dash of white to lemon yellow. Or, according to the Standard Dictionary, 58% of white, 24% of yellow, and 18% of green. Originally English, it should be a very pale yellow tint.
Rich yellow: Orange chrome, 1 part; white, 6 parts
Straw color: White, 5 parts; lemon yellow, 2 parts; vermilion, a drop or two
Sulphur yellow: Lemon chrome, 1 part; white, 1 part

BLUES

Azure blue: White, 35 parts; ultramarine blue (medium), 1 part
Bird's egg blue: Add ultramarine blue to white until a tolerably intense blue is reached; then give a dash of light chrome green.
Brunswick blue: Made in three shades; popular in some areas.
Cerulean blue: White, colored with ultramarine blue
Changeable blue: Prussian blue
Cobalt blue: A fine pale blue, and a most beautiful panel color.
Grass blue: White, 6 parts; emerald green, 2 parts; Prussian blue, 1 part
Ocean blue: White, 15 parts; Prussian blue, 1 part; raw sienna, 2 parts
Ultramarine blue: Three shades, light, dark, and medium

GREENS

Bottle green: Dutch pink and Prussian blue, glazed with yellow lake; or medium chrome green, 5 parts; drop black, 1 part
Brilliant green: Paris green, 4 parts; chrome green, 1 part
Bronze green: Chrome green, 5 parts; burnt umber, 1 part; black, 1 part
Grass green: Yellow, 3 parts; Prussian blue, 1 part
Marine green: White, 30 parts; chrome green, 1 part
Milori green (or green lake): Resembles chrome green when dry, but is a much richer color when mixed up. A fine panel color for business wagons, with good covering power.
Nile green (or body green): Milori green, Prussian blue, and black, mixed to the desired shade, and glazed over with yellow lake.
Olive green: Golden ochre, 5 parts; coach black, 1 part
Pea green: White, 5 parts; chrome green, 1 part
Quaker green: Chrome yellow, 5 parts; Prussian blue, 2 parts; vermilion, 1 part
Sage green: White, 60 parts; light chrome green, 2 parts; raw umber, 1 part
Scheele's green: Paris green
Tea green: Made of blue chrome green and raw umber. A striking panel color for business wagons.
Willow green: White, 5 parts; verdigris, 1 part

The greens form a class of colors extensively employed in the painting of all classes of vehicles. There are two orders of green, namely, cold and warm. In cold greens, blue or black predominates; the warm greens contain an excess of yellow. As a class, the greens contrast with reds and colors containing reds, and harmonize with colors having blue or yellow in their composition.

BROWNS

Amber brown: Burnt sienna, 4 parts; medium chrome yellow, 5 parts; burnt umber, 8 parts
Bismark brown: Dutch pink, burnt umber and lake. Or, with a mixture of burnt umber, 2 parts; white lead, 1 part, make a ground, over which put a coating of burnt sienna, and then glaze with carmine, 1½ parts; crimson lake, 1 part; gold bronze, 1 part. An English vermilion makes a base over which the glazing makes a considerably lighter brown.
Burnt sienna: A fine, warm, reddish brown, if the sienna be of good quality. A very close imitation of Bismark brown.
Chestnut brown: Red, 2 parts; chrome yellow, 2 parts; black, 1 part
Chocolate color: A little carmine added to burnt umber.
Coffee brown: Yellow ochre, 2 parts; burnt sienna, 1 part; burnt umber, 5 parts
Dark brown: Indian red, 5 parts; Prussian blue, 1 part
Indian brown: Indian red, 1 part; yellow ochre, 1 part; lampblack, 1 part
Japan brown: Black japan, to which is added a little vermilion.
Olive brown: Burnt umber, 3 parts; lemon yellow, 1 part
Orange brown: Orange chrome, 2 parts; burnt sienna, 3 parts
Seal brown: Burnt umber, 4 parts; golden ochre, 1 part

Tan brown: Yellow, 2 parts; raw umber, 1 part; burnt sienna, 5 parts

Umbers: A class of natural earths, affording varying shades of brown, the Cypress mines yielding rich, warm, olive colors. Calcined, this umber reaches a positive violet shade. Burnt umber used alone or in connection with red and black, gives a very striking panel color for business vehicles.

Vandyke brown: A product of natural deposits of brown color. Vandyke brown is a warm color of a reddish hue, and is permanent. Most of the Vandyke browns with which the carriage painter is familiar are made, however, from black, red, and yellow.

MISCELLANEOUS

Blue-black: Ivory black, 15 parts; Prussian blue, 1 part

Burgundy: A bright lake given a small percentage of asphaltum

Cane color: White and ochre shaded with black

Chamoline (so named because it resembles wet chamois skin): White, 5 parts; raw sienna, 3 parts; lemon chrome, 1 part

Copper color: Yellow, 2 parts; red, 1 part; black, 1 part

Dove color: Medium chrome yellow, 1 part; blue, 1 part; white, 4 parts; vermilion, 2 parts

Drab color: Burnt umber, 1 part; white, 9 parts

Fawn color: White and ochre with a bit of vermilion

French gray: White, tinted with ivory black, warmed with a pinch of vermilion

Lavender: White, 15 parts; mauve lake, 1 part; rose madder, 1 part

Leather color: Burnt sienna, 2 parts; burnt umber, 1 part; a little white added

Light gray: White, 9 parts; black, 1 part; blue, 1 part

Lilac: Blue, 1 part; carmine, 4 parts; white, 3 parts

London smoke: Red, 1 part; umber (burnt), 2 parts; white, 1 part

Maroon: Carmine, 3 parts; yellow, 2 parts. Or crimson lake and burnt umber

Medium gray: White, 8 parts; black, 2 parts

Normal gray: White, black and purple; or simply white and black

Peach blossom color: White, 8 parts; blue, 1 part; red, 1 part; yellow, 1 part

Pearl gray: White, black, and blue

Plum color: White, 2 parts; red, 1 part; blue, 1 part; or white, 2 parts; blue, 2 parts; red, 1 part

Salmon color: White, 5 parts; burnt umber, 1 part, yellow, 1 part

Silver color: White, indigo, and black

Snuff color: Yellow, 4 parts; Vandyke brown, 2 parts

True lead color: White, 8 parts; blue, 1 part; black, 1 part

Wine color: Ultramarine blue, 2 parts; carmine, 3 parts

COMPASS — An instrument for describing circles, having two slender legs, joined at the top by a rivet upon which they move; also called dividers. (C)

COMPRESSED BAND HUB — A hub having bands set under pressure into grooves turned in the wood. (Y)

CONCEALED HINGE — A coach hinge so constructed that when the door is closed, no part of the hinge is visible. (O)

CONCORD GEAR — A running gear of a carriage employing the Concord spring, and three reaches. (Y)

CONCORD SPRING — A half-elliptic side-spring, which characterizes the Concord buggy. (h)

COOM — Dirty refuse matter, as that which collects in the boxes of carriage wheels; coal dust, etc. (C)

COOMED-UP — A wheel on which the grease has gone dry and stiff, and other foreign matter has collected, is said to be coomed-up. (O)

COPAL — A resinous substance flowing spontaneously from two trees found in the East Indies and in Central or South America; it is used in the manufacture of most kinds of varnish. There are a variety of grades on the market, the principal ones being Zanzibar, Benguela, North coast, Angola, and Kowrie. (C)

COPAL VARNISH — A varnish made of gum copal, linseed oil, and turpentine. (C)

COPSE — (British) Same as *clevis*, for attaching or suspending parts. (O)

CORDUROY — A thick cotton fabric, corded or ribbed on the surface, used to some extent for trimming carriages. (C & T)

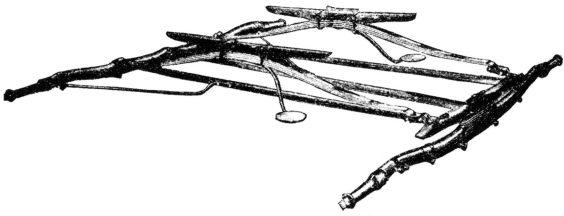

Concord gear. (Cray Bros., Cleveland, Ohio, 1910.)

CORK-LEDGE — A strip of cork, fastened to the footboard of a carriage, to afford a firm hold for the driver's feet. (h)

CORNER BRACE — A metal strap bent at an angle, screwed, riveted, or bolted in place to reinforce two pieces of wood coming together at an angle. (O)

CORNER IRON — An L-shaped iron plate, screwed, riveted, or bolted over the junction of two wooden members. (K)

CORNER PILLAR — The pillar that forms the corner support to a carriage body. (C)

COTELINE — The commercial name given to a woven fabric of silk; it is of French origin, and is much used for trimming fine closed carriages. (C & T)

COTTEREL — Same as *king-bolt*. (h)

COUNTERSINK — (*v.t.*) To form, by drilling or otherwise, an angular depression for the reception of a bolt or screw head. (*n.*) The tool with its cutting end beveled to the same angle as the underside of a screw head, used to form the countersink. (C)

COUNTERSUNK BOLT — A bolt with the underside of the head set at an angle greater than a right angle to the body of the bolt. It is used when the bolt head is to be set in level with the surface. (C)

COUPE-PILLAR — When the coach was cut down, or coup'd, to give a tasteful finish to the front corners, a gracefully swept pillar was devised, which has since been known as the coupe-pillar, and applied to a variety of other vehicles. Illustrations of coupe-pillars may be found as far back as 1771 (Roubo, plate 182). See *pillar*. (h)

COUPLING — The name given to the hinged metal fastenings used to attach shafts, poles, etc. (C)

COUPLING POLE — The *reach* or *perch* of a lumber wagon, or of other types of heavy wagons. (O)

COUPLING POLE PIN — An iron pin of about three-fourths inch diameter, used to pin the rear hounds to the coupling pole. (O)

CRAB — A metal socket having a loop on either side, used on the end of a coach pole to receive the pole straps. (C)

CRADLE — A skeleton platform placed behind a carriage body to hold luggage. (h)

CRANE, or CRANE-NECK — Cranes had their origin in the use of the double-perch of the Berlins. As the double-perch allowed less room for the fore-wheels to turn, the perches were bent up to allow the wheels to turn or *lock* under. The accidental resemblance of this curve to a crane's or swan's neck led to perches of this type being called *crane-necks* or *swan-necks* (see Felton, plate 7). They were also made with a neck in both the fore and hind part of the perch, in which case they were called *double-bowed cranes*. Again, they were made in the form of a single perch, with two forked crane-necks at the front (Roubo, plate 210, fig. 2). The crane-neck itself (*i.e.*, the bent or bowed part) was usually iron, while the perch was of wood, or of wood and iron. See *perch*. (h)

CRAWL — Said of varnish when it wrinkles on the surface after having been spread over the paint. (C)

CREST — A painted ornament, bearing a coat of arms, heraldic device, initials, etc. (O)

CREST PANEL — A narrow panel under a door window, on which a crest or monogram is painted. (O)

CRIMSON — A deep red color (carmine). (C)

333

CROSS-BAR — (1.) Any piece of wood placed transversely in a carriage body or gearing. See *bar*, *horn bar*, *spring bar*, etc. (C)   (2.) On a one-horse vehicle it is the piece that is fixed between the shafts, to which the whiffletree is pivoted. (O)

CROSS STRAPS — See *back cross straps*.

C-SPRING — A curved spring, of C-shape (and thence the name), made up of several overlapping plates or leaves. The C-spring, originally called a *scroll spring*, is the natural successor and the latest development of a series (the S-spring and whip spring being the preceding forms) which supplanted the primitive wooden pillar or post attached to the axle, from which braces extended to the coach body. They were introduced shortly before 1800. C-springs are often used in conjunction with elliptic springs, when the combination is called *double-suspension*, which arrangement, almost without exception, demands the use of a perch. So called *dumb C-springs* are sometimes introduced with elliptics to give the appearance of style; these are simply iron scrolls, otherwise useless. (h & C)

CURLED HAIR — The hair of animals, twisted into a rope and afterwards picked out loose; used for stuffing cushions, etc. (T)

CURTAIN — A leather or cloth piece, used to enclose carriages, attached in such a manner that it can be removed or rolled up. (T)

CURTAIN FASTENER — A device that takes the place of a knob and knob eyelet for securing curtains. It is

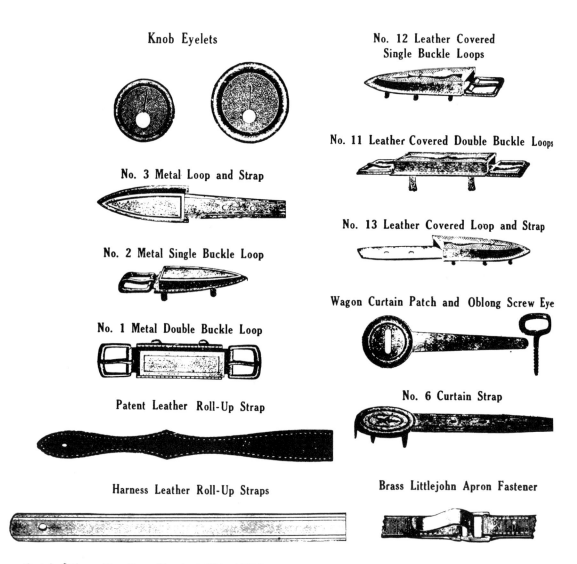

Curtain fittings. (Cray Bros., Cleveland, Ohio, 1910.)

slightly oval in shape and has a swiveling top, which turns 90° and keeps the eyelet from slipping over the fastener. It was developed late in the carriage era, and was much used on early automobiles. (Y, B, & G)

CURTAIN-FRAME — (1.) A pair of metal frames, joined by means of screws or other devices, used for holding a small glass in a curtain. (C & T) (2.) The wooden frame to which the leather curtains of a covered carriage are attached. (h)

CURTAIN HOLDER — An ornamental strap, attached to the top of a carriage, used as a support to curtains when rolled up. (T)

CURTAIN LOOP — A small box or housing made of leather or metal, and attached to the curtain or quarter for securing a billet or chaser. It is also known as a *case loop*. More specifically, the loop containing the buckle is known as the *buckle loop*, and the one containing the strap is known as the *strap loop*. (T & G)

CURTAIN PATCH — Similar to a *knob eyelet*, but having a slot in the center of the leather, which slips over an eyelet, and a strap which protrudes beyond the ring and passes through the eyelet. It is generally used on a wagon curtain. (Y & B)

CURTAIN ROLLER — A small tubular roller, used inside a coach as a support for a small curtain. They are generally made with a coiled spring in the barrel which tightens when the curtain is drawn down; a small ratchet wheel prevents it from flying back. (C)

CURTAIN STRAP — A short leather strap attached to a curtain or quarter by means of a *curtain loop*. The curtain strap fastens into the buckle of the opposite loop, securing the curtain. (Y)

CURTUER — A projection on the end of an axle-bed, originally intended as a scraper to prevent the accumulation of dirt upon the back end of the hub. They are now little used, and then only as an ornament. (C) Also called a *dub*.

CUSHION — A case made of cloth or leather, and stuffed with some soft and elastic material. It is laid on a seat board for the comfort of passengers. (C & h)

CUSHION STICK — A triangular strip of wood fastened to the front edge of a seat frame, used to keep the cushion in place. (C)

CUSHION STRAP — Same as *seat strap*.

CUT-UNDER — That portion of the body which is cut away so as to allow the front wheels to pass under without striking the body. (C)

D

DAMASK — A linen fabric with raised figures, woven in imitation of damask silk; used for slip linings in carriages. (C & T)

DAMASSE — A kind of linen manufactured in Flanders; an imitation of damask, used by carriagemakers for slip linings. (C)

DAMMAR — A resinous gum used for the manufacture of white varnish. (C)

DASH — An iron frame, covered with leather and fastened in an upright position on the front of a carriage body as a protection against the mud and water thrown by the horses' feet. It is also called a *dasher*. When applied to a sleigh, it is sometimes called a *splash-board*. Also see *footboard*. (C, T, & h)

DASHBOARD — A board used in place of an iron and leather dash, such as the dashboard on a buckboard, or certain types of spring wagons. (C)

DASHER — See *dash*.

DASH FOOT — An iron fitting used to attach the dash to a carriage. (I)

DASH FRAME — An iron frame which forms the outline of a dash, and supports the leather covering. (T)

DASH LAMP — See *lamp*.

DASH MOLDING — A metal molding applied to the top edge of a dash, both for the purpose of appearance and preservation. (K & T)

DASH RAIL — An extra rail placed outside of the covered rails of a dash frame. Also an iron rail used on a dashboard. (T & C)

DASH ROD — An iron rod having a turned collar or other ornament, used as an extra rod on iron dashes. A variety of dash rail. (C)

DAUMONT, A LA — Driven by outriders or jockeys, the carriage having no coach box or other driving seat, a small box called a *coffre* being substituted upon the front gearing. Four horses are employed, with two jockeys, and generally two footmen behind. This method of driving was introduced in France under Napoleon I, by an eccentric nobleman, the Duke de Daumont. (h)

DEAD COLOR — A color mixed with turpentine or other substances that cause it to dry without any luster. (C)

DEAL — A soft wood, generally fir or pine, that is sometimes used in place of whitewood, but inferior to it. (C)

DECALOMANIA — (1.) The process by which a picture or design on a specially prepared paper is transferred and permanently fixed to another surface, such as a carriage panel. (2.) Also, the picture or design prepared for the above operation. This was also called a *transfer ornament*.

According to the March 1, 1881, number of *The Hub*, this process, as applied to vehicles in the United States, was developed by Charles Palm, of New York City, and offered to the trade initially by the firm of Palm and Fechteler of that city. This eliminated the need for a heraldic artist in the paint shop, making possible such ornamentation at but a small cost. (C & O)

DEE — An iron ring having one side flattened like the letter D, used as a loop for a stay or other strap. (C)

DENNETT SPRING — The Dennett spring more properly belongs to the chaise mode of suspension, in contradistinction from the gig suspension. The early chairs, whiskies, and chaises described by Felton all have their springs under the body; while the gig class, including the curricle, cabriolet and gig proper, have the coach suspension by C-springs and thoroughbraces. Both methods have been found to be defective, and a great mechanical difficulty has been experienced in the various efforts to arrive at a perfect suspension of two-wheeled vehicles. The difficulties to overcome are these: to balance the weight of the body and passengers so as not to put too much load on the horse (this is the defect of the cariole and volante); to prevent the liability of the passengers being thrown out, in case the horse stumbles (the defect of the Stanhope and

Dennett); to reduce the excessive weight of the ironwork (the defect of the Tilbury); and to obviate swaying motion (the defect of the gig class), and unequal motion (the special defect of the Dennett). The Dennett spring was the result of one of these attempts; invented in England about 1825, it consists of three springs, two half-elliptics at the sides, and one cross spring behind. The Dennett spring, a type of platform spring, is like the *mail spring*, but has only one cross spring. They are also called three-quarter springs. (h & FR50)

DESILVER — To remove silver by the use of acids from any article that has been plated. (C)

DICKEY-SEAT — (1.) A front driving seat made distinct from the body of a carriage, being either framed to it, or suspended separately on iron loops. The driving seat to that class of vehicles where a coachman is employed. Members of the coach family usually have dickey-seats, while those of the phaeton family, being owner-driven, do not. See *boot*, and *driving seat*. (h)   (2.) Formerly the servant's seat behind the body. (C)

DIP — See *swing*.

DIRT-IRON or DIRT-BOARD — A British term for a projection extending over the hub from the end of the axle-bed, serving the same purpose as a *sand-band*. (O)

DISH — The technical term applied to what an ordinary observer would term the outward cant of that part of a wheel above the axle, and which would be more exactly defined as the deflection of the upper portion of the wheel from the perpendicular plane in which the lower portion travels. The object of dishing is this: the tendency of all wheels is to collapse inwardly, and by dishing them outwardly, thus forming a trusslike bearing, this tendency is counteracted. Two other incidental results are used to advantage by modern builders. First, by dishing wheels, mud is thrown away from the body. Also, the carriage can be built with a greater width of body to a given width of track. (h)

DISH-STAFF — A wooden staff, screwed or held against the end of a hub, with an iron screw or stud passing through the free end, which serves to adjust or guide the angle or dish at which the spokes are to be driven. Also called a *driving-sett*, or *face-sett*. (h)

DISTEMPER — A peculiar sort of painting, wherein the colors are mixed with water, glue, egg yolk, etc. (C)

DODGE — A technical term applied to setting back from a given line each alternate mortise in a hub, to give a wider bracing to the spokes at this point. Also called *stagger*. (h)

DOESKIN — Enameled cloth finished to imitate dash leather, the back resembling leather in color. Also the leather made from the skin of does, and sometimes from the skin of sheep and lambs. (T)

DOG — Various tools bear the name *dog*, such as *spoke dog* — this being a tool with a pivoted hook, used to draw two spokes together when fitting a felloe to the spoke points. Another is the *tire dog*, which is a specially shaped tool used to help force the hot tire over the wooden rim of a wheel. (O)

DOG-STICK — Same as *drag-staff*. (O)

DOOR — That part of a frame of a carriage that is hung on hinges, and made to open and close, affording ingress for passengers and luggage. (h)

DOOR CAM — A device for holding the window in a carriage door at any height, and to prevent rattling. (T)

DOOR CREST — The upper panel of the door. (C)

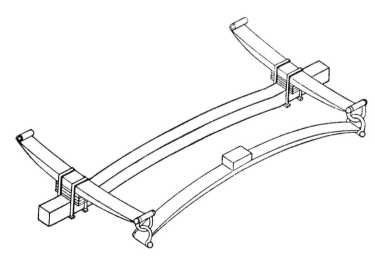

Dennett, or three-quarter springs. (From Fairman Rogers: *Manual of Caoching,* 1901.)

DOOR HANDLE — A metal handle used for turning the catch of the door lock and opening or closing the door. See also *handle*. (C & T)

DOOR HEAD — An extra bar fitted against the top rail between the standing pillars. (C)

DOOR LINING — A thin panel nailed against the inside of the door, extending from the guard rail to the bottom; it acts as a support to the trimming, or on some cheaper vehicles is simply painted, and that part of the trimming is omitted. (C)

DOOR MOLDING — A wide flat molding fastened on the sides of a door to hide the opening between it and the body pillars. (C)

DOOR PILLAR — The pillar that forms the side of a door; one is known as the hinge pillar, and the other as the lock pillar. (C)

DOOR RABBET — The rabbet made in the lock pillar of the door to act as a guard against the door shutting in too far. (C)

DOOR-STILE — The transverse frame-piece at the middle of a carriage door, the top edge of which is rabbeted out, forming a recess for holding the glass-frame when it is elevated. (h & C)

DOOR-STOP — A band of lace, leather, or metal used to prevent the carriage door from opening beyond a certain point. Also a small piece of metal or wood attached inside a carriage body at the corners of a door, to serve as a stop when the door is closed. (O & T)

DORMEUSE — A boot which can sometimes be formed on a post chaise, by means of a removable front panel, to lengthen the body of the vehicle for sleeping purposes. (A)

DOTED — Said of timber when it has turned white and become brash. (C)

DOUBLE CHECK LOOP — A double loop attached to the perch, which receives the two check-straps. Also called a *perch loop*. (K)

DOUBLE PERCH CARRIAGE — A carriage-part having two perches framed into it. (C)

DOUBLE-SUSPENSION — Suspension both front and rear upon a combination of C and elliptic springs. (h)

DOUBLETREE — Same as *evener*. Also see *swingletree*.

DOUBLETREE BLOCK — A block of wood bolted to the tops of the hounds, to raise the doubletree to a higher level. (O)

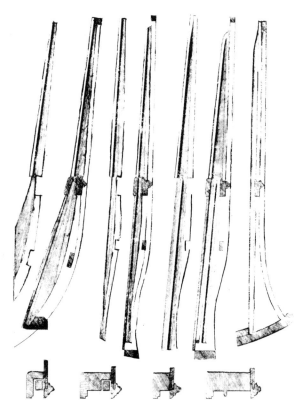

Door and body pillars, showing the construction of sliding glasses. Figure 7 shows the design of the coupe-pillar. (From Roubo: *L'Art du Menuisier-Carrossier,* volume 17, 1771.)

DOUBLETREE CLEVIS — A clevis for attaching singletrees to the doubletree. Some are a combination fitting, with a stay chain hook formed on the rear. (I)
DOUBLETREE CLIP — A clevislike fitting having either a hook or loop for the attachment of the swingletrees or stay chains, or both, to the doubletree. (Y)
DOUBLETREE COUPLINGS — A pair of plates, usually of malleable iron, on which the doubletree swivels. Generally the surfaces of the plates are formed so as to fit into one another, increasing the bearing surface and lessening the strain on the doubletree pin. Also called *evener plates*. (I & K)
DOUBLETREE HASP — Same as *hammer-strap*.
DOUBLETREE PIN — The heavy iron pin on which the doubletree swivels. (O)
DOUBLETREE PLATE — An iron plate nailed or screwed to the underside of a doubletree to receive the wear caused by the swiveling action of the doubletree. (K & Y)
DOUBLETREE STAPLE — A staple with threaded shanks, secured to a doubletree with nuts; in reality, a U-bolt. Either stay chains or swingletrees may be attached at this point. (O)
DOUBLE WHIFFLETREE — Same as *doubletree*.
DOVETAIL — A common form of mortise and tenon, so called from its fanlike shape.
DOVETAIL CATCH — A catch made of two pieces of metal, one piece having a wedge, or dovetail-shaped projection placed diagonally across the center, while the other piece has a mortise of the same shape into which the wedge fits when the door is closed. It is used to prevent a door from sagging. (C)
DOWEL-BOUND — The condition existing in a wheel in which the dowel holes are not drilled deeply enough, so that the dowel keeps the felloes apart, rather than allowing them to draw tightly together. (O)
DOWEL CATCH — A half round metal dowel attached to a plate. It is used on coach doors to prevent the bottom from springing out when only two hinges are used. The dowel is attached to the face side of the hinge pillar of the door, near the bottom. When the door is closed, the dowel enters the standing hinge pillar, holding the door pillar firmly in place. (C)
DOWEL PIN — A pin of wood or metal, used for joining two pieces of timber where there is no tenon or other fastening. (C)
DRAB — A dull, brownish yellow or gray color. (C)
DRAFT — (1.) The amount of power needed to move vehicles. (2.) The moving of loads by drawing or pulling. (3.) A team of animals, together with the load they draw. (4.) The bevel given to a pattern for a casting, so that it may be removed from the sand without injury to the mold. (C)
DRAG-CHAIN — A variety of brake, consisting of a strong chain with a hook at one end, which, being attached to the hind wheel of a vehicle while descending a hill, prevents its revolving, and thus retards motion. Also the chain by which a *shoe* is attached to a vehicle. (h)
DRAG-FRONT — A heavy, full-paneled driver's seat, generally without a wheel-house. (h)
DRAGON'S BLOOD — A brownish-red resinous gum, used for coloring varnishes, etc. (C)
DRAGON TONGUE — A peculiarly shaped iron attached to the end of a whiffletree to hold the end of a trace. (C)
DRAG-STAFF — A straight bar of wood or iron attached to the underpart of a vehicle, which in ascending hills is allowed to drag behind. Upon the stoppage of the horse, it presses against the ground and thus prevents the vehicle from rolling backward. It is also called a *mountain-point*, *mountain-check*, and *dog-stick*. (h)
DRAUGHT — Same as *draft*.
DRAW-BAR — Same as *splinter-bar*. (FR 22)
DRAW BOLT — One of the bolts used to secure a pole between the hounds. (O)
DRAWBORE — A hole bored in a tenon in such a position as to cause the mortise pin to draw the shoulder snug to the face of the mortise. (C)
DRAW-IRON — An iron stay extending from the front axle-bed or transom to the draw-bar (or *splinter bar*), used to couple the draw-bar to the front gearing. Also called a *wheel-iron*. (h)
DRAW-ROD — Same as *brake-rod*, except that the term here may be used only when the rods draw the beam tight, while the term, *brake rod*, must be applied to the variety that pushes the brake beam. (O)
DRAYEL — A British term applied to a stout iron staple driven into the front end of each shaft, for the attachment of traces for an additional horse. (O)
DRIER — A substance prepared from asphaltum, or a metallic oxide which, when mixed with paints, causes them to dry quickly. (C)
DRIP-MOLDING — A three-cornered or concave molding fastened over the windows and doors of a carriage body to prevent the water from dropping inside of the carriage. (C)1

DRIP-TUBE — A small metal tube inserted in the floor of a carriage, to allow rain water that blows into the vehicle to drip out underneath. (O)

DRIVING BAND — A temporary band of iron, which is driven upon a hub to prevent it from splitting during the operation of driving the spokes. On many wheels, particularly on wagons, these bands are permanent. (h)

DRIVING CUSHION — A cushion having a high back and a low front, used on certain styles of vehicles when the driver is compelled to sit with his legs nearly in the same position as when standing. (T)

DRIVING RAIL — A rail, generally plated, attached to the top of a dash, serving as a support for the reins. Also called a *rein rail* or *dash rail*. (O)

DRIVING-SEAT — An elevated front seat especially intended for the driver. See *boot* and *dickey-seat*. (h)

DRIVING-SETT — See *dish-staff*.

DROP-BOTTOM — A sunken bottom to a carriage body to give sufficient leg room without making the side of the body too deep and heavy. (C)

DROP-BOX — (or DROP-SEAT BOX) — A small box suspended underneath the seat of a carriage or wagon for the accommodation of small objects. (O)

DROP-FRONT — That portion of a body in front of the seat that falls below the line of the bottomside, the object being to give increased leg-room to the driver. (h)

DROP HANDLE — A coach handle hinged on the bolt so as to allow it to hang down when not in use. (C & T)

DROP LAKE — A red or purple pigment which, in its marketable form, is in small globules. (C)

DROP-LIGHT — A coach light that drops into the framework of the body. (C)

DROPPER — Same as *spring-shackle*. (H)

DROP TONGUE — A wagon tongue used with two pairs of hounds, front hounds and tongue hounds, which construction permits the tongue and its hounds to drop forward when the wagon is not hitched up. (Y)

DRUG-BAT — A British term meaning the same as *shoe*. It was also called a *drag-bat, drug-shoe,* or *skid-pan*. (O)

DRUGGET — A coarse woven cloth or felt, printed in colors on one side; it is used for lining lap robes, etc. (C)

DRUNKEN WHEEL — A carriage wheel which wobbles when revolving on the axle-arm. The cause can be a box that is not set true, or spokes that are not set true. (C & h)

D-SHACKLE — An iron loop, with bolt through the ends, forming a D, used on top of C-springs to keep the leather braces in position. (h)

DUB — Same as *curtuer*.

DULEDGE — A wooden peg which joins the felloes of a gun carriage. (C)

DUMB SPRING — A British term meaning an iron plate, of the same shape as the half-elliptic spring to which it is attached. It is used only on perch carriages with C and under springs. (O)

DUMMY-BOARD — A small board on the back of a coach for a footman. Also called a *page-board*. (h)

DUMP STICK — A slender bar of wood or iron which slips into the eyes on the shafts of a dump cart, to hold the front of the body down on the shafts. When the stick is removed, the cart dumps. (O)

DUPLEX SPRING — A type of double-sweep end spring, generally used with drop center axles on light carriages. (Y, B, & G)

DUSTER — An extra panel or molding around the top edge of drop-front bodies. (C)

DUST HOOD — A covering of leather or fabric used to cover a carriage top when it is down, and protect it from dust. (Y)

DUTCH METAL — An alloy of eleven parts copper and two parts zinc, rolled or beaten out into thin sheets, used for cheap gilding. It is sometimes called Dutch gold, or Dutch leaf. (C)

D WHEEL — A fifth-wheel of a carriage having but little more than a half circle. (C)

Drop tongue used on a Studebaker mountain wagon. (Studebaker Bros. Mfg. Co., Catalog No. 222, 1903.)

E

EAR — A projection forged to or connected with any part of a carriage, having a hole drilled through for the purpose of receiving a bolt. (C)

EAR CUSHION — A small cushion stuffed with curled hair, which is hung in the back corner of a coach to lean the head against. (C)

ECCENTRIC — A wheel or disk having its axis to one side of center, used to give a motion precisely like that of a crank. (C)

ELBOW RAIL — A light iron rail attached to the end of a carriage seat. Also used synonymously with *arm-rail*. (C)

ELBOW REST — A metal arm piece with a leather- or fabric-covered pad or cushion on top, affixed to the end of a seat, and upon which the arm rests; used in place of a wooden arm on open-top carriages. (C)

ELBOW SPRING — A steel or wooden spring, straight or slightly curved, and pivoted to a loop at one or both ends. A single elbow spring, pivoted at one end only, may be pivoted to another single elbow spring. The elbow spring was the forerunner of the elliptic spring. See *grasshopper spring*. (F)

ELBOW SWEEP — A short sharp sweep, such as now used on corner pillars of the most fashionable carriages. (C)

ELECTRO GILDING — The process of depositing gold upon white or yellow metal by means of electricity. (C)

ELECTROPLATE — (1.) (*v.t.*) To cover iron or other metal with a film of silver, gold, nickel, etc., by electrolysis. (2.) (*n.*) The carriage hardware plated by this process. (C)

ELEVATING COLLAR — A wood or metal spool, generally turned ornamentally, used as a spacer and support between other parts that must not come together. (I)

ELLIPSE — An oblong or oval figure, the perimeter of which is a regular curve. The term *oval* is most commonly used by the trade to designate this shape when it is applied to windows, molded ornaments, etc. (C)

ELLIPTIC SPRING — A carriage spring made up of two sets of overlapping steel plates or leaves hinged together at the ends by bolts, the general form being an acute ellipse — hence the name. Elliptic springs were invented by Obadiah Elliot of England in 1804; they had an important influence on the mechanical development of carriages, first permitting them to be built without perches. A half-elliptic spring is one-half of an elliptic spring, cut transversely. A three-quarter-elliptic is one having a complete bottom half, and only half of the upper section. A double elliptic is one having two sections above, side by side, and one heavier section below, joined between the two upper springs. Some manufacturers term this last named spring as a triple elliptic. (h, Y, & O)

ELM — A wood used in the carriage-building trade, particularly for hubs. (C)

EMBOSS — To finish the surface with raised work of ornamental figures, flowers, etc.; formerly a popular way of ornamenting leather boots to buggies. (C)

EMERALD — A rich green color like that of the emerald. (C)

EMERALD GREEN — A very durable pigment of a vivid light green color, made from the arsenite of copper. (C)

EMERY — A granular dark-colored corundum extensively used for grinding and polishing metals. (C)

EMERY BELT — A leather belt, the face of which is covered with a coating of emery and glue, used for polishing metal surfaces; a belt called an emery belt is also used for rounding wood; it is made in the same manner as an emery belt, except that a fine, sharp white sand is used in place of emery. (C)

ENAMEL — (1.) (*n.*) A paint, usually glossy, that flows out to a smooth hard coat. (2.) (*v.t.*) To form a glossy surface upon cloth or other material. (3.) (*v.i.*) Said of varnish when it crawls, and presents a wavy appearance on the surface. (C)

ENAMEL CLOTH — Cotton or linen cloth painted with a preparation, the component parts of which are linseed oil, lampblack, and turpentine. It is used for aprons, and on cheap work for curtains, the covering for calash tops, etc. It is intended as an imitation of enamel leather. (C & T)

ENAMEL LEATHER — Leather having one side colored and varnished; the face is covered with a preparation much like that used on enamel cloth. Several coats are used; these are baked in a hot oven, the skin being stretched upon a frame; after the necessary number of coats have been put on, the leather is put through a grainer and afterward cross-worked by hand, which gives to it the rough grained surface that distinguishes it from patent leather. Enamel leather is made exclusively from the grain or outside split of the hide. Other types of leather are also made of the grain split, but they are all japanned upon the split side, while enamel leather is japanned on the grain. It is extensively used for the covering to calash-tops, etc. (C & T)

END-BAR — The cross-bar that forms the end to the framework of a body. (C)

END BOARD — The back end of a wagon body, hinged or otherwise, attached to the bottom in such a way that it can be dropped or removed for convenience in loading and unloading. It is held in place by spring catches, chains, a link and pin, etc. Also called a *tail board*, *tail gate*, or *end gate*. (C & h)

END-BOARD NUT — One of the nuts used to secure the end board of a wagon. They are made either as double-wing nuts, or with a single projection on one side, enabling the nuts to be secured very tightly without the use of a wrench. The same type of nut is also employed to secure one of the chains which supports the side boards of some wagons. (K)

END CLIP — A fitting at the end of a singletree. It consists of a thin, broad strap in the rear, changing into a loop in front, to which is attached the singletree hook. (K & Y)

END-EXTENSION BOARDS — Boards which can be applied at will to the top of wagon ends (in conjunction with *side-extension boards*), to increase the depth of a wagon box.

END GATE — Same as *end board*. (Y)

END-GATE CHAIN — A chain for securing an end gate in either a closed or horizontal position.

END-GATE FASTENER — One of the many devices used to hold the end gate of a wagon in a closed position. (Y)

END-GATE HINGES — Hinges used on the end gate of a wagon, allowing the gate to drop for loading and unloading. (Y)

ENGILD — To gild; to brighten. (C)

EQUIVALENT CARRIAGE GEAR — A patented type of front carriage gear manufactured by Bartholomew & Co., of New Haven, Connecticut. It was so named because it was said to be the equivalent of the platform and perch, doing the work of both. It consists of an ordinary elliptic spring across the front of the vehicle, supplemented by a series of parallel motion flat leaf springs, just behind and at right angles to the center of the elliptic spring. (O)

EVENER — The movable front cross-bar of a pair-horse vehicle, secured to the futchells by a center bolt (on which it swivels); the singletrees are attached to this part. It is also called a *doubletree*. Compare with *splinter-bar*. (h & C)

EVENER DROP STAY BRACE — A fitting attached under a dropped evener to brace it from the rear. (I)

EVENER DROP SUPPORTS — A pair of couplings that drop the evener below the part of the gearing to which it is attached. (I)

EVENER PLATES — See *doubletree couplings*. (K)

EXBED — A term frequently used in England in referring to an axle-bed. (O)

EXCELSIOR — The trade name of a preparation of wood, cut into fine fibers by machinery, curled and dried. It is used for stuffing cushions, etc. (T)

EXTENSION STAKE — A removable stake which slides into a socket on the regular stake of a wagon. (O)

EXTENSION TOP — A type of falling top used on a two-seated vehicle, wherein the bow sockets of the front, or extended portion of the top, are attached to the front bow socket of the rear section when the top is to be put down. (O)

EYELET — A wide-flanged metal ring, made in two sections that are clinched or screwed together over a piece of leather or fabric, to form a reinforced eye or hole in the material. (T)

F

FACE-SETT — See *dish-staff*.

FAIR LEATHER — Same as *russet leather*.

FALL — An apron attached to the front edge of the seat or cushion, suspended between the point of attachment and the bottom of the carriage body. (T)

FALLING TOP — A carriage top that can be put down. (O)

FALSE BOLSTER — An additional piece of timber bolted to the top of a regular bolster, to raise the load or provide an additional wearing surface. These are generally used on lumber and pipe-line wagons, and were sometimes used on the rear bolster of a Conestoga gear, to raise the rear end of the body. (O)

FALSE-BOX — Same as *sand band*. (O)

FALSE LINING — An adjustable lining, made of linen or other thin fabric, used as a protection to the finer materials covering such things as cushions, squabs, etc.; also called a *slip lining*. (C & T)

FAWN COLOR — Lead color slightly tinted with brown. (C)

FEED BOX — A long narrow trough generally carried by means of chains across the rear end gate of a wagon. When used to feed the animals it was usually taken from the carrying position and placed on the tongue of the wagon. The feed box was standard equipment on such wagons as Conestogas and army wagons. (S226)

FEED BOX CHAIN — One of the two chains secured to the rear portion of a wagon, for the purpose of carrying the feed box. (O)

FELLOE — A sawed or bent segment of the rim of a wheel. The bent ones now used are generally termed *rims*, while those sawed out in short segments are called *sawed felloes*. The term was originally spelled *felly*, and even now is generally pronounced as if so spelled. The bent circle attached to the futchells is also called a felloe. (C & h)

FELLOE-BAND — A metal band, the outside of which is of the same shape as the ends of the felloe; it is put on as a ferrule, and acts as a support to the joint of the felloes (or fellies). (C)

FELLOE BOUND — Same as *rim bound*.

FELLOE-COUPLING — The name applied to the various types of patented metal devices which are used in place of a wooden dowel to hold felloes in alignment with one another. (h)

FELLOE-PLATE — A small plate of wrought or malleable iron, used for the same purpose as a felloe-band, but held in place by bolts which pass through and tie together the tire, felloe, and plate. (C)

FELLY — The earlier spelling of *felloe*.

FELT — A coarse cloth of untwisted fibers, made compact by rolling and pressing; when printed in high colors it is used by the carriagemakers for linings to aprons, lap robes, etc. (C & T)

FENCE — The raised part of a glass-rest, intended to keep the glass frame in position when elevated, and to prevent rain from blowing in under the frame. It is also called a *guard*. (h & C)

FENDER — (1.) The bent piece framed to the ends of the beams of light sleighs, originally to protect the side of the sleigh from injury. (2.) An iron frame, covered with leather, attached to a carriage body by stays, and extending over the wheels, its purpose being to protect the carriage and its occupants from mud. A thin panel of wood or a piece of sheet metal, bent to the required shape, is sometimes substituted. It is also called a *wing*. (C & h)

FERRULE — A metal band put around tool handles, singletree ends, etc., to prevent them from splitting. (C)

FIFTH-WHEEL — A horizontal metal circle, or section of a circle, placed between the body and front axle of a carriage, and generally connected with both by means of a king-bolt, its purpose being to allow the axle to turn laterally and change the line of motion without disturbing the balance of the body. Most fifth-wheels

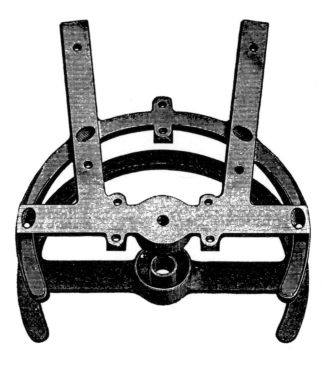

Fifth-wheel for double perch gear. (The Eberhard Mfg. Co., Catalog No. 8, 1915, Cleveland.)

consist of two full-circle plates of iron, revolving one upon the other; in some, one metal plate revolves upon a wooden segment; there are also *half-circle* and *elliptical* fifth-wheels, whose names are self-explanatory, (h & C)

FIFTH-WHEEL STOP — A spur projecting from the edge of one of the plates of a fifth-wheel, or attached to the futchells, the purpose being to prevent the fifth-wheel from revolving beyond that point, and in turn to prevent the front wheels from striking the body or perch. (h)

FILLET — A little rule or reglet of gold leaf drawn over moldings, etc. (C)

FILLING — A coarse paint used as a preparatory coat to panels of carriages. (C)

FIRMER — A broad, thin chisel, such as is used by bodymakers. (C)

FLAKE WHITE — The purest white lead, in flakes or scales. (C)

FLAPPER — A folding support, two of which are used to steady a movable window in the back of a landau, when the window is in its raised position. (O)

FLARE — The cant given to the sides or ends of a carriage body or seat. (C)

FLARE BOARD — An extra board attached to the top of a wagon box to increase the carrying capacity. It is so named because it flares outward at the top. (Y)

FLARE BOARD IRON — The iron bracket which supports a flare board. (Y)

FLOAT — (1.) A steel instrument having long teeth filed across the face, at the same angle as those of a rip saw; a type of rasp. It is used for cutting down the uneven surfaces of wood. (C)   (2.) A British term for a *side-extension* board. (O)

FLUTE MOLDING — A molding having one or both edges concaved. (C)

FOLDING SEAT — (1.) A seat hung on a special hinge, on a truck or dray, made so that the seat can be turned over and out of the way while the vehicle is being loaded. (I)   (2.) Generally, any seat that folds out of the way when not in use. (O)

FOLDING STEP — Iron carriage steps so made that they can be folded up when not in use, the motion generally being automatic and governed by the movement of the door. During the period when carriage wheels were six feet or more in diameter and footmen were literally necessary to assist ladies in alighting, folding steps were first devised. The original folding steps consisted of at least four or five steps, which folded one within another and finally folded into the coach, resting on the floor. (h & C)

FOLDING TOP — A carriage top that can be put down. (O)

FOOTBOARD — (1.) Originally a platform for the footman, afterward used as a rack for the support of baggage, and so constructed as to be closed up when not in use, presenting a carved or finished face; also called a *pole-guard*.   (2.) Also the dash or footrest of a coach or other carriage, where no upright dash is used; a *toe board*. (T & C)

FOOTBOARD HANDLE — A handle attached to a footboard to assist the driver in getting to his seat. (K)

FOOTBOARD LEDGE — A strip of wood fastened to the footboard of a carriage, as a brace for the driver's feet. (h)

FOOTMAN — A servant or page who accompanies a vehicle, occupying a seat beside the driver or in a rumble, or standing upon a page-board behind the carriage. Footmen were formerly used in considerable numbers, as many as five or six sometimes accompanying a coach. Their office was more than mere ornament. At one time *running footmen* were used. (h)

FOOTMAN CUSHION — According to Felton, this was a wooden frame covered with stout leather and stuffed, to ease and elevate the servant behind the carriage. (C)

FOOTMAN-HOLDER — An ornamental pendant, attached to the back end of a coach body at the top corners, originally designed as a holder by which the footman could steady himself. The only use of these straps in this country is as an ornament, and they were already going out of style even for this purpose by the 1870s. Also called a *page-holder*. (C, T, & h)

FOOTMAN-HOLDER LOOP — A metal loop to which a footman-holder is attached. (K)

FOOTMAN'S STANDARD — See *standard*.

FOOTMAN-STAND — Same as *dummy-board*.

FOOTMAN-STEP — An iron step affixed to the hind gearing of a vehicle, to assist the footman or page in mounting. (h & C)

FOOT MUFF — A muff made of furs as a receptacle for the feet in cold weather. (C & T)

FOOT WARMER — Same as *carriage heater*.

FORE CARRIAGE — The framework of a no-perch carriage, which rests upon the fore axle. (C)

FORE-GATHER — Same as *fore way*. (O)

FORE TRANSOM — The timber used in place of the front bolster upon carriages that are made without fifth-wheels. It is held to its place by the king-bolt, the springs being fastened to either end. (C)

FORE-WAY — A term used in England, meaning *gather*. (O)

FORGE — An elevated platform of brick or iron, having an open chimney, beneath which is a small fire bed for heating metals. (C)

FORGER — The head blacksmith who superintends the handling and forging of the iron at the forge; also called the fireman. (C)

FORM — A pattern or shape over which felloes or other bent timber is drawn to give it the required shape. (C)

FOXY — A yellowish or reddish brown color used in shading. (C)

FRENCH-BOOT — A style of coach boot; the risers to the seat and the toe board brackets are bent in one piece, the back of the seat being supported by carved or ornamented iron legs. (C)

FRENCH RULE — The application of practical geometry in the construction of carriage bodies, as practiced in France. It was brought to the United States by foreign workmen, who were taught it by Professor Henri Zablot, the originator, and Albert Dupont, and afterwards published in Mr. Brice Thomas's journal, called *Le Guide du Carrossier*, from which it was translated into English by the late Professor John D. Gribbon, the first instructor in the Carriage Builders' National Association, Technical School for Carriage Draftsmen. By this rule the bodymaker obtains the points by which he draws the correct side-sweep for the different pieces of the framework, the turn-under and side-swell being given by the operation of right lines drawn over the side elevation and cant of the body. To the bodymaker it is invaluable in dressing up his stock on a scientific plan, and to the designer it is necessary to lay out and properly prepare working drafts. Though difficult to learn, when once understood, it lightens the burden of the work.

The operation of French rule is nearly identical to that of *square rule*, as explained in volume 18 of the *Carriage Monthly*, page 115.

"It is generally understood in the body-shop that the proper meaning of the French rule is when the different pieces of the body are inclined and contracted according to the turn-under and round of body to save material, while by the square rule it is understood that all the pieces of the body should have vertical faces, although sometimes contracted. In giving our drafts in the Monthly we make no distinction between the two, laying them out according to the amount of round and turn-under of the body, and constructing the lines in the most simple manner. If convenient, the pieces are laid out square as on a buggy body, but when contraction is necessary it is much easier to construct the draft according to the necessary contraction, and dress the pieces accordingly. There is decidely less labor when done in this way than by first laying it out square and then cutting the wood for the contracted lines for the inside and outside surfaces afterwards. This holds good for pieces which are inclined, or inclined and contracted. We make no distinction between the French and square rules, simply calling it 'practical geometry in the construction of carriage bodies.' " (O & h)

FRINGE — An ornamental border, having a narrow heading, the remainder being formed of loose or twisted thread pendants. The technical names of fringe are bullion, festoon, and carpet. (C & T)

FROG — (1.) An ornament of silk or worsted, woven to match the lace used. (2.) A silk cord, attached to the inside of a carriage door, to hold the glass-string or lace when the glass-frame is raised. (T & h)

FRONT-PILLAR — An upright piece of timber at the front corner of a body, to which the frameworks of the side and front are attached. (C)

FRONT-RAIL — One of the pieces of timber framed across the front end of a body. (C)

FRONT WASHER — See *axle washer*.

FULCRUM, BRAKE — The point on which the brake lever pivots. (O)

FURCHELL — A varient spelling of *futchell*.

FUSTIAN — A kind of coarse twilled cotton stuff, including corduroy, velveteen, etc., used to some extent for trimming light carriages. (C)

FUTCHELL — One of several crooked timbers that are framed through the front axle-bed, the back ends of which support the sway bar or the outer edge of the lower part of the fifth-wheel. The front ends are finished in different ways: on some carriages they are cut off even with the bed; on others they support the splinter-bar, and on others they are brought nearly together to form a socket for the pole. In the latter case they are generally called *hounds*. There are both outside futchells and inside futchells. The former run from the sway bar to the ends of the splinter-bar, supporting the ends of that part. The inside futchells are more nearly like ordinary hounds, forming a socket for the pole. (C, h, & FR)

FUTCHELL BAR — A term sometimes applied to a *sway bar*.

FUTCHELL BED — A stout piece of timber bolted to the tops of the futchells of a platform gear vehicle. It approximates the sand board of a wagon, and the king-bolt passes through the center of it. (O)

G

GAMMON (or GAMMON-BOARD) — A name sometimes applied to a roof-seat of a coach. The term originated in England, where in 1788 a Mr. Gammon was instrumental in securing the passage of an Act of Parliament which prohibited coaches from carrying more than six persons on roof seats. (FR)

GARDEN-SEAT — A style of seat used on the upper deck of a double-decked omnibus, wherein all seats face forward, with a central aisle. These seats, introduced about 1880, were used on the later types of busses. (O)

GARNISH-RAIL — The inside rail, either top or bottom, enclosing a carriage glass. (h)

GATHER — A technical term for the forward set of an axle, meaning that the ends of the axle-arms are bent forward, away from the centerline of the axle. This is done to compensate for the outward cant of the wheels which causes the wheels to turn outward and run hard against the axle-nuts or linch pins. By *gathering* the axle-arms, the wheels toe-in slightly, and consequently run against the shoulders of the axle-arms. In England, the term *fore-way* is sometimes used to denote *gather*. (C)

GAUGE — (1.) A tool having an adjustable head, used for marking timber to a uniform thickness or width. The double, or mortise gauge, has one stationary spur and another attached to a movable bar, for the purpose of gauging both sides of mortise or tenon at one time. (2.) In some localities the track of a vehicle is called the gauge. (C)

GEAR — See *gearing*.

GEARING — The underpart or running part of a vehicle, including the axles, springs, fifth-wheel, beds, futchells, wheels, etc. By English builders, and according to ancient usage, called simply *the carriage*. (H)

GENET ROBE — A carriage robe made from cats' skins. They are colored to imitate the fur of the African genet. (C)

GERMAN SILVER — A silver-white alloy composed of copper, zinc, and nickel, used to some extent for carriage mountings, such as handles, lamp parts, etc. It is also called *nickel silver*. (O)

GILT — Gold laid upon the surface of baser metals. The showy gilt ornaments on carriages and harness are of some base metal covered with a thin film of gold by the electroplating process. (C)

GIMP — A woven ornamental braid of silver, gold, and silk or worsted thread. (T)

GIMP-COVERED NAIL — A trimming nail, having the head covered with gimp. (T)

GIMP TACK — A small round-headed tack used to tack gimp in place. Also called a *Hungarian nail*. (K)

GLASS — Synonymous with *window* or *light*. (h)

GLASS FRAME — A wooden frame for the support of glass in the movable windows of a carriage. (T & C)

GLASS FRAME HOLDERS — A pair of iron beds hinged to the top of a Landau door for the purpose of supporting the glass when it is lifted out of the body of the door, obviating the necessity of lowering the glass frame before the door can be opened. The use of these holders has done much toward giving the Landau and kindred carriages their present popularity, as it has rendered the use of the upper framework of the door unnecessary. (C & h)

GLASS FRAME LIFTER — A strip of lace or leather, the lower end of which is attached to the lower bar of a glass frame, by which it is lifted out of the bed in the door or body. (T & C)

GLASS-HOLDER — See *holder*.

GLASS-REST — That portion of a middle door bar that supports a window frame in the raised position. (O)

GLASS ROLLER — One of the small rollers that are let into the bottom of a glass frame to reduce the friction when the frame is sliding. (C)

GLASS-STRING — A strip of broadlace or leather attached to a window frame, by which the frame is raised or lowered. Also called a *glass frame lifter*. (h & T)

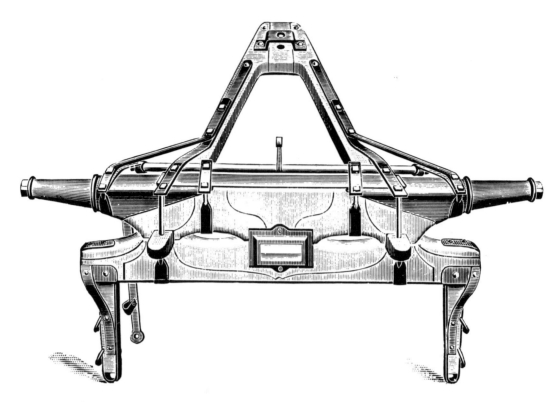

Bottom view of the rear section of Studebaker's steel-skein mountain gear, used on both farm and freight wagons. (Studebaker Bros. Mfg. Co., Catalog No. 222, 1903.)

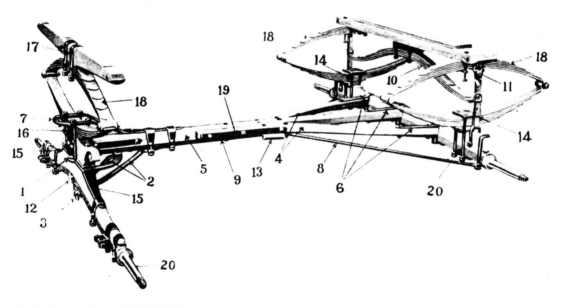

Studebaker wood-hound DELIVERY WAGON gear. The names Studebaker applied to the various parts: 1, Forged D. & T. plates, male and female; 2, Double clipped forged drop stays; 3, Norway split head king bolt; 4, Bent wood hounds, long jaws; 5, Tapered reach; 6, Norway forged reach and hound clips; 7, Norway double spring brace and pole loop; 8, Iron stay braces; 9, Iron reach plate; 10, Check spring; 11, Check strap; 12, Swelled center axle; 13, Rolled steel rub iron; 14, Norway shrunk spring clips, bolts and plates; 15, Extension D plates; 16, Iron head block plate; 17, Norway forged spring bar clips; 18, Oil tempered steel springs; 19, Hound jaw plates; and 20, Concord steel axles, increased arms. (The Studebaker Corp., Cat. No. 601, 1912.)

GLASS-STRING SLIDE — See *slide.*

GLAZE — (*v.t.*) To apply a transparent or semitransparent color over another color, to modify the effect. (O)

GLAZED LEATHER — A term sometimes used to describe *patent leather.*

GOLD LEAF — Gold beaten into a very thin leaf or sheet, used by painters for gilding. (C)

GOLD SIZE — A composition upon which gold leaf is laid in gilding. (C)

GOOSENECK IRON — A fitting which is secured to a seat-iron for the attachment of the bow sockets of a falling top. The front end of a shifting rail, made for the same purpose as above, is likewise called a gooseneck. (Y & G)

GRAIN DASH — The trade name given to a plain glazed patent leather, the hair side of which is unfinished, the japan being put upon the flesh side. (T)

GRAIN TANK WAGON BOX — A special wagon box intended for hauling grain. The lower part is like an ordinary wagon box, above which the sides flare, as a wagon with flare boards, and then the sides become vertical again above the flare boards. (O)

GRASSHOPPER SPRING — A name given at an early day to what was otherwise known as the *double elbow-spring.* In Felton's time the term was mainly applied to chaise springs, and gave rise to the variety of chaise known as the *grasshopper* (see Felton, vol. 2, page 117 and plate VII). It consisted simply of a relatively straight steel spring, with loops at the ends to support the body above it. (See Felton, vol. 1, page 76 and plate IX, fig. 6.) (h)

GROMMET — A small, round, metal eyelet that is put into fabric, such as a wagon cover, to prevent the fabric from being torn by a rope passed through it. (Y)

GROUND — The surface upon which figures or objects are painted; the principal color used. (C)

GUARD — A thin strip of wood on the middle door bar or other bar that supports a window frame. It is used to keep the frame in its place when it is raised, and also to keep rain from blowing in under the frame. Also called a *fence.* (C & h)

GUARD STRING — Silk cord, used for holding up lace and tassels, attached to doors or windows. (T)

GUDGEON — A metal pin fixed in the end of a wooden shaft, serving as the spindle on which the shaft turns. An iron band often binds the end of the wooden part, as a reinforcement, before the gudgeon is driven. (O)

GUG — A small metal stud fixed on the inner side of a clip. It imbeds itself in the wood as the clip is drawn up tight, providing security against the sidewise movement of the clip. (O)

GUM — A vegetable secretion in the juices of many plants, which hardens when exposed to the air. Some types of gum are used in the manufacture of varnish. (C)

GYPSY TOP — A buggy top, the sides of which are partially closed by triangular curtains, knobbed or otherwise secured to the back bows and the back quarter. The term is also applied to a variety of canopy top generally without fringe, having roll-up curtains which enable the vehicle to be enclosed. However, the term is generally applied to the first described top, and if reserved for that type confusion will be avoided. (T)

H

HAIR PENCIL — A brush made of hair, used for striping. (C)
HALF-ELLIPTIC SPRING — See *elliptic spring*.
HALF-PATENT AXLE — The trade name of a style of axle having a nut on the front end, while the collar at the back end is sunk in level with the end of the box. (C)
HALF-SPRING — Same as *half-elliptic spring*. (O)
HALF-STIFF TONGUE — A wagon tongue constructed just as a stiff tongue, but installed with specially formed plates on the forward ends of the hounds, so as to allow a slight vertical movement of the front of the tongue. (O)
HAMMER — An iron or steel tool used for driving nails, pounding iron, etc. (C)
HAMMER, WAGON — Same as *hammer-pin*. (O)
HAMMERCLOTH — An ornamental valance to a dickey-seat. It is made of cloth or bear skin and lined with buckram or other heavy material, then laid up in ornamental folds and embellished with lace, gimp, fringe, and tassels. Its use is limited in this country to a few styles of carriages, such as coaches, chariots, loop landaus, and bretts; in Europe it is the fashionable trimming for the dickey-seats on the carriages of the court and nobility. The English word is by some supposed to be corrupted from *hamper cloth*, a cloth covering to conceal the hamper for tools, provisions, clothing, etc., formerly carried under the driver's seat. Another theory is that the original hammercloth served to cover the box containing implements, which in the frequent breakdowns incident to the early days of coaching, formed an indispensable part of the driver's equipment. Both theories, however, are purely conjectural. (C & h)
HAMMERCLOTH-STANDARD — See *standard*. (h)
HAMMER-PIN — A doubletree pin with a head in the form of a hammer, thus having the additional capacity to serve as a tool as well as a pin. The end of the pin is generally chisel-shaped, to serve as a prying bar for such purposes as lifting the linch pins from the axles. (O)
HAMMER-STRAP — A metal strap, one end of which hooks onto the rear end of a wagon pole, while the other end goes over the doubletree, and the hammer-pin passes through it, for support, before it passes through the doubletree. Also called a *doubletree hasp*. (K)
HANDLE — A fitting attached to some part of a carriage to assist a passenger or driver in mounting or descending from a vehicle; also a fitting by which a door is opened or closed. The former is sometimes known as an *ascending handle*, while the latter may be of two types: the *door handle*, outside and attached to the lock, and the *pull-to handle*, inside and unconnected with the lock. Handles may be of metal, leather, ivory, or other material. Also see *drop handle*. (C, T, & h)
HANDLE NUT — A nut having a handlelike projection, so that it can be put on tightly without the use of a wrench. Greater leverage can be applied than with a *wing* or *thumb nut*.
HANGER, BODY — An iron fitting which serves the same purpose as a body loop. The term is frequently applied to the type of fitting used on vehicles having elliptic end springs. (Y & B)
HANGER, BRAKE — One of the connecting links which attach a swinging brake beam to the hound bar above. (O)
HANG OFF — A technical phrase designating the act of ironing and attaching the body to the carriage part. (C)
HARD STOPPING — An English term, equivalent to *rough stuff*. (O)
HARNESS LEATHER — As the name implies, this is leather used in the manufacture of harness. It is also employed to a lesser extent in carriage trimming, where it is used in exposed positions, in most cases bearing

a heavy strain. Harness leather is known in the trade as *long harness, weight, trimmed,* and *backs. Long harness* is that in which the hides are tanned and curried without being trimmed. *Weight* stock is much the same as long stock, though in some localities the shanks, etc. are shortened and trimmed. *Trimmed* harness is that in which the legs and thinner portions of the bellies are trimmed off before being curried. The best quality stock, *backs*, are strips eighteen to twenty-two inches wide, measuring from the backbone line. (T)

HAWN – Same as *hound*. (K)

HEAD – Synonymous with *top* (which term is more common in the United States), meaning the upper portion of a carriage, above the body proper, including the roof or upper quarters, enclosing or covering the body. (h)

HEAD BLOCK – A small block of wood into which the perch is mortised, the center of which forms the rest for the front spring. The lower side is bolted to the top circle of the fifth-wheel. It forms the stationary point upon which the forward axle turns. (C)

HEAD BLOCK PLATE – A plate used to secure the perch to the head block. Together with the bed plate underneath, it also serves as a bearing surface around the king-bolt. It frequently serves also as a clip tie. (K & I)

HEAD-BOARD – The front end board of a wagon. It may open, or it may be fixed in place. The term is also sometimes used synonymously with *top rail*. (O)

HEAD LIGHT – A narrow strip of glass fixed into a door head. (C)

HEAD LINING – The lining to the underside of a carriage top. (C)

HEAD PLATE – The strip of iron generally attached to the front half of a landau or landaulet top, or head, for the purpose of covering the joint and excluding rain or dust when the top is closed. (h)

HEARSE MOUNTINGS – Any of the decorative objects mounted on a hearse body, such as rails, urns, angels, crosses, plumes, etc. (K)

HEEL BAR – A bar of wood placed upon a footboard, on which the driver may rest his feet. (C)

HEEL BOARD – A board or panel used to enclose the space under a carriage seat. (C & h)

HICKORY – An American tree of the genus *Hicoria*. There are several species: the shagbark, the pignut and the swamp hickory. The shellbark or white hickory is one of the most valuable timbers for carriage work, particularly for spokes. (C)

HIND CARRIAGE – That part of the gearing to which the hind wheels are attached. (C)

HINGE – A jointed metal fitting on which a door, end gate, or other swinging part turns. (O)

HINGE PILLAR – The pillar of either a body or door into which the hinges are set; generally it is the hind door pillar. (C)

HOLD BACK – An iron loop or hook attached to the underside of a thill, into which a breeching strap of a harness is buckled for the purpose of checking the motion of a carriage when going down a hill. Also the similar fitting on the underside of a pole, against which the ring of the neck yoke bears. (C & Y)

HOLDER – A strip of broadlace, or stitched leather, attached to a carriage body, either within or without, and variously employed: 1. To lift the windows, when it is known as a *glass-holder*; 2. To support a passenger's arm, in which case it is generally padded inside and attached to the inside of a door pillar, and known as the *arm-holder* or *armrest*; 3. As a support for the footman, known as the *footman-holder* or *page-holder*. (h, C, & T)

HOLDER TASSEL – An ornamental pendant, made flat, two and one-half inches wide to agree with the *holder* which is generally made of broadlace. These have netted heads and heavy bullion fringe. (T)

HOLLOW AUGER – A tool for cutting a round tenon on the end of a piece of timber, such as the point of a spoke. (O)

HOOD – An extra projection attached to the front bow or bar of a carriage or wagon top, extending forward as an additional protection from rain or sun. It generally consists of an iron frame with a leather covering, and is frequently made to hinge back against the bow when not in use. It is also known as a *storm hood*. The name is also sometimes applied synonymously with *head* and *top*, but is more properly reserved for the first usage. (C, T, & h)

HOOP – The name given in some areas to the tire of a wheel. (C)

HOOPED WHEEL – A carriage wheel having the hoop or tire set. (C)

HOOP STICK – The original name given to a carriage bow. (C)

HORN AND BUMPER – A fitting attached to the rear end of the reach of a trail freight wagon. A clevis on the trail tongue slipped over the horn, to guide the following wagon, and the end of the tongue butted against the bumper, to prevent the wagon bodies from coming together. Attachment of the wagons to one another was not by means of the horns, but by short heavy chains on the underside of the trail tongues, which in turn attached to a long chain running under the wagons. (S226)

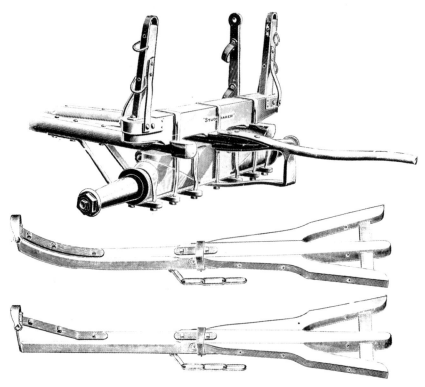

Upper, Horn and bumper for TRAIL FREIGHT WAGONS; *center,* bent Trail tongue; *Lower,* straight Trail tongue. (Studebaker Bros. Mfg. Co., Catalog No. 222, 1903.)

HORN BAR — The ornamental cross-bar tying the ends of the half circle fifth-wheel. The term is also applied to a crooked *draw-bar* or *doubletree.* (C)

HORN-CASE — A receptacle for the coach-horn, generally made of leather, and attached to the rear roof-seat iron on the off side. (FR)

HORN CATCH — A staple fixed to the bottom of the hinge pillar of a coach door when but two hinges are used. Its purpose is to keep the bottom of the door in its proper position. (C)

HOUND — One of the crooked timbers framed through the front axle to receive the pole, the back ends supporting the fifth-wheel or sway bar. On a drop tongue gear there are two pairs of hounds in front, the forward pair being called *tongue hounds*, and the pair behind are called *front hounds*. The tongue hounds are fastened rigidly to the tongue, but pivot in the front hounds, so that the assembly can swing downward. On heavy wagons hounds are framed into the back axle, and act as braces to the reach; these are sometimes called *reach braces*. Rear hounds are called *open crotch hounds* if they do not come together in front, and the coupling pole, or reach, passes between them. If they come together and the pole passes underneath them, they are called *closed crotch hounds.* (O)

HOUND BAR — A stout wooden bar bolted to the tops of the rear hounds, to which the hangers for a swinging brake beam are attached. (O)

HOUND PIN — An iron pin with a large, flat, vertical head, passing through a hound just behind the axle, as an additional support to prevent the hound from being drawn through the axletree. (O)

HUB — The turned block forming the center of a wheel, into which the spokes are mortised. Metal hubs, or wooden hubs with metal shells are sometimes substituted, and there are many patented varieties. Sometimes called a *nave.* (C & h)

HUB BAND — A metal band which encircles a hub, both as a protection against splitting, and as an ornament. (C)

HUB CAP — A metal covering which fits over the open end of a hub, to keep out dirt, and to present a more finished appearance. (K)

HUB RUNNER — Same as *runner attachment.* (O)

HUG (of a wheel) — (*n.*) The portion of a hub that bears against the collar or shoulder of an axle-arm. (*v.i.*) Said of a wheel when the axle is properly gathered, in which case it *hugs* the axle-collar. (C & h)

HUNGARIAN NAIL — A small round-headed tack used to tack gimp in place; a *gimp tack.* (K)

HUNGARY GREEN — Same as *Saxon green.*

I

IMPERIAL — (1.) A roof seat of a coach, omnibus, or railway car. (2.) A box for luggage intended to be carried on the roof of a coach. (h)

INTERLINE — To draw a fine line through the center of a broad stripe; to split with a fine line. (C)

IRON — This word is frequently used in conjunction with a great number of other words, such as *seat iron*, *fender iron*, etc., and means approximately that the named part is shod with, braced with, or supported by this part. Used as a verb, it means to make and attach the irons of a carriage. (O)

IRONSMITH — A worker in iron; a blacksmith. (C)

IRONWORK — The portion of a carriage made of iron or steel. (C)

IVORY BLACK — A black made from the bones of animals, the best of which is made from ivory shavings burned to a black coal in a closed crucible, and afterwards ground fine; when purified it is the best known black. (C)

IVORY HEAD SCREW — A wood screw having a bore and thread cut into the head, into which an extra screw is placed having a round ivory head. (C)

IVORY NAIL — A nail made of brass wire, having an ivory head. (C)

IVORY TRIMMING — An inside ornament used on the interior of a carriage body, such as a handle, card box, slide, etc. These parts may be made entirely or partially of ivory, either white or colored. (T)

J

JACK — (1.) A small windlass which receives the end of a leather suspension brace after it has passed around the back of a spring. By means of a wrench, or winch handle, the jack may be wound up or let down so as to shorten or lengthen the brace. (2.) An upright iron support at the corner of a running gear, which receives and supports a thoroughbrace. (3.) A mechanical device for raising a carriage wheel a short distance off the ground, used when greasing the axle, making repairs, etc. (4) Also see *shaft-jack*. (A, C, & h)

JACK OF SWEEPS — See *scroll*.

JAGGING — The roughing of smooth metal surface, such as a step tread, by raising sharp points by means of a triangular punch. (O)

JAPAN — A peculiar varnish prepared from asphaltum and oil, used as a drier for paints. The word originated from the fact that the use of varnish or lacquer was first learned from the Japanese. There are two distinct varieties: brown japan, a liquid drier, used to mix with paints; and black japan, a black varnish, often used on carriages as a substitute for a color coat. (C & h)

JAPAN GOLD-SIZE — A liquid drier for paints, similar in character to japan, but lighter colored, and having greater siccative power. (h)

JAPANNED LEATHER — Same as *enamel leather*. (C)

JAPANNING — The glossy black finish given to leather or metals. (h)

JARVEY — A colloquial English term applied to a hack-driver. The following vivid picture of the typical *jarvey*, from Sidney's *Book of the Horse*, merits quoting here. He says: "The sexagenarian of 1873 can remember when the only conveyance for hire in the streets of London was a rickety, creaking, generally dirty vehicle, drawn by a pair of miserable, broken-winded screws, driven by a very ancient 'jarvey,' in a much patched, never cleaned, many-caped outer garment, who descended very slowly from his perch, opened the door, and unrolled the steps with grating clang, and drove away with much cracking of whip, and some oaths, at the rate of about four miles an hour." (h)

JEW'S PITCH — Asphaltum. (C)

JOCKEY — The rider of a race horse. Also see *Daumont, A la*. (O)

JOINT — One of the hinged props or braces that support a bow top when raised. Also called a *top joint*. (C)

JOINT BREAKER — Same as *top lever*. (I)

JOINT END — The trade name for a small piece of iron having a milled eye on one end. It is a readymade end, used when making up a joint. (C)

JOINT PROP — Same as *top prop*. (C)

JOURNAL — The portion of a rotating shaft or spindle which turns in a bearing. (O)

JOURNAL BOX — A box or bearing for a journal. The term is frequently used to describe the bearing in which a *brake roller bar* turns. (Y)

JUMP SEAT — A type of seat used on carriages, by means of which a vehicle may be made to carry several additional passengers. These seats are usually fitted with iron legs, which move upon stationary bolts. In changing the seats they are lifted, and jumped backwards or forwards to their places. When not in use they may be folded out of sight. (C & h)

KEY — A pin or strip of metal inserted into a hole or slot to hold parts together, or to keep a nut from turning off a bolt. Some are split keys, such as cotter keys, and others are wedge-shaped, in order that they may be driven tightly into place. This latter type of key may also be made of wood. (O)

KEYHOLE — The small hole in the end of a whiffletree, used in conjunction with the keyhole strap to keep a trace from slipping off of the whiffletree. (Y)

KEYHOLE STRAP — A small tapered strap tacked near the end of a whiffletree. It passes through the keyhole at the end of the tree after the trace has been attached, and keeps the trace in place. (Y)

KEYSTAFF — The wooden staff slipped into the staples at the rear ends of the shafts of a dump cart, to secure the front of the body to the shafts. A quarter turn, and an endwise movement of the staff permits the body to tip over backward. (O)

KING-BOLT — The bolt that connects the front axle to the rest of the carriage-part, and the pivot upon which the axle turns. They formerly passed through the axle, but now are clipped to it so as not to weaken the axle, with the exception of heavy vehicles which still have the bolt going through the axle. Also called a *king-pin*. (C)

KING-BOLT KEY — An iron pin fixed through the lower end of a king-bolt, below the king-bolt nut, for the purpose of securing the nut. Keys are also used without a nut, the key alone preventing the king-bolt from working out. (h)

KING-BOLT NUT — A nut screwed onto the lower end of a king-bolt to keep the bolt in place. (h)

KING-PIN — See *king-bolt*.

KNEE — One of the uprights that connect the beams and runners of a sleigh. (C)

KNEE BOOT — A wooden covering for the legs of passengers on outside seats, now used only on the Hansom cab. It has been replaced by the leather or enameled cloth *apron*, to which the term *knee boot* is still sometimes applied. The apron is also called a *knee-flap*. (C & h)

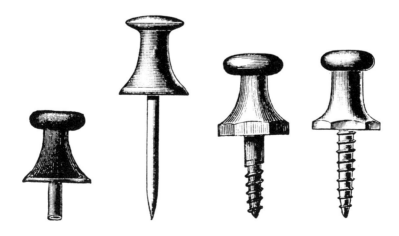

Carraige knobs were riveted, driven, or screwed into place. (Cray Bros., Cleveland, Ohio, 1910.)

KNEE BOOT STRAP — One of the straps used to keep the knee boot in its place. (C)
KNEE CHECK — A strap attached to the side of a knee boot. (C)
KNEE-FLAP — See *knee boot*.
KNIFEBOARD-SEAT — A style of seat used on the upper deck of a double-decked omnibus, wherein the seats are long benches, placed back to back, and running from the front to the back of the bus. These seats, introduced about 1849, were used on the earlier types of busses. (O)
KNOB — An ornamental stud of ivory or metal, used for securing curtains. (T)
KNOB EYELET — A small, circular, leather patch, with a hole and slit, fastened to a curtain with a metal ring as a reinforcement to the knob hole. These slip over the carriage knobs to secure the curtains. Also called a *knob patch*. (T & Y)
KNOB PATCH — Same as *knob eyelet*. (T)
KNUCKLE — A double bolted joint, the bolts of the opposite arms crossing each other at right angles, used to couple the side and cross springs on combination spring carriages; a type of spring shackle. (C)

L

LAC — A resinous substance produced by the *coccus lacca*, a scale-shaped insect, the female of which exudes from its body this resinous substance; when melted and reduced to thin sheets it is called shellac. It is the principal ingredient in lacquers. (C)

LACE — A woven close web, with a loop surface, plain or of ornamental figures, used for trimming carriages. Coach laces are of three types: *broad lace*, *seaming lace*, and *pasting lace* (see these types for further descriptions). Originally most laces had a silk face, but by the last quarter of the nineteenth century, most had worsted faces. (T)

LACE PLATE — A perforated metal or ivory plate, used for securing the end of a piece of lace in position. (T)

LACE TACK — A fine tack having a smooth round head. (C)

LACQUER — A yellowish varnish made by dissolving shellac in alcohol, and colored with gamboge, saffron, etc. (C)

LADDER — A light framework of side rails and rungs, arranged over the sides of a wagon body somewhat like flare boards. These are used chiefly on wagons intended for hauling hay. Sometimes known as *hay ladders*. (O)

LAKE — A deep red color derived from cochineal (the finest dark red known), and from madder; that known as madder carmine is nearly equal to cochineal in the brilliancy of its color and durability. (C)

LAMP — A carriage lantern, attached to the seat, boot, body, or dash. A lamp attached to a boot or quarter is termed a *coach lamp*; one attached to a dash is termed a *dash lamp*; and any inside lamp is called a *reading lamp*. The parts are: *body*, the frame which supports the glass; *stem*, or *barrel*, the long tube forming the bottom; *collar*, the ornamented projection at the top of the stem; *tip*, the bottom ornament of the stem; *panel*, the raised part on the top, to which the head is attached; bottom and top *heads*, the respective parts of the head, which serve to ornament and ventilate the lamp; *neck*, the part which joins the head to the panel; *reflector*, the highly polished concave metal parts in the back and door; *candle cap*, the cap which covers the top of the tube in the stem; *spring*, the coiled wire in the stem which supports the candle. The various carriages use lamps with glass sizes as follows: buggies and pony phaetons, about three inches; two- and four-passenger phaetons, three and one-half inches to four inches; Victorias, mail phaetons and T carts, five inches; rockaways, broughams, and coupes, five and one-half inches to six inches; landaus and Berlins, six inches to seven inches; hearses, six inches to seven and one-half inches, and five inches to five and one-half inches for a child's hearse. (T & h)

LAMP BARREL — The tube at the lower end of a carriage lamp, into which the oil or candle is placed. It is sometimes added for appearance only. Also called a *lamp stem*. (h)

LAMP BLACK — The soot collected by holding a plate over the flame of a lamp or candle; that prepared for the market is obtained by burning various woods. It is the cheapest black used, but is well adapted to toning mixed colors. (C)

LAMP PROP — An iron brace attached to a carriage body, and used as a support to the lamp. (C)

LAMP SOCKET — A ring or short collar attached to the lamp prop for the insertion of the lamp barrel; used to support a lamp. (C)

LAMP SPRING — A coiled wire spring, encased in muslin and placed inside the lamp barrel, for the purpose of keeping the candle up to the top of the cylinder until it is consumed. (T)

LAMP STEM — Same as *lamp barrel*.

LANCE WOOD — A light, tough, elastic wood found in the West Indies. It is used to a considerable extent for making the shafts to gigs, trotting sulkies, etc. (C)

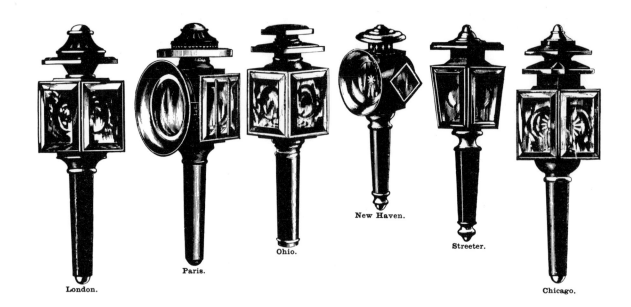

A variety of carriage lamps from an advertisement in *The Hub*, June, 1882.

LANDAU TOP HOOK — One of the specially constructed metal hooks used to hold tight the joint of division in a landau roof-rail. (K)
LANDAU TOP LEATHER — A grade of enameled leather made from large, heavy hides, of even texture and fine grain. The process of manufacture is like that of other top leathers. This class of leather is used for heavy tops, such as landaus, cabriolets, doctors' phaetons, etc. (T)
LAP BLANKET (or ROBE) — A carriage or sleigh blanket used as a covering for the lap. They are made in bright colors and are considered indispensable; the finest of these is the worsted afghan. (C)
LAZY-BACK — A light and temporary seat back which can be attached to a carriage seat when needed. Some are attached permanently to the seat, some attach to the shifting rail, and some fold down when not in use. They generally have a padded back rest about four inches wide. (O)
LAZY-BOARD — A small board projecting from under the left side of a wagon (such as a Conestoga or other freight wagon), near the center, on which the driver could ride, and from which he could operate both the jerk line and the brake. This board was generally constructed in such a manner that it could be pushed back under the wagon body when not in use. (O)
LEAD-BARS — A three-part unit fixed to the pole-hook for the attachment of the leaders (leading pair of horses). The unit consists of a *main-bar*, resembling an over-sized singletree, to which is attached a pair of *single-bars*, or *lead-singletrees*. The main-bar is also known as a *leader doubletree*. It serves the same purpose as the *spreader bar* and chains, though the construction differs. (S226 & FR)
LEAD COLOR — A mixture of lamp black or indigo and white lead; the color used for priming coats. (C)
LEADER DOUBLETREE — A term sometimes applied to a *main lead-bar*. (O)
LEAD MOLDING — A T-shaped molding cast from lead, used to hide the joints on door pillars, etc. (C)
LEAF — One of the layers of steel making up a steel spring. They are also called *plates*. (C)
LEMON YELLOW — A color made by mixing realgar and orpiment. (C)
LEVELING — The act of glazing over bodies after the rough stuff has been rubbed down, in order to produce a perfect surface. (C)
LEVER LOCK — A spring bolt let into the edge of a Landau door to prevent the door from being opened while the glass is up. When released from the pressure of the glass frame, the bolt moves into a hole in the opposite pillar, and acts as a lock. (O)
LIFTER — Same as *glass-string*. (h)
LIGHT — Any window in a standing or falling top carriage; known as a *door light*, *front*, *back*, or *side light*, according to its location. (C)
LIMBER — (1.) The forepart of a gun carriage, converting the gun carriage into a four-wheeled vehicle, for moving

the gun from place to place. It consists of two wheels, an axle, and a pole or shafts. (O)   (2.) Also an old term for a *shaft*. (H)

LINCH PIN — A metal pin used to prevent the wheel from running off the arm of the axle. (C)

LINE OF BEAUTY — A curved line representing an elongated form of the letter S. (C)

LINING — The general term for the inside trimmings of carriages. (T)

LINING NAIL — A small iron nail that has an extra cap, plated or japanned, over the head. (C)

LINSEED OIL — Oil obtained by pressure, from flax seed. Used in painting. (C)

LIST — Same as *selvage*.

LITHARGE — A yellowish red lead oxide, used in drying oils.

LIVERY — The dress of a coachman. (C)

LIVERY STABLE — A stable where horses and carriages are kept for hire. (C)

LOCK — (1.) (*n.*) A mechanical device made for the purpose of keeping a carriage door closed.   (2.) (*n.*) A device for locking a wheel when going down a hill.   (3.) (*v.t.*) To chain or otherwise fasten a wheel so that it cannot revolve.   (4.) (*v.i.*) *Lock* and *locking* are terms much used in the nineteenth century, particularly in England, meaning *turn*, and *turning*. The application was not to the turning of the wheel on the axle, but to the turning of the front axle and wheels around the king-bolt. In the case of an equirotal carriage, the terms applied to all four wheels, turning about the central pivot point. The lock of a vehicle is defined as follows: *quarter-lock*, the straight sides of the vehicle restrict the movement of the wheels; *half-lock*, a small section of each side is indented, to permit greater movement of the wheels; *three-quarter lock*, the wheels may pass under the body, so that only the perch restricts their movement; *full-lock*, the wheels may turn completely under the body with no restriction whatever. (A, C, & O)

LOCK CHAIN — A chain secured to some part of a vehicle, and passing around the felloe of a rear wheel, for the purpose of preventing the wheel from turning. (K)

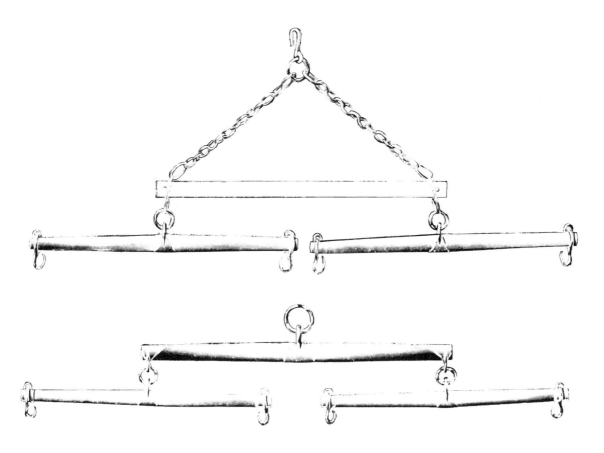

Upper, Lead singletree with chains and spreader bar; *lower,* lead bars or leader doubletrees. (Studebaker Bros. Mfg. Co., Catalog No. 222, 1903.)

LOCK CHAIN HANGER — An iron fitting to which the center ring of a lock chain is attached on the body or gear. (O)

LOCK CHAIN HOOK — A hook on the side of a wagon, to which the long end of the lock chain is attached when the chain is not in use. (O)

LOCKING PLATES — (1.) The original name for *wear irons*. (2.) In England the term is often applied to *bed plates* or *transom plates*. (C)

LOCKING-STOP — A projection on the fifth-wheel to prevent it from rotating beyond a certain point. (h)

LOCKING-WHEEL — A British term, meaning *fifth-wheel*. (O)

LOCK PILLAR — The upright piece of the framework in a carriage body against which the door closes, and which receives the bolt of the lock. (h)

LONG BED — A term used by the trade to designate an axle-arm with a long bed section about twenty-eight inches long. A long bed is convenient for new work, as the two bed sections are approximately the correct length to make an axle with only one weld. (C & Y)

LONG STAPLE — Same as *cart shaft rod*.

LOOP — See *curtain loop* and *body loop*. (T & h)

LOOP HEAD NUT — A nut similar in purpose to a double wing nut, only the head is formed like a loop. These do not catch readily in clothing, and they are stronger than wing nuts. (K)

LOWER QUARTER — The quarter panel below the arm-rail or belt. (C)

LUCENE — A light volatile oil used in carriage lamps. Receptacles for this oil are filled with cotton, which is thoroughly saturated with the oil. The heat from a small wick-filled tube leading from the container vaporizes the oil and causes it to burn as long as any oil is in the cotton. (T)

LUG — (1.) A piece of iron welded upon a body plate so as to rest on the top of a rocker or bottomside. It acts the double part of a support to the plate, and as a foundation for the head of the bolt that is used for attaching the bottom plates to the body, and also for making the connection where combination springs are used. (2.) Generally, any small projection from another iron part. (C)

LUGGAGE BOARD — Same as *opera board*. (h)

LUGGAGE BOOT — A rack attached to the back part of a stage or other public vehicle for the purpose of carrying baggage. The top is covered with a large apron of leather or duck. (C)

LUGGAGE IRON — The iron frame or support of a luggage boot. (C)

LUNETTE — (Ordnance) An iron ring at the end of the trail of a field-piece, which is placed over the pintle of the limber in limbering up the gun. The term is also applied to the hole through an iron plate on the underside of the stock of a seige-piece, into which the pintle of the limber passes when the piece is limbered. (W)

LUNETTON — A smaller sort of *lunette*. (W)

M

MADDER — A plant, the roots of which are much used in the manufacture of red dyes and pigments. (C)

MADDER CARMINE — A brilliant red pigment made from madder. It is next to cochineal carmine in brilliancy and durability. (C)

MAIL BOOT — The English name for what is known as the full or high boot; a coach boot, the pillars and framework of which are covered with a panel, such as the boot on the front of an Abbot-Downing passenger wagon. (C)

MAIL PATENT — The trade name for an axle without any nut or thread on the arm. The wheel is kept on by means of bolts which pass through the hub, and into a large plate back of the collar on the axle-arm. It is also called the *bolt patent axle*, and the *mail axle*. (C)

MAIL SPRING — A type of platform spring, consisting of four half-elliptic springs forming a quadrangle, arranged in parallel pairs and shackled together at the ends. By them the body is placed at two removes from the concussion. They were so named because they were first used on early mail coaches. They are not easy without a great weight on them, and their use is therefore confined chiefly to heavy vehicles such as mail phaetons, stage coaches, express wagons, trucks, etc. They are also called *telegraph springs*. (h & FR)

MAIN-BAR — See *lead-bars*.

MAIN BRACE — Long straps of heavy leather stitched together, and used to suspend a body on a gearing having *whip*, *S*, or *C-springs*. If the straps connect the body and gearing of a vehicle, they are *main braces*; if the straps extend from gearing to gearing, with the body resting upon them, they are then termed *thorough-braces*. (C, T, & h)

MALLEABLE CASTING — An iron casting that has been annealed. Following Seth Boyden's 1831 patents on the manufacture of this type of casting, carriage builders gradually began to employ such castings in the construction of vehicles. By the latter part of the horse-drawn era, many carriage irons that had formerly been made by forging were being made of malleable iron, resulting in a considerable cost reduction in the more common vehicles. (C & O)

MALLEABLE IRON — Cast iron made from a variety of pig iron which is suitable for conversion into a crude form of wrought iron without subsequent fusion. In England the term is frequently applied to *wrought iron*. (O & C)

MANHEIM GOLD — A composition metal made of three parts copper, one part zinc, and a small quantity of tin. (C)

MAROON — Brownish crimson; claret color. (C)

MAROON LAKE — A deep lake prepared from madder, and prized for its transparency and durability. (C)

MASSICOT — The commercial name of calcined lead. It is lead in its first change after being combined with the oxygen of the atmosphere. Heated a little, it is converted into red lead; greater heat produces brown oxide. It is a good light yellow, and is very serviceable for mixing with blue for making greens. (C)

MASTIC — A yellowish white resin, which exudes from a small tree that grows on the coast of the Mediterranean Sea. (C)

MASTIC VARNISH — A transparent varnish made by dissolving mastic in spirits of turpentine. Its principal use is for varnishing oil pictures. (C)

MAT — A thick pad, used as an extra covering for the floor or footboard of a carriage. (T)

MAUL STICK — Same as *moll stick*.

MAUVE — A delicate purple obtained from aniline, a product of coal tar. (C)

MAZARINE — A deep blue color. (C)

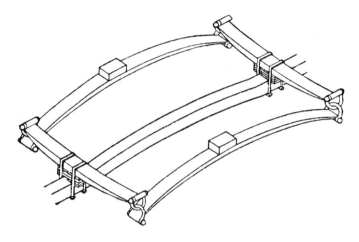

Mail, or telegraph spring. This drawing is abbreviated so that the separate spring leaves do not show. (From Fairman Rogers: *Manual of Caoching*, 1901.)

MEASURING WHEEL — Same as *traveler*.
MEGILP — A gelatinous compound of linseed oil and mastic varnish, used as a vehicle for colors. (C)
MEXICAN-QUARTER — An Americanism introduced about the year 1850, by Wood Bros., of New York, and applied to the quarter panel of a coach, where two curves intersect. See *step-piece*. (h)
MIDDLE PILLAR — The center post of a shifting front; the lock pillar of a curtain carriage having but three pillars on a side. (C)
MIDDLE RAIL — The middle or arm piece of a paneled body; also called a *belt rail*. (C)
MILORI GREEN — A popular green somewhat resembling chrome green, but richer in color. Also called *green lake*. (O)
MINIUM — A red pigment made by reducing common lead by calcining to an oxide or litharge, which is afterward put into a hot furnace and stirred until it assumes a red color; red lead. (C)
MOCK GOLD — A composition metal composed of sixteen parts copper, seven parts platinum, and one part zinc. (C)
MOCK IRON — An alloy composed of nine parts lead, two parts antimony, and one part bismuth. This alloy expands on cooling, and is used to fill small defects in iron castings; also called *expanding alloy*. (C)
MOCK JOINT — A top joint without a hinge, used on panel-quarter coaches, and intended merely for ornament. (T & h)
MOCK PLATINUM — An alloy composed of eight parts brass and five parts zinc. Much like white brass. (C)
MOLDING — (1.) A projecting or depressed ornament, placed upon a carriage body or gearing to relieve it of a monotonous appearance, or for the purpose of hiding joints, or dividing off panels into smaller sections to make them appear lighter. (C)   (2.) A metal bead, beveled or half round, filled with soft metal which supports tacks, whereby the molding is secured to the body. It is much used as an ornamental border around leather tops. (T)
MOLDING FINISHER — A piece of metal used to finish the ends of metal moldings, made similar to the molding, but with a more ornamental form. (T)
MOLDING TOOL — A tool having two cutter-bits shaped like the bit of a molding plane, but placed so as to work right and left. It is used for working out moldings on bodies where they cannot be cut until the work is put together. (C)
MOLE SKIN — The technical name of varnished cloth, finished to imitate dash leather, the underside of which is colored to imitate plain leather. (T)
MOLL STICK — An implement used by carriage painters, intended to steady the hand while using the pencil in ornamenting. It is made of a straight piece of wood, with a knob on one end, made of some soft substance, covered with leather. The knobbed end is rested lightly on the work, and the other end is held in the left hand, thus rendering the stick a support to the right hand. (O)
MONKEY-BOARD — The small shelf on the rear of an omnibus, on which the conductor stood. (O)
MONOGRAM — The initial letters of a person's name ingeniously interwoven so as to compose an ornament, and

yet retain the character of each letter. This device is extensively used on harness and carriages. (C)

MOROCCO – Fine grained leather used in carriate trimming, made from goatskin and finished without oil or grease. The color is imparted by means of dye, applied with a brush. Sheepskin is also made into morocco, but this type is unsuited to carriage trimming. Some fine trimming leathers, closely resembling morocco, are made from the grain split of fine cow hides, dyed various colors, finely grained, and given a dull finish. These are far superior to sheepskin or low grades of morocco. (T)

MORTISE – A recess cut into a piece of timber, made to receive another piece called a tenon, commonly square or having square corners. In some areas a round hole made with an auger, intended to receive a round tenon or dowel, is called a mortise. (C)

MOSAIC GOLD – An alloy composed of fifty-two parts zinc and forty-five parts copper. (C)

MOSS – A fine plant gathered in the Southern forests. When cured it is of a dark brown color and has a thread-like appearance; being quite elastic it is extensively used for stuffing to carriage trimmings. (C)

MOUNTAIN-POINT – A device which trails loosely at the back end of a vehicle, to prevent the backward motion of the vehicle when ascending mountains. The point, or points of this part are shod with heavy irons, which are forced into the ground if the motion of the vehicle is reversed, thus serving as a bar to hold the entire weight of the carriage and check all further backward motion. It is also called a *mountain-check,* and a *drag-staff.* (C)

MULLER – A hard, fine stone, egg-shaped, the larger end being flattened. It is used by painters for grinding paint on a grind stone. (C)

MUNTZ METAL – A composition containing sixty parts of copper and forty of zinc. (C)

NACARAT — A pale red color with a cast of orange. (C)

NAIL — A slender piece of metal, having a head of metal or other material, used both as an ornament and a fastener. See nails in illustration, page 401, A, cloth covered; B, japanned and silver tufting; C, ivory head. (T)

NAME PLATE — A plain or ornamental metal plate bearing the name of the builder of a carriage. Placed in some conspicuous position, it serves the double purpose of advertisement and ornament. Frequently, name plates are found on some rear portion of a carriage. (C)

NAPE — A piece of wood used to support the pole of a wagon or cart. It is sometimes hinged to the underside of the pole, and when not in use is fastened up against the underside by means of a strap. (C)

NAPLES YELLOW — A beautiful pigment having a rich golden hue; being more unctuous than other yellows it readily combines with other colors, and is considered by oil painters to be the best yellow known. It was formerly supposed to be obtained from the lava of Mount Vesuvius, but was later analyzed by a French chemist and found to be composed of ceruse, alum, sal-ammoniac, and antimony. In grinding it is necessary to avoid using steel or stone, as they will change the color. Instead, use an ivory spatula. (C)

NAVE — Synonymous with *hub*. The term is not much used in this country, but is commonly employed in England. (C)

NEAP — The pole and futchells of a wagon, cart, or sled, when made from a single piece, generally a sapling which has been split at the rear. Also called a *spear*. (O)

NEAR SIDE — The left-hand side in driving. In England it means the opposite side, as the driver when walking is on the right-hand side of his team. (C)

NECK — That portion of a coach or other heavy body that is between the front corner pillar, and the boot or rest for the driver's seat. (C)

NECK YOKE — A bar attached to the front of a pole that is not fitted with a *crab*, which, through the breast straps, attaches the horses to the front end of the pole. (O)

NECK YOKE RING — A metal or leather ring attached to the center of a neck yoke, for the purpose of securing the yoke to the pole. (K)

NETTING — Cords crossed beneath the head lining of a carriage, as an ornament, but formerly as a support for light packages. (T & C)

NEUTRAL TINT — A gray pigment composed of blue, red, and yellow in various proportions. (C)

NIB (or NEB) — A term used in England for the shaft or pole of a wagon. (C & O)

NICKEL — A grayish-white metal of considerable luster, very malleable and ductile. It is much used for electro-plating carriage and harness mountings. (C)

NICKEL PLATING — The thin film of nickel applied to other metals by electrolysis. (C)

NICKEL SILVER — See *German silver*.

NOSE — The portion of a hound that extends forward of the front axle, which with its mate supports the pole. Also called a *nose-piece*. (C & h)

NOSE PLATE — The plate that is attached to the lower sides of the nose-pieces, near the ends; its purpose is to tie the hounds together and support the pole. (C)

NUNTER BAR, or NUNTER — (1.) One of the short pieces of timber that connect the upper axle-bed with the rear transom. (2.) One of the small cross-bars framed into the front transom, to help keep it in place. (3.) A small cross-bar, mortised into any two wooden members to hold them apart. (C, h, & O)

NUT — A small, square piece of iron or steel, with internal threads, screwed onto the end of a bolt to keep the latter in place, and to secure the parts through which the bolt passes. (h)

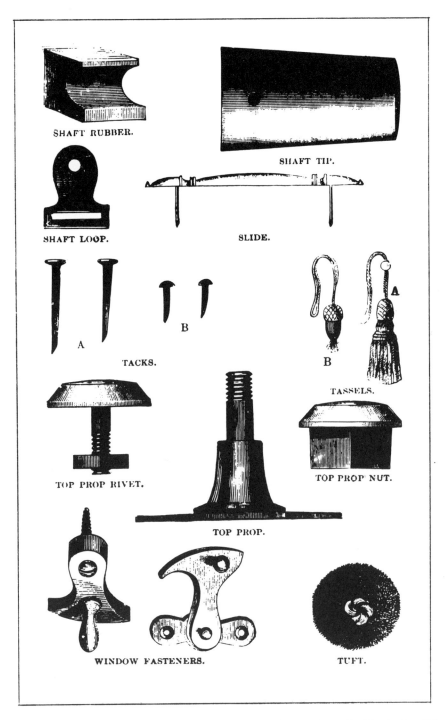

Trimmings and hardware used in the trimming of carriages. (From William N. Fitzgerald: *The Carriage Trimmers' Manual and Guide Book,* 1881.)

NUTCRACKER SPRING — An English term applied to a style of spring combination, now obsolete, consisting of a pair of half-elliptics shackled to a pair of *dumb springs*, the whole forming a quadrangle. It is similar in arrangement to the *mail spring,* except that the latter has four true half-elliptic springs. See Bridges Adams's *English Pleasure Carriages,* page 125. (h)

NUT WASHER — See *axle washer.*

OAK — A tree of the genus *Quercus*, of which there are many species. White oak is the most valuable to carriage builders, being extensively used for gearing parts and body framings. (C)

OCHER — A variety of fine clay containing iron ore, used as a pigment in paints. The most common colors are yellow and red. (C)

ODOMETER — A device attached to the wheel or axle of a carriage to measure the distance traveled. (C)

OFF SIDE — The right-hand side in driving. In England it means the opposite side, as the driver when walking is on the right-hand side of his team. (C)

OGEE — A surface having a concave and a convex section run together. The curve of this surface approximates the so-called *line of beauty*, or an elongated S curve. The abbreviation is O.G. (C & h)

OGEE-QUARTER — A quarter panel in which the ogee sweep is introduced. (h)

OIL CARPET — Same as *oilcloth*.

OILCLOTH — Heavy cloth, covered with numerous coats of paint, presenting an ornamental surface. It is somewhat similar to linoleum and during the first half of the nineteenth century was much used for covering the floors of carriages. It is also called *oil carpet* and *painted carpet*. (T & O)

OIL COLOR — Color made by grinding dry coloring matter in oil. (C)

OIL SKIN — Fine linen saturated with linseed oil to make it waterproof. At one time it was extensively used by carriagemakers as an extra cover for dickey seats, and in the South it was used as a covering for carriage panels to protect the paint from the weather. A patented article also called *oil skin* is made of woolen cloth dressed with oil, and is said to be as impregnable to water as the linen. (C & T)

OLD'S WIRE-BOUND WHEEL — A patented wheel, the hub of which is strengthened by being bound about with wire. (h)

OLIVE COLOR — A brownish green color prepared by the mixing of black and a little blue with yellow; yellow pink with a little verdigris and lampblack; or violet and green mixed in equal quantities. (C)

OPEN LINK — A stout link of chain, left open with the ends overlapping, used to attach a singletree to a doubletree. They may also be used to repair a broken chain, in which case the ends are hammered together. (O)

OPERA BOARD — A board affixed to the back part of a coach, hinged in such a manner as to act as a footboard for servants when let down. The underside, which is exposed when the board is in the raised position, is ornamented with a molding and generally with a carved device. The opera board, when raised, served as a protection to the rear panel of the body from the pole of a carriage following too closely in the rear. By 1871 they were used only on public vehicles, and then often for the purpose of carrying luggage. (C & h)

OREIDE — A yellow composition metal, consisting of seventy-three parts copper, twelve and three-tenths parts zinc, four and four-tenths parts magnesia, two and five-tenths parts sal ammoniac, six and five tenths parts cream of tarter, and one and three-tenths parts quick lime. (C)

ORNAMENT — A wood or metal device of ornamental form affixed to some part of a carriage; also a painted device on a crest panel, etc. (C)

ORPIMENT — Yellow arsenic. A pigment composed of fifty-eight parts arsenic and forty-two parts sulfur. It is also called king's yellow. (C)

ORPIN — Orpiment. (C)

ORSEDEW — Dutch gold. This is an alloy of copper and zinc, beaten into sheets and used like gold leaf. (C & O)

OSCILLATORS, SLEIGH — Hingelike fittings used to attach *bobs* to a sleigh, allowing the two units to swing up and down, to facilitate the passage of the vehicle over uneven terrain. These are not necessary on two-passenger sleighs, and are generally used on four-passenger sleighs, and larger varieties, or on freight-carrying sleds. (I)

OUTRIDER — A mounted servant, whose duty it is to procede a royal equipage, clear the way, and herald its approach. (h)

OUTRIGGER HINGE — A projecting hinge put on the lower part of a carriage door when there is much turn-under on a body, so as to allow the door to open square, and allow a step cover to be fitted on if required. (O)

OUTSIDE JOINT BREAKER — Same as *top lever*. (O)

OVAL LIGHT — A carriage window of oval shape. (C)

OXBOW — A bent wooden piece by which the ox yoke is secured to the neck of the ox. (O)

OXBOW PIN — A wooden or metal pin which holds an oxbow in place in the yoke. Some patent types of pins are also to be found, having mechanical latches. (O)

OXFORD BOUNDER — An English term applied to an unusually high dog cart wheel. (h)

OX YOKE — A stout wooden beam, curved to fit the back of the neck of an ox, against which the animal exerts force to move a vehicle. Single yokes have an iron fitting at both ends of the yoke for attachment to the vehicle, while double yokes have a single ring in the center for the same purpose. A bent wooden bow prevents the yoke from falling from the animals' necks. (O)

P

PACKFONG — A variant spelling of *paktong*. (C)

PAD — That part of a carriage step on which the foot is placed. (h)

PAGE-BOARD — Same as *dummy-board*.

PAGE-HOLDER — Same as *footman-holder*.

PAINTED CARPET — Same as *oilcloth*.

PAINT STICK — One of the strips of wood attached to a body in the paint shop, the ends projecting so that the body can be set up on the side without marring the paint. (C)

PAKTONG — An alloy resembling *German silver*. Sometimes spelled *packfong*. (C)

PALETTE — A thin oval-shaped board, made of hardwood, with a thumb hole and slight indenture in one end, whereby it can be easily held in the hand. It is used by painters for holding and mixing pigments. (C)

PALETTE KNIFE — A flat knife without a cutting edge, generally made of steel and highly tempered, extremely thin and flexible, used by painters when mixing or grinding colors on a stone or palette. For colors that contain arsenic in their composition, ivory palettes should be used, as metal has a tendency to change the color of the pigment. (C)

PALLET — A flat brush used to take up, apply, and maneuver gold leaves. (C)

PANEL — (*n.*) One of the thin boards forming the outside covering to the framework of a body; also a raised or depressed surface on the body, the borders of which are finished with a molding. (*v.t.*) To partition off with moldings, or by striping. (C)

PANEL KNIFE — A drawknife having adjustable bits or cutters of different widths, or a stationary cutter with a narrow edge. (C)

PANEL SAW — A short, fine-toothed saw of the same form as an ordinary handsaw. (C)

PANEL-STUFF — Boards in the rough, adapted for carriage panels, generally sawed out from the log one-half inch or five-eighths inch thick. (h)

PARTITION FRONT — A partition, either stationary or adjustable, between the outside seat and the inside seat of a rockaway body. When made adjustable, it is also called a *shifting front*. (C)

PASTE — A mixture of flour and water, forming an adhesive substance used for uniting cloth and leather when trimming. (T)

PASTING LACE — A narrow lace used as a finishing edge or border in trimming carriages. It has one unornamented edge, or selvage, which is tacked to the edge of the trimming, and the ornamented face of the lace is then turned over and pasted over the tacks. (T, C, & h)

PATENT LEATHER — Leather having one or both sides covered with coats of color which are baked on. It is distinguished from enamel leather by having a smooth surface. It is sometimes called *glazed leather*. (C)

PATENT WHEEL — A term applied to any wheel in which some particular patent is involved. Specifically, it is usually applied to one of the many types of wheels having a patent hub, these being designed to provide a stronger hub, and one that will not loosen rapidly through use. Most such hubs are reinforced with a patented malleable iron sleeve, or band. While there have been many such wheels patented and marketed, the two most notable are the Sarven and Warner wheels. (O)

PATENT YELLOW — A beautiful yellow made by mixing sea salt and litharge in proportions of two and one respectively, moistening them and leaving them together for twenty-four hours. The product is then washed, filtered and evaporated, by which means a soda is obtained; this, upon heating, changes in color to yellow. (C)

PATTERN — (1.) A thin board cut to the shape of the outline of some part of a body, or other piece of woodwork. (2.) An exact likeness of an article to be cast from metal. This pattern makes an impression in molding sand, into which cavity the molten metal is poured. (C)

PAULIN — From the word *tarpaulin*, meaning a wagon cover. This term is used to a large extent by the army. (O)

PEARL GRAY — A color made by mixing white lead, indigo, and a very slight amount of black. (C)

PEARL WHITE — A white powder precipitated from the nitrate of bismuth by a solution of sea salt; a pigment of white lead slightly blued. (C & O)

PENCIL — A small brush made of fine hair confined in the barrel of a quill, or a small tin tube, and placed on a wooden handle when used. It is used for striping and ornamenting carriages. (C)

PERCH — A piece of timber, or a bar of iron or steel which connects the front and hind parts of a carriage gearing. Synonymous with *reach*. In the primitive coach the perch consisted of a heavy timber or several pieces of wood bolted together, and was used to connect and brace the fore and hind axles. Later, when the improvement was introduced of suspending coach bodies at the corners by means of straps from four posts or pillars, the necessity of great size and strength in the perch increased. Although subsequent improvements in transforming these corner posts or pillars into springs tended to diminish this necessity, the numerous appendages such as footmen's boards, etc., together with the length and weight of the vehicle, still prevented the employment of a lighter perch. The substitution of an iron perch was an effort to reduce the weight of this part, but it was not until the invention of elliptic springs by Obadiah Elliot in 1804 that this became practical.

The developments of the past half-century (1830-1880) in the application of iron and steel to carriage building have resulted in the attainment of an exceedingly light and graceful perch, and the improvement known as platform suspension has led to an admirable substitution for the perch. This consists of two continuous iron plates, running from the boot, along the neck and rockers to the back quarter, where the body is attached to the rear platform springs. The connection between the front and hind gearing is thus established in a most compact and complete manner, the function of the perch being fulfilled by the combined services of the body and neck framing, and the fore and hind springs. The object formerly sought is thus more than realized.

One phase of the mechanical development of the perch was due to the exigencies of locking. Various devices have been restored, from time to time, to avoid the collision of the wheels with the perch in locking. Among these was the curving or bending up of the perch at the point where the wheels would strike, which resulted in the various modifications of the perch known as the *crane-neck* or *swan-neck* (which see). Another phase of development was due to the effort to remove the perch from danger of contact with the coach body. The high wheels of early carriages made access difficult, and to avoid this problem the perch was either dipped (a more recent instance of which is found in the Dorsay), or made double, as in the case of a coupe described by Roubo (page 562, plate 202) under the name of the *inversable*. This use of double perches allowed the body to hang between the two, but, on the other hand, renewed the difficulty of locking, and compelled entrance from the rear. All of these obstacles have been happily overcome in the admirable combination known as platform suspension without perch. (h)

PERCH-BED — An extra bed bolted to the back bed, into which the three perches of a style of side spring wagon are fixed. (C)

PERCH-BOLT — Same as *king-bolt*. There is likely to be confusion between the terms *perch-bolt*, *perch-pin*, and *reach-pin*, since all three could appear to mean the same pin. Generally *perch-bolt* and *king-bolt* are synonymous while *reach-pin* and *perch-pin* are synonymous with *coupling pole pin*. (C & O)

PERCH-BRACE — A brace, generally of iron, running from the rear axle and meeting the perch at an angle. *Hounds*, or *perch-wings* are sometimes called *reach braces*. (K & O)

PERCH CARRIAGE — A gearing in which the perch is used, as distinguished from a platform or no-perch carriage. (C)

PERCH-CLIP — A patented clip for attaching the perch plate to the perch, used in place of bolts. (C)

PERCH-COUPLING — The name given to a clip and eye coupling used to join the perch to the rear axle. (C)

PERCH-HOOP — The hoop or band used to unite and bind together the hounds and perch of a gearing. (C)

PERCH-IRON, REAR — A fitting for attaching a perch to the rear axle, the hindmost portion serving as a *clip tie*. (I)

PERCH-LOOP — A double loop attached to the perch, which receives the two check-straps. Also called a *double check loop*. (K)

PERCH-PIN — Same as *reach-pin*.

PERCH-PLATE — An iron plate bolted to the underside of the perch; also the side plates screwed to the perch. (C)

PERCH-STAY — (1.) A metal arm extending from a perch to the hind axle, with which it connected, its purpose being to strengthen and stiffen the connection. (2.) In some areas the term is synonymous with *perch-brace*. When it is called a brace, then the term *stay* is applied to an additional reinforcement which extends from the perch to the brace, at approximately a right angle, to further stiffen the gearing. (h & K)

PERCH, SWAN-NECK (or CRANE-NECK) — One of the early forms of the perch, where, to avoid collision with the front wheels, the perch was bent upward in a curve, suggesting the name. See *perch*. (h)

PERCH-WING — One of the two pieces of timber extending from near the center of a wooden perch, and spreading out at an angle, over and beyond the rear axle. Synonymous with *hound*. Sometimes called a *perch-brace*. (h)

PICKING OUT — A technical term applied by carriage painters to fine lines or stripes. (h)

PIECING OUT — A term applied to the welding of a piece of iron to the axle arms, to give the axle bed the required length. This term may be generally applied to the extending of any piece of iron or wood, by welding or splicing. (C)

PILLAR — The general name for any piece of upright framework in a carriage body. (C)

PILLOW — (British) This part corresponds with the front bolster of an American wagon, while the part corresponding with the sand-board of an American wagon is called the bolster on an English wagon. (O)

PILLOW SPRING — A coiled wire spring, used for upholstering backs and cushions of carriages. (T)

PIN — A slender piece of metal or wood, used to attach or secure different parts of a carriage. See *linch pin, hound pin*, etc. (C)

PINCHBECK — An alloy of copper and zinc in proportions of one pound of copper to three ounces of zinc. It is used to imitate gold. (C)

PINTLE — The upright pivot pin on a limber, over which the lunette is placed in limbering up a gun. The pintle serves the same purpose as a *king-bolt*. (O)

PIPE BOX — The long iron box lining a hub, into which the arm of an axle is inserted. They are now more generally used than any other. The *short box* for which they are a substitute is seldom used, except for wooden axles. (C)

PLANK SIDE — The side of a carriage body that has been worked to the desired sweep from a thick plank. (C)

PLATE — (1.) (*v.t.*) To strengthen certain wood parts of a carriage with strips of iron or steel. (2.) (*v.t.*) To cover certain metal parts with gold, silver, nickel, etc. (3.) (*n.*) A strip of metal used to plate wooden parts. (4.) (*n.*) The leaf of a spring. (C & h)

PLATFORM BAR — One of the transverse wooden bars, generally three in number, which together with the upper half of the fifth-wheel make up the unit of a platform-spring vehicle known as the chair. See *chair*. (O)

PLATFORM CARRIAGE — A carriage part employing platform springs, and in most instances, no perch, although a *half platform* does often have a perch. There is some discrepancy in the use of the term *full platform* and *half platform*. Most manufacturers used the term *full platform* when both ends of the carriage had platform springs, and *half platform* when one end, usually the front, had one or two elliptic springs. A few manufacturers, however, meant by *half platform* that the two side springs were joined by one cross spring, and by *full platform* that the side springs were joined by two cross springs. This last arrangement is more often called a *mail spring*. Truck platform springs are generally different, having no cross springs. (C, Y, & S226)

PLATFORM SPRING — A combination of springs applied to a carriage, designed to bring the body as near as possible to the axle. It consists of two side springs, crossing the axle at right angles, joined by a cross spring. On passenger carriages, a platform spring usually consists of two springs resembling three-quarter elliptics, attached to the body by means of *pump handles*, and connected at their forward ends by a cross spring. These springs, sometimes known as *French platform springs*, are used only on the rear axle of passenger vehicles. On freight wagons, omnibuses, and related heavy commercial types, platform springs are used both front and rear, though the side springs, joined by a cross spring at their rear ends, are similar to half-elliptics. *Mail springs*, having two cross springs, are by some manufacturers termed *full platform springs*. Truck springs, though they have no cross springs, are sometimes called platform springs, though they are more often called *truck springs*. The ends of platform springs may be joined with *barrel shackles*, or with loose shackles, which are D-shaped. Also see *Dennett spring*. (Y & h)

PLATING — (1.) The covering of iron or other metal with gold, silver, nickel, etc. (2.) Technically, the strengthening of the timbers of a carriage by means of an iron plate. (C)

PLAY — The movement, regular or irregular, of a wheel upon the arm of an axle. (C)

PLOW — A tool for cutting grooves, working in the same manner as the groove plane, except that the guard is adjustable. (C)

PLUG — A circular piece of wood, cut across the grain, and used to cover the head of a screw or other fastening device. (C)

PLUG CUTTER — A tool for cutting plugs. Some are made in the form of a gouge, the end being sharp, while others have an adjustable cutter bit. (C)

PLUME — A feather ornament used on horses. Also used on hearses and sleighs. (T)

PLUSH — A bright colored fabric, having a long velvet nap on one side, popular for trimming sleighs. (C)

POINT BAND — A hub band used on the front of a hub. (Y & O)

POLE — A long piece of timber connected with the futchells of a carriage, extending between the wheel horses and used as a lever to guide the carriage. Also called a *tongue*. (C & h)

POLE-BRIDGE — The iron passing over the front end of the futchells, to keep the pole from lifting. (h)

POLE-CAP — An iron fitting for the tip of a pole, made with straps which are fitted to the sides of the pole and terminate in front with a ring, to which the pole-chains are attached. When the ring is vertical, the hold back is frequently made integral with the cap. (Y)

POLE-CHAIN — One of the two chains attached to the pole-ring, and extending to the breast-chain or breast-strap, for the purpose of guiding and holding back a vehicle. (K)

POLE-CIRCLE — A bent piece of wood, to the center of which the rear end of a pole is secured, and to the ends of which the pole-eyes are attached. This unit may be used to convert a light, shafted vehicle to a two-horse carriage. (K & Y)

POLE-COLLAR — An iron band fastened to the draw-bar, through which the pole passes and is thereby kept in place. (h)

POLE-COUPLING — One of the two metal fittings which secure a pole to the axle. The shackle part, consisting of a pair of lugs, fastens to the axle, and the pole-eye is bolted to the pole circle. These are identical to *shaft-couplings*. (K)

POLE-CRAB — The metal socket at the front end of a pole, having a loop on either side to receive the pole-straps. It sometimes has a hook in front. (h)

POLE-EYE — The portion of the pole-coupling through which the bolt passes to secure the pole to the carriage. (C)

POLE-GUARD — Same as *footboard* and *opera board*.

POLE-HOOK — An iron or steel hook attached to either the pole-crab or to the front of the pole, used to hold the lead-bars when four horses are employed. Also called a *swan-neck*. (h)

POLE-HOUND — Same as *tongue-hound*. One of the two hounds attached to the back end of a drop-tongue, or pole. (O)

POLE-PAD — A cushion of leather, stuffed with hair or other material, attached to each side of a carriage pole, to prevent injury to the horses. Also a cup-shaped pad made to fit over the end of the pole of an army wagon, to prevent injury to the horses immediately in front of the wheelers. (T & O)

POLE-PIN — The long iron pin passed through the futchells and the rear end of the pole, to keep the latter from being drawn out. (h)

POLE-PIN-STRAP (or CAP) — The small piece of leather passed through an eye in the end of the pole-pin, to keep it in place. (h)

POLE-PLATES — (1.) Plates of iron screwed to the sides of the pole at the rear end to protect it from wear from the futchells.   (2.) The pair of plates upon which the evener works. (C)

POLE-RING — A ring attached to the end of a pole, to receive the pole-chains; it is part of the *pole-cap*. Also, one of the rings attached to the pole-crab, to receive a pole-chain, when the latter is used in place of a pole-strap. (C & h)

POLE-SCREEN — A protective screen at the rear of a pole, designed so as to prevent small stones from being thrown against the vehicle or passengers by the horses' hooves. (O)

POLE-SOCKET — The iron socket frequently attached to the transom of the front gearing, into which the rear end of the pole is inserted. (h)

POLE-STOP — An iron hook passed down through the rear end of the pole between the futchells, and in front of a small bar, to keep the pole from being forced back. (h)

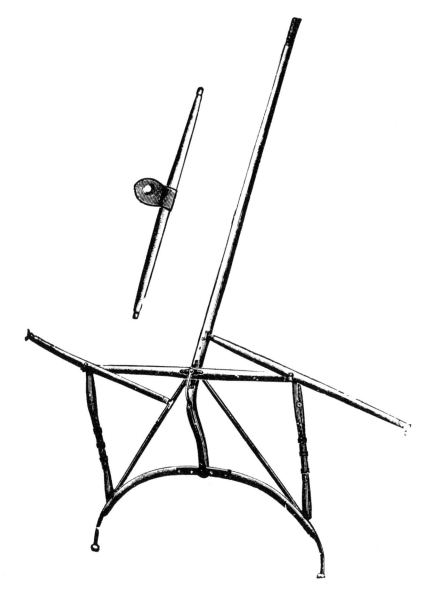

Pole attached to pole circle, with whiffletrees and neck yoke, used for the conversion of a single-horse vehicle into a pair-horse vehicle. (Cray Bros., Cleveland, Ohio, 1910.)

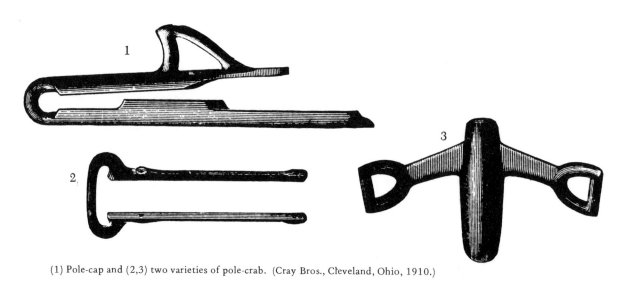

(1) Pole-cap and (2,3) two varieties of pole-crab. (Cray Bros., Cleveland, Ohio, 1910.)

POLE-STRAP — One of the two leather straps attached to a pole-crab, and serving the same purpose as a pole-chain. (O)
POLE-TIP — A metal socket fitted to the front end of a pole, having a flange on the lower side which extends about halfway around it, against which the ring of the neck-yoke bears when in use. (C)
POLE-YOKE — A short yoke made of iron, attached to the end of a pole by a bolt. It is used in place of a neck-yoke. (C)
PONGEE — An inferior kind of Indian silk. (C)
PORTLAND STONE COLOR — A yellowish drab color, made by mixing umber, yellow ochre, and white lead. (C)
POST-HORN — See *coach-horn*.
POSTILION — A mounted driver. When a vehicle is turned out in the style known as *a la Daumont*, it has no coach-box or driver's seat, but is driven by jockeys mounted upon horses, who are known as postilions. The employment of postilions is said to have been first suggested by the divulgence of a state secret of the Duke of Orleans by his coachman, the proximity of the driving-seat having permitted him to overhear the conversation of his master. (h)
POST-MASTER — A master of a post where post chaises were supplied to travelers. See Felton, volume 2, page 51. (h)
PRINCE'S METAL — An alloy that imitates gold, made in the proportion of about three ounces of copper to one ounce of zinc. It is also called Prince Rupert's metal. (C)
PROP — Same as *top prop*. (h)
PROP-BLOCK — A block of rubber or leather which fits onto the bow rest of a falling top carriage, providing a rest for the bow sockets. (Y)
PROP-REST (or PROP-IRON) — Same as *top prop*. (O)
PROP-WASHER — A leather washer on a bow rest, used on either end of the prop-block. (h & Y)
PRUSSIAN BLUE — Any of several complex cyanogen compounds of ferrous and ferric iron; specifically, a dark blue amorphous substance having a coppery luster, obtained by adding a solution of potassium ferrocyanide to a ferric salt. Used in dyeing and painting. Also known as *Williamson's blue*. (O)
PULL-TO HANDLE — See *handle*. (T)
PUMICE STONE — A substance ejected from volcanoes, used for rubbing down *rough stuff*. It is of various colors, but the gray is most highly prized by painters as it possesses the finest grain and is free from any foreign substance. (C)
PUMP HANDLE — A bar of wood or iron attached to a bottomside or rocker of a coach or other body at the back end, acting as a support for the rear of the body. They were originally used to support the footman's board, the carved ends being used as handles to assist the footman in mounting. Also called a *break*. (C)
PUNCHEON — Synonymous with *nunter-bar*. (h)
PUTTY — A cement made of whiting or ceruse and linseed oil beaten or kneaded together until it becomes quite stiff. It is used to fill small nail holes, blemishes, etc., and is sometimes colored by lampblack, umber, or other coloring matter. (C)

QUARTER — (1.) One of the sections of a body between the door and corner pillars, designated as the upper and lower, back and front quarters. (C)   (2.) The technical name given to that portion of the back of a carriage top on either side of the curtain.   (3.) The side piece of a close leather top, or the side section of an ordinary falling top. (T & O)

QUARTER LIGHT — A window, either stationary or movable, in an upper quarter of a coach body. (C)

QUARTER PANEL — A thin panel used to cover the framework of a quarter of a coach body. (C)

QUARTER SQUAB — The quilted trimming which forms the lining to a quarter of a body above the belt-rail. (T)

QUEEN BOLT — A term sometimes applied to a *draw bolt*. (O)

R

RABBET — The groove cut in the edge of a board, into which the tongue of another board is inserted; also a groove cut into one board or beam, into which another board is fitted, making a longitudinal joint. (C)

RAILING — An iron rod around the top of a seat, dash, etc. (C)

RAILING LEATHER — Thin patent leather, used for covering seat rails, etc. (T)

RAISER — Same as *riser*. (h)

RATTAN MOLDING — A molding made from the core of the rattan. (C)

RAVE — (1.) A framework of rails or boards added to the side of a cart or wagon to enable the carrying of a greater load. It differs from a side board in that it is generally horizontal, extending over the wheels, or sloping slightly upward, like a flare board. In some instances however, it may be vertical, in which case it is much like a side board. This usage of the term is common in England. (2.) A horizontal rail in the side of a wagon or cart body; also one of the vertical side pieces mortised through the horizontal rave. The side of a Conestoga body is an example of this construction. This usage of the term is more common in the United States. Also see *ribbed body*. (3.) See *bob sleigh* (*gearing nomenclature*). (O)

REACH — Synonymous with *perch*. Also see *bob sleigh* (*gearing nomenclature*). (C & H)

REACH BRACE — This term is sometimes applied to a *hound*.

REACH COUPLING PLATE — An iron plate bolted to the front ends of the rear hounds, and having a hole in the center to receive the reach pin before it passes through the reach or coupling pole. Generally two such plates are used, one above and one below the hounds. (K & Y)

REACH-PIN — An iron pin of about three-fourths inch in diameter, used to pin the rear hounds to the reach, or coupling pole. (Y)

READING LAMP — A small lamp used to a limited extent inside a carriage. It may be made either oblong or oval, the long side being the front; there is no glass on the outside. When used, the doors, which are double and hinged to the sides, are opened; the doors are lined with mirrors, and are very convenient as toilet glasses. See *lamp*. (T)

REALGAR — A combination of sulphur and arsenic having a brilliant red color; red orpiment. Used as a pigment. (C)

REAVE — A variant spelling of *rave*.

RED VARNISH — A varnish made from Dutch sealing wax. (C)

REGULATING BOW-SPRING — An adjustable type of spring invented about 1835 by W. B. Adams. It consisted of a single leaf spring with adjusting screws at the ends which secured a stout cord. By adjusting the screws, the cord drew the spring into a shape resembling an archer's bow, and the spring was adjustable for both light and heavy loads. When installed in the carriage, the cord was on the underside. (A)

REIN HOOK — A metal hook attached to a carriage, on which reins may be hung when not in use. (K & I)

REIN RAIL — A rail, generally plated, attached to the top of a dash, serving as a support for the reins, and protecting the leather at the top of the dash. Also called a *driving rail*. (O)

REP — A silk or cotton fabric having a surface appearing as if made of small cords, presenting a ribbed effect. It is used to a limited extent for trimming sleighs and buggies. (C)

REST-STICK — Same as *moll stick*. (O)

RIBBED BODY — A body with sides and ends which are framed up with light strips, the panel being placed inside so as to show the framework. Express wagons and grocer's wagons are examples of this construction. Such bodies are sometimes called *raved bodies*. (C)

RIDGE POLE — A thin wooden strip bolted or tied underneath the tops of wagon bows, at right angles to them,

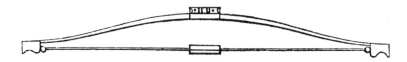

Adams's regulating bow-spring. (From W. B. Adams: *English Pleasure Carriages*, 1837.)

    to aid in supporting the bows and cover. (O)

RIM — The outer framework of a wheel. The term is sometimes applied to a tire, but is more properly applied to the wooden part. The bent half circles that form the outer framework are called rims, and the sawed segments make up a rim, but individually they are generally called *felloes*. (C)

RIM BAND — A metal band encircling the end of a hub; a hub band. They are so-called because they are constructed with a raised rim, or molding, around one edge. (Y, B & G)

RIM BOUND — Said of a wheel when the felloes are too long, preventing them from settling down upon the shoulders of the round tenons of the spokes. (C)

RISER — (1.) A piece of wood or an iron stay placed under the seat of a carriage to raise it. (2.) A piece of material placed under trimming to lift it, so that it appears to be embossed. Also spelled *raiser*. (h)

RIVED SPOKE — A spoke made from timber that has been split, instead of being sawed. (C)

RIVET — A pin of soft iron or other metal used for uniting wood or metal. The point of the rivet is spread out by hammering, so as to form a head which prevents it from being withdrawn. (C)

RIVET, WAGON BOX — A rivet with a large thin head, slightly oval-shaped like a carriage bolt head, used to rivet strap bolts to wagon boxes. (G & Y)

ROBIN — A round band of iron, covered with leather or India-rubber, used to connect the side and cross springs. It is the English equivalent of *knuckle*. (O)

ROCKER — A thin piece of timber, wider than the bottomside, to the inside of which it is fastened, for the purpose of giving greater strength, and to increase the foot room in a carriage; it is also called a *sunken bottom*. The bottom boards are secured to the rockers. (C)

ROCKER-PLATE — A long plate of iron or steel, fitted and fastened to the inside of a rocker, to stiffen and strengthen the body. (h) The term is sometimes used to refer to bolster plates, though this usage confuses the term. (Y, I, & G)

ROLLER — (1.) A revolving cylinder, such as a log, used to move heavy loads over the ground; the forerunner of the wheel. (2.) One of the small wheels in a metal frame which is attached to the underside of a sliding glass frame to ease its movement. (3.) Same as *curtain roller*. (4.) See *bob sleigh* (*gearing nomenclature*). (T & O)

ROLLER-BOLT — A large-headed bolt, surrounded by a wooden roller, attached vertically to a splinter-bar. A trace is looped to this fitting. (h)

ROLLER, GLASS STRING — A roller over which a glass string may slide when a carriage window is being lowered or raised. (O)

ROLLER-SCOTCH — (British) A small roller serving the same purpose as a *mountain-point*. It is attached to a vehicle in such a manner that it rolls along immediately behind a rear wheel, where it can immediately scotch the wheel if the vehicle starts to move backwards. (O)

ROLL-UP STRAP — Same as *curtain holder*.

ROOF — The covering to a standing-top carriage. (C)

ROOF-BOARD — One of the boards used as a covering to the frame of a roof. (C)

ROOF-CURVE — One of the slightly curved pieces of framework that crosses the top of a carriage from side to side, forming the support for the roof-boards. (C)

ROOF-MOLDING — A triangular or half-round molding that is nailed against the top and end-rails to cover the edge of the leather or cloth that is used as a covering for the roof. (C)

ROOF-RAIL — Same as *top rail*. (h)

ROOF-SEAT — A seat on the roof of a coach, for carrying additional passengers. (FR)

ROOF-SLAT — One of the slats supporting the covering of a standing-top roof, differing from roof-boards by having intervals between them. (O)

ROSE LAKE — A rich tint, prepared from lac and madder, precipitated on an earthy basis; rose madder. (C)

ROTONDE — A round-cornered seat of a phaeton, T-cart, or cabriolet. (h)

ROTTEN STONE — A soft stone used for giving a fine polish to metals. It is also called Tripoli, after the country from which it was formerly brought. (C)

ROUGH LOCK — A short section of very heavy chain which passes around the felloe of a locked wheel, and slides along under the wheel, for the purpose of retarding a vehicle on a steep grade, particularly when the road is muddy. (O)

ROUGH STUFF — A coarse paint made of about six parts yellow ochre to one of white lead, and mixed with varnish, brown japan, and a very small quantity of raw oil. It is used as a leveling paint, being put on quite thick, and when dry is rubbed down with pumice stone and water until the surface is perfectly level and smooth. (C)

ROWEN — Short fine grass used in stuffing cushions and other parts of upholstery. It is cheap stuffing, not of the best quality. An advantage over excelsior is that it will not pack so hard. (T)

RUBBER — (1.) An elastic material made from the milky juice of certain tropical plants. It is used for making carriage tires, and for smaller parts, such as prop blocks, bow separators, etc.   (2.) The term is also applied to either the entire brake mechanism of a vehicle, or to the brake blocks. (O)

RUBBER CLOTH — Cloth that has one side covered with India rubber; it is used in place of leather for aprons, tops, and curtains of cheap carriages. The heaviest kind is known as *rubber duck*. (T)

RUBBER DUCK — See *rubber cloth*.

RUBBER TIRES — In a continuing effort to improve both the durability of vehicles and the quality of the ride, a few carriage builders considered the application of rubber to the rims of the wheels. At least as early as 1835 a French inventor worked with the idea. A decade later, in 1845, Robert William Thompson was granted an English patent on a pneumatic carriage-tire made of rubberized fabric and leather, and the following year worked toward the development of a solid rubber tire. He thus showed that both ideas were workable, and, in fact, several carriage builders applied the pneumatic tires to a few broughams, yet the effort was premature, and, consequently, soon forgotten.

In both Europe and America inventors worked on the problem of rubber tires, attempting to produce a durable tire that would stay on the wheel. In 1856 the Boston Belting Co. made a few solid tires according to the design of George Souther and George Miller, but nothing came of the effort. Thompson became active again in 1861, at which time he was fitting solid tires to steam-propelled road tractors. In 1866 rubber tires were observed in use in Berlin by Channing M. Britton, who later became head of Brewster & Co., in New York. At the time, Britton was a student at the University of Berlin, where one of his fellow students was the Earl of Shrewsbury and Talbot, who later would play an important part in advancing the use of rubber tires in London. During the 1870s a number of patents were granted on solid tires for bicycles.

By 1885 solid rubber tires were coming into limited use in London, following the W.H. Carmont patents of 1881 and 1883, and in that year the Earl of Shrewsbury and Talbot introduced to the streets of London a fleet of Forder-built Hansom cabs equipped with solid rubber tires. Reportedly having an average life of five months, the tires were a success and were immediately popular with cab patrons, but unpopular among other cab companies who were forced to go to the expense of fitting rubber tires to their cabs because of popular demand. A number of the better private carriages in London were likewise equipped with rubber tires by 1889.

In America, Channing Britton introduced solid tires through Brewster & Co. a short time after his friend had fitted them to his cabs in London. Besides Brewster, only Healy & Co., of New York, and C.P. Kimball & Co., of Chicago, are known to have used rubber tires during the 1880s. However, in 1890, advertisements of other firms offering solid tires began to appear in the publications of the carriage trade. The use of rubber tires continued to be extremely limited during the early years of the 1890s and even though a number of the better carriages shown at the World's Columbian Exposition in Chicago in 1893 had rubber tires, this feature was not rapidly accepted.

Solid tires were becoming more popular by 1895, and were given a considerable boost by the 1896 patent of Arthur W. Grant. Grant's tire, held on by longitudinal wires imbedded in the rubber, lay in a rolled steel channel with flaring sides that prevented the loss of resiliency through compression of the rubber, this having been a serious problem with the earlier tires that were held on by the tight compression of the channel. Another important tire patent was that of James A. Swinehart in 1901. This tire was secured in the channel by two side wires, which were forced on over the ends of a series of short transverse wires imbedded in the rubber, and had the advantages of the Grant tire, plus the additional advantages of easier application, and less likelihood of the rubber suffering from wire-cutting. The Grant tire became known as the Kelly-Springfield and the Swinehart as the Firestone, the latter tire also having been made by the B.F. Goodrich Co. In addition to these two popular tires, a great many other patented varieties were available.

In an effort to improve the effectiveness of solid rubber tires, another type known as the cushion tire was made. This was a solid tire with air spaces (the air was not under pressure, so flat tires were not possible), these spaces allowing a greater flexing and compression of the rubber. The spaces came in many shapes, some being one or more holes running longitudinally through the tire, the retaining wires being in these oversized holes. The holes were also put in the sides of some tires, while still others had angular sections cut out of the surface. As with solids, there were many varieties, though the most common was a round tire with a single large longitudinal channel in which rested the single retaining wire. The tire rested in a steel channel as did the more common solid tire, and the rubber was sufficiently heavy to support the weight of the vehicle without assistance from air pressure.

Prices of solid rubber tires were about $25 to $30 per set of four for a light vehicle just after 1900, and gradually decreased by about $10 toward the end of the decade. Cushion rubber tires frequently cost about $5 more per set.

The third type of rubber tire is the pneumatic, or one supported by air pressure. As previously mentioned, a tire of this type was invented by Thompson in 1845. Following that, there was no further development until the 1888 and 1889 patents of the Irish veterinary surgeon John Boyd Dunlop. Oddly, Dunlop learned in 1890 that because of the Thompson patent, his patents were invalid, yet his tires were successful, and soon underwent a period of rapid development, with the result that the pneumatic tire quickly became popular on the newly developed safety bicycle. During 1891 and 1892, the pneumatic tire came into use in America too, on most safety bicycles.

In 1892 a number of racing sulkies were fitted with bicycle wheels and pneumatic tires. One famous trotter, Nancy Hanks, attracted much attention to the pneumatic tire with her record-breaking performances on the track, thereby introducing this tire once again to horse-drawn vehicles. Thus, the use of pneumatic tires was practically simultaneous with the use of solid and cushion tires, though the former were used generally only on a few lighter vehicles, such as sulkies and bike wagons, while the heavier carriages employed solid or cushion tires.

The bike wagon was one of the last of the innovations of the horse-drawn era. Essentially a runabout, it was distinguished by rubber tires (of any of the three types), and generally by ball-bearing hubs and wire wheels, while some even featured tubular steel running gears. Such features as these were soon to be adopted by the embryonic automobile industry, so that rubber tires were not developed for the automobile, but were simply borrowed from the carriage industry. Remnants of several of these early rubber tires, once mounted on certain pioneer automobiles of the 1890s, survive in the collections of the Smithsonian Institution.

RUB IRON — Same as *wear iron*.

RUG — A coarse fabric mat with a long nap, used as a floor covering. Some are made of animal skins, with the wool or fur left on. (T)

RUMBLE — A footman's seat attached to the back part of certain carriages, used in place of the footman's board. In the case of a ladies' phaeton, the horses are sometimes driven from this seat. It was originally called a *booby-boot*. (C & h)

RUNNER — One of several curved strips upon which a sled or sleigh slides. (C)

RUNNER ATTACHMENT — A sled runner made to fit onto an axle-arm of a wheeled vehicle, a set of which converts said vehicle into a sled or sleigh for winter use. (K & Y)

RUNNER-BEAK — A small metal figure, generally the head of a bird, mounted on the front tip of a sleigh runner, above the dash, as an ornament. (O)

RUNNING FOOTMAN — See *footman*.

RUNNING-GEAR — Same as *gearing*.

RUSSET — A color produced by mixing two parts of red with one each of yellow and blue. (C)

RUSSET LEATHER — Leather which is generally made of large spread cowhides, as these produce a finer grained and softer leather than steer or other hides. This leather is sometimes bleached quite light, and colored with a yellowish-brown stain, producing what is known as cuir or dorado color. (T)

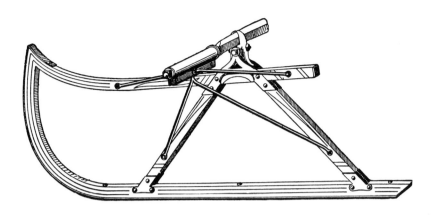

Wyeth's Runner Attachments were offered in 1910 by Cray Brothers, of Cleveland, Ohio, for the conversion of a carriage to a sleigh.

S

SADDLE — See *bob sleigh (gearing nomenclature)*.
SADDLE-BAR — Same as *back-bar*.
SADDLE-CLIP — A patent clip with a long flange projecting from either side of the top bar, so as to form a bed for a spring which crosses an axle at right angles. Another variety has two clips and a loose top plate, and binds down a spring that is parallel to the axle. (C & K)
SAFETY BRACE — An extra brace used on a carriage in connection with a main brace, as a protection against accident in the event that a main brace breaks. (C)
SALISBURY BOOT — A boot of curved form which superseded the square boot to some extent in the late eighteenth century. First used on post coaches in England, it carried the tools necessary to the coach in its early days. Its name was derived from Salisbury, the town where it was first made. The introduction of the brougham, and the consequent framing of the driver's seat and the boot into the body, making the boot simply a light paneled framework supporting the driver's seat, resulted in dispensing with the necessity for a boot so far as actual use was concerned. By 1871 this type of boot was little used, except on the coaches of the aristocracy. Hence the term Salisbury was eventually applied to a mere relic of its original form, namely, a box covered with leather, or a paneled box, painted, placed on the front gearing under the driver's seat. (C & h)
SAMSON — A type of clamp used at an early date to draw two felloes tightly together while a strake was being nailed to them. (O)
SAND-BAND — (1.) A metal band having narrow prongs that can be driven into wood, or a flanged band with holes to admit screws or nails. It is secured to the back end of a hub to prevent sand from working into the back end of the axle box, and it is used mainly with wooden or other large axles.
(2.) A sheet-iron band attached to a carriage axle, and extending over the back of a hub, to prevent sand from working into the axle-box. (C & h)
SAND-BOARD — A stout piece of timber bolted to the tops of the front hounds directly over the axle, and extending the full length of the axle between hubs. The bolster of the wagon rests on the sand-board, and the king-bolt passes through both parts. (C)
SAND-BOARD PROP — One of the small spacers placed between the sand-board and the axletree, on either side of the king-bolt.
SAND-BOLSTER — Same as *sand-board*. (O)
SARVEN WHEEL — One of the best and most popular patent wheels; patented in 1857 by James D. Sarven. It employs a wooden hub, over which two malleable iron shells or bands are pressed, and riveted together against the spokes. The spokes are so designed that the shoulders meet outside the hub, so as to form a wooden arch, which adds to their support.
SASH — The window frame enclosing a glass of a window or door. (h)
SASH-HOLDER — A lace or leather strip fastened to the bottom of the window frame, the loose end being finished off with a piece of fringe or other ornament, and hanging out in such a manner as to be of easy access. It is used to raise and lower a carriage window. Same as *glass-holder*. (C)
SASH-LIFTER — An ornamental knob or handle, attached to a window-sash as a handle for lifting. (T)
SATIN — A glossy fabric, originally made only of silk, but now sometimes of wool, woven in a satin weave. Some cheaper grades are cotton backed. It has a thick, close texture, and overshot woof, and is used to a limited extent for trimming closed carriages. A heavy quality is much used for trimming fine clarences, etc. (T & C)

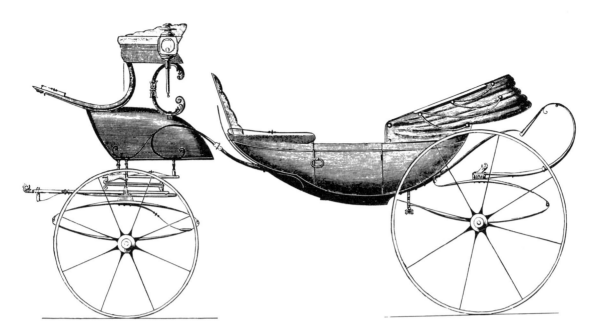

A CALECHE equipped with a Salisbury boot. (Lawrence, Bradley & Pardee, New Haven, Connecticut, 1862.)

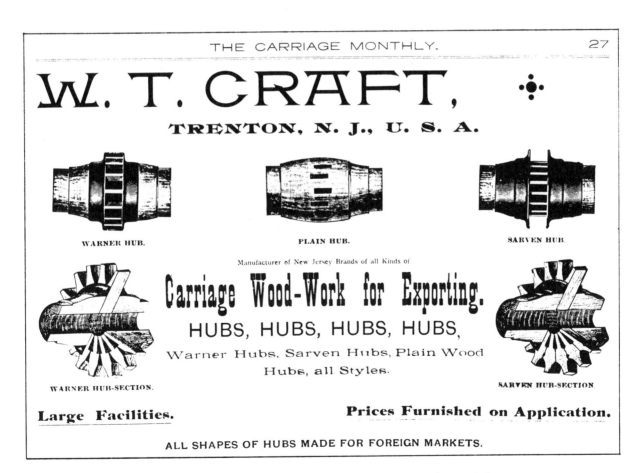

This *Carriage Monthly* ad of April 1891, shows cross sections of both the Warner and Sarven hubs.

SAXON BLUE — A beautiful blue compound bearing close resemblance to ultramarine; it will not bear oil without changing its color, and is therefore only used for enameling surfaces. (C)

SAXON GREEN — A carbonate of copper found in a natural state in the mountains of Saxony and Hungary. It is a pale hue, also known as *Hungary green*. (C)

SCHEELE'S GREEN — A green pigment first discovered by the chemist from whom it takes its name. It is prepared from arsenite of copper, and grinds well in oil. (C)

SCHWEINFURT GREEN — A green pigment superior to Scheele's green, obtained by dissolving verdigris in pure vinegar, and adding to it a solution of white arsenic. (C)

SCREW BOLT — A bolt with a heavy sharp thread, made to be used without a nut in the same manner as a wood screw. (C)

SCREWING-UP — Turning up and tightening the loose nuts of a carriage. (h)

SCRIM — (1.) A light, coarse, loosely woven fabric of linen or cotton. (2.) The plural, *scrims*, refers to the thin canvas glued to the inside of panels to give them increased strength and stiffness. (C)

SCROLL — (1.) An ornamental finish of spiral form on the ends of spring bars, perches, head blocks, pump handles, etc. (2.) A wooden pattern of curved shape, used by carriage draftsmen in filling up the sweeps in drawings. It is commonly called the *jack of sweeps*. (h)

SCROLL SPRING — (1.) A steel spring having one or both ends bent in the form of a scroll. *Grasshopper springs* are sometimes called scroll springs. (2.) *C-springs* were originally called scroll springs. (C & K)

SEAMING CORD — A heavy cord, used as the body of a welt for seaming up cushions, etc. (T)

SEAMING LACE — A narrow lace used for covering cord and joining seams. It has two unornamented edges, or selvages, which are turned around the cord and laid together, where they are stitched to the edges of the cloth to be joined. (T, C & h)

SEAT — That part of a carriage upon which a passenger sits. As technically applied to carriages, the term *seat* describes a *single sitting*, as distinguished from *seat-board* which may afford two or more sittings. A sulky is thus an example of a one-seat vehicle, while a buggy is a two-seat vehicle. (h)

SEAT-ARM — A rail or wide strip placed above the end of a seat, as a rest for the arm. (C)

SEAT-BACK — (1.) The stuffed cushion which supports the backs of passengers in a vehicle. (2.) Also an extra back screwed down on the back of a seat and used in connection with the seat for the purpose of making a closed, trimmed back. (C)

SEAT-BOARD — (1.) One of the boards forming the seat bottom on which the cushion rests. (C) (2.) Technically, a carriage seat affording two or more sittings. (h)

SEAT-BOX — A small box hung on the underside of a seat frame, the seat-boards forming the lid. It is used to carry small parcels, side curtains, or any small articles needed with a carriage. (C)

SEAT-BRACE — Same as *seat-iron*. (I)

SEAT, CHILD'S — A supplemental seat, generally hinged to the front of a carriage body, so that it can be closed when not in use. The origin of this style of seat is much earlier than is usually supposed, as Roubo refers to one in 1771 (page 459). (h)

SEAT-FALL — Same as *fall*. (T)

SEAT-FASTENER — A device used with a removable seat, such as that on a spring wagon, to hold the seat in place when it is in use. Also called a *seat lock*. (Y & I)

SEAT-FRAME — A hard wood frame to which the ends and back of a seat are attached. The frame supports the seat-boards, or in the absence of the latter, a canework seat. (C & h)

SEAT-HANDLE — A metal handle attached to the end of a seat to assist passengers in getting in and out of a vehicle. (K)

SEAT-HOOK — One of the metal hooks which fasten over the top edge of a wagon box to hold the support on which the seat springs are mounted. (Y)

SEATING — The manner in which seats are arranged in a vehicle. The most common method of seating is by two separate, transverse seat-boards, where all four passengers face forward, known as the *phaeton method* of seating; a four-passenger box wagon and a surrey are examples of this class. A second method is by two similiar transverse seat-boards, with the forward one removed to the front of the body so that the passengers occupying that seat sit facing the passengers in the rear seat, known as the *vis-a-vis method*; this characterizes all coaches, landaus, six-passenger rockaways, and the *vis-a-vis phaeton*. A third and less common method is by two transverse seat-boards placed together at the center of the body, so that the front and rear passengers sit back to back, known as the *dos-a-dos method*; the seats of several styles of dog carts are arranged in this manner. A fourth method is by two longitudinal seat-boards, placed together along the center of the body, so that the passengers sit back to back, with their feet over the wheels; this method is confined to the Irish jaunting-car, and is therefore known as the *jaunting-car method*. This seating arrangement offers the advantages of easy draught (the burden being equally distributed over the gearing), and of easy access and egress. The fifth method resembles the jaunting-car method in that it also consists of two longitudinal seat-boards, and possesses the same advantages of equal weight distribution and of offering convenient entrance and exit. In this instance, however, the seat-boards are removed to the extreme sides of the vehicle, so that the passengers face one another; this is known as the *wagonette method*. Any vehicle characterized by this arrangement of passengers is technically a true wagonette, and a familiar example is the omnibus, which has amply demonstrated the advantages of the method in connection with the public service. (h)

SEAT-IRON – (1.) A corner iron or other iron used on a seat to strengthen it. (2.) An iron stay, either plain or ornamented, used to support the driver's seat. (C & h)

SEAT-LOCK – Same as *seat-fastener*. (I)

SEAT-RAIL – (1.) A bent or sawed stick forming the top of a stick or spindle seat. (2.) A strip framed into the standing pillars of a heavy body to support the seat-boards. (3.) The iron rail around the top of a seat. (C & h)

SEAT-RISER – One of the wood or metal supports which elevate a seat above the body of a wagon. (O)

SEAT-RISER, PENNSYLVANIA – A riser on a farm wagon, made T-shaped, the lower part being bolted to the wagon box while the upper part, or T, supports a seat spring. (O)

SEAT-RISER, ST. LOUIS – A riser on a farm wagon, made in the form of a side board which runs about half the length of the wagon. (O)

SEAT-ROLL – (1.) A cord, or a strip of wood covered with leather, cloth or lace, and secured to the front edge of a seat frame to keep the cushion from slipping off the seat; it serves the same purpose as a cushion stick. (2.) The cushion around the rail of a Stanhope gig or phaeton. (C,T, & h)

SEAT-SPINDLE – One of the turned sticks used to support the seat rail, filling the space between the rail and the bottom. (C)

SEAT-STRAP – A strap used to keep a seat cushion in place. It may be made of leather, cloth, or lace, and attached to the cushion from below, while the other end is attached to the seat frame by a knob or other device. In some instances it is not attached to the cushion, but only to the frame, passing over the top of the cushion. (T,C, & h)

SELVAGE – The edge of woven fabrics, consisting either of one or more stronger cords, or a narrow border, usually of different weave from the body, serving to strengthen the fabric and to prevent warp threads from fraying. The selvage edge is called *fast* when it is enclosed by all or part of the picks, and *not fast* when the filling threads are cut off at the edge of the fabrics after every pick; the selvage of such fabrics, usually split goods, consists either of *leno,* or in the cheapest grades the fabric is simply sized along the edges to prevent fraying. Also called *list*. (O)

SET – A technical term, meaning the downward and forward inclination of axle-arms. The downward set is called *swing*, and the forward set is called *gather*. The former is necessitated by the dish of the wheel, the axle-arm being turned down slightly to cause the bottom spoke to stand at right angles to the ground. *Gather* is consequently required to compensate for the tendency of the outwardly canted wheels to turn outward. An additional reason for *gather* is that the actual bearing point of the axle, when the vehicle is running, is not on the bottom of the axle-arm, but slightly ahead of this point, which, with a tapered arm, causes the wheel to run against the nut or linch-pin. The proper amount of *gather* reverses this tendency, so that the wheel runs against the shoulder. A properly set axle will allow the wheels to run with the least amount of friction, with a slight tendency to run against the shoulder to relieve strain on whatever device secures the wheel on the arm. (O)

SHACKLE – (1.) An iron staple which serves to receive a leather suspension brace on the spring or gearing of a vehicle. (2.) The part used to couple springs together, or to couple a spring to the gearing. There are many varieties, such as D-shackles, barrel shackles, single and double shackles, etc. (3.) See also *shaft coupling*. (A & h)

SHACKLE, BARREL – A type of spring shackle used to fasten together the ends of platform springs. They are made with two holes, at right angles to each other, to receive the shackle bolts. Ordinary shackle links join the barrel shackle to the end of the spring. (Y)

SHACKLE-EYE — The hole through which the bolt of a shackle passes. (h)
SHACKLE-JACK — A device for pressing the eye and shackle of shaft-couplings into alignment, for the insertion of the bolt. These are necessary where antirattlers are used, to compress the spring or rubber. (K)
SHACKLE-LINK — A short plate having two holes, a pair of which are used to join the end of a spring to a vehicle or to another spring. (Y)
SHADING — (Painting) The filling in of an outline so that it represents the effects of light and shade, such as a painted letter which is shaded in one or more colors to give it a three-dimensional effect. (C)
SHAFT — One of the two wooden poles attached to the gearing of a carriage (to the axle, futchells, springs, etc.), between which the horse is secured. Also called a *thill*. (h)
SHAFT-BAR — The cross-bar that joins the two shafts at the back ends. It is frequently called a *cross-bar*. (C)
SHAFT-BRACE — A straight bar of iron underneath a shaft, running from a point near the shaft eye to a point near the junction of the shaft and cross-bar. Frequently the lower part of the brace is formed into the shaft-eye. (I)
SHAFT-COUPLING — The pair of metal fittings which secure one of the shafts to the vehicle. The shackle part, consisting of a pair of lugs, fastens to the axle or to some other part of the gearing, and the shaft-eye is bolted to the end of the shaft. Also called a *shaft-jack*. (h)
SHAFT-COUPLING ANTIRATTLER — A device to prevent worn shaft-couplings from rattling. It is generally a spring or a piece of rubber which keeps the shaft-eye tight against the bolt. (K)
SHAFT-HORSE — The horse of a tandem team that works between the shafts. (C)
SHAFT-JACK — Same as *shaft-coupling*. (K & h)
SHAFT-LOOP — A metal loop through which the shaft-strap passes. (T)
SHAFT-ROD — Same as *cart shaft rod*.
SHAFT-RUBBER — A block of vulcanized rubber, used to prevent shafts from rattling at the coupling. Same as *antirattler*. (T)
SHAFT-SOCKET — A long, tubular steel fitting pushed on over the end of a shaft. Some were finished to resemble leather, and were put on new shafts in place of leather in order to make a more rigid shaft. In other instances these were used only to repair broken shafts, and were pushed on over the break. (K)
SHAFT-SPRINGS — Patented springs which keep the weight of the shafts off of the horse's back, and also hold the shafts at any desired position when not in use. (Y)
SHAFT-STRAP — (1.) A leather strap nailed around a shaft a short distance in front of the cross-bar, through which a trace passes for support. (K) (2.) A leather strap attached to the underside of a shaft, near the cross-bar and at the shaft-coupling. (T)
SHAFT-T — The T-shaped iron that is bolted over the juncture of the shaft and cross-bar to reinforce that joint. (C)
SHAFT-TIP — A thimble of iron or other metal that is pushed onto the end of a shaft as an ornament and for protecting the wood. (C & T)
SHAM-DOOR — An English term, meaning the pillar of a gig or Stanhope, let into the bottomside by a stump tenon. It is also called an *anglet*. (O)
SHAMMY — A kind of leather originally prepared from the skin of the chamois. Sheepskins are more commonly used for this leather at the present time (1871). It is very soft and pliable, and readily absorbs water. Carriage painters use this skin to wipe water from carriages after washing. (C)
SHEAR SPRING — A carriage spring so named because of its shearlike action, used where a C-spring cannot be used because of insufficient space between the body and the gear. The lower half, which is placed on the outside of the upper half of this spring, rests on the perch (the perches are doubled), and the two halves are hinged together by an iron rod which extends across to the opposite side of the carriage and also serves as the hinge pin for the mating shear spring. An iron strap extends from one upper half to the other, to steady the springs. (O)
SHELLAC — (1.) The resin lac spread in thin sheets after being melted and strained. (2.) Lac, dissolved in alcohol and used as a wood filler or a finish. (C)
SHELL-BODY — The body of a carriage or sleigh having the sides and end worked out with a series of convex moldings, to imitate a scallop shell. (C)
SHELL-HANDLE — A T-shaped handle for a carriage door, the handpiece of which is in the shape of an elongated oval.
SHELLING — A technical term relating to hubs, meaning the clipping of the wood fiber around the hub bands,

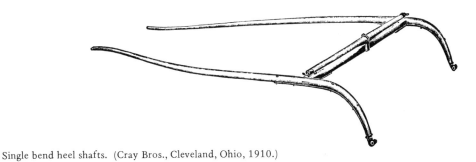

Single bend heel shafts. (Cray Bros., Cleveland, Ohio, 1910.)

causing the wood to shell off. (h)

SHIFTING BAR — An arrangement on the front of a sleigh which allows the shafts to be moved off-center, enabling the horse to walk in one of the ruts created by the passage of earlier vehicles. (I)

SHIFTING FRONT — A movable glass partition used on rockaways between the front and middle seat. (C)

SHIFTING QUARTER — A squab, shaped to fit the upper quarter of an open standing-top carriage, to convert the vehicle to a closed carriage for winter use. The outside of the squab is covered with leather, to which straps are attached for holding it in position. (C & T)

SHIFTING-RAIL — An iron rail, extending around the top of a carriage seat, to which the top is attached. It is so constructed that it can be removed from the seat without detaching the top from the rail. (T)

SHIFTING-RAIL CLIP — A clevis-shaped fitting which fits around a shifting-rail, and receives the bolt which secures the rail to the eye of the seat-iron. (Y)

SHIFTING-RAIL EYE — The eye on the seat of a carriage where the shifting-rail is attached. (I)

SHIFTING-SEAT — A movable seat, so constructed that the seating arrangement of a vehicle may be altered. (O)

SHIFTING-SEAT BODY — A body having two seats so constructed that they can be shifted toward the back or front with but little effort. (C)

SHOE — (1.) An iron or wooden box placed under the wheel of a vehicle, and attached to the vehicle by a chain, to prevent the wheel from rolling when going down a hill. It is used when a road is too steep or slippery to depend on the brake alone. The shoe consists of the sole, two cheek pieces, and the neck. At the forward end of the neck is a stout ring, in which a chain is attached to hold it to the carriage. (2.) The iron or steel strap fastened to the bottom of a sleigh runner. (C)

SHORT BED — A term used by the trade to designate an axle-arm with a short bed section about eight inches long. These are convenient for old work, as the two short beds can be welded to an old axle-bed, the arms of which have worn out. (Y)

SHORT BOX — A short iron box lining a hub, into which the axle-arm is inserted. Short boxes are generally used only with wooden axles, and two are inserted in each hub, whereas only one is needed when a *pipe box* is used. (O)

SIDE BAR — A variety of wooden side-spring, of American invention, applied principally to road wagons. This suspension is the result of an effort to secure compactness, and to relieve the horse from the play of the elliptic suspension, which it has sometimes supplanted where speed is the requisite. It commonly consists of two elastic wooden bars, close to and parallel with the body, to which the latter is directly attached; the ends of the side-springs are connected with two half-springs which rest on the axles. (h)

SIDE BOARD — Same as *side-extension board*.

SIDE BRACE — An additional support of metal or wood, generally the former, running from the end of a crossbar to the side of a wagon box. (Y)

SIDE-EXTENSION BOARD — A vertical board which can be applied at will to the top of a wagon box, to increase the depth of the box. Often called simply, a *side board*. (O)

SILL — One of the main bottom timbers of a body into which the various pillars are framed; the bottomside. The thills of a cart are sometimes called the sills. (C)

SILL PLATE — A metal plate with an antiskid design worked in the surface, attached to the top of a sill at that point where a passenger steps into a vehicle. Its purpose is to provide a firm footing for the passenger and protect the sill from wear. (I)

SILVER PLATING — The coating of iron or other metal with silver, either by close plating or electrolysis. (C)
SILVER SOLDER — A silver alloy solder used to unite metals. Hard silver solder is composed of four parts silver to one of copper; soft silver solder is two parts silver to one part brass. Both types require a much higher heat than soft solder, and they are much more durable. See also *solder*. (C)
SINGLE-BAR — See *lead-bars*.
SINGLE-PLATE SPRING — A spring, elliptic or half-elliptic, composed of one leaf, instead of several overlapping leaves or plates. (h)
SINGLETREE — Same as *swingletree*, of which it is a corruption.
SINGLETREE CLIP — See *center clip* and *end clip*.
SINGLETREE FERRULE — An end fitting for a singletree, having an integral hook or eye. If the eye is integral rather than the hook, then a separate hook hangs in the eye. (K & Y)
SINGLETREE HOOK — An iron hook attached to the clip or ferrule at the end of a singletree, to which a trace is attached. These hooks are made either with a small eye to attach to an end clip or a singletree strap, or with a large eye to go around a ferrule. (Y)
SINGLETREE STAPLE — A staple with threaded shanks, secured to the center of a singletree with nuts; in reality, a U-bolt. By means of an open link this fitting attaches to the doubletree, or it may be attached directly to a doubletree clevis. (Y)
SINGLETREE STRAP — An iron strap bent U-shaped, and rounded in the center, bolted or riveted to the end of a singletree in place of an end clip or ferrule. The ends of the strap extend along the front and back of the singletree, where they are secured in place. (Y)
SIZE — A glutinous material used to stiffen or fill the pores of fabrics, leathers, etc. It was originally made by boiling in water pieces of parchment, or the skins of animals or fins of fishes, and evaporating to a proper consistency. Colored sizes are prepared by grinding colors in thick drying oil, or by using the parchment size in place of the oil. (C)
SKEIN — A strap of iron let into the top and bottom of a wooden axle to protect it from wear, and to give support to the linch-pin or axle-nut. (C)
SKELETON BOOT — A carriage boot which, instead of being paneled, is composed of iron stays. The original purpose of the skeleton boot was that it should be removable, so that the vehicle might be driven either from the boot or *a la Daumont*. It is the skeleton boot which chiefly distinguishes the Victoria from the cabriolet. (h)
SKELETON RUMBLE — A rumble which is supported by iron stays, instead of being paneled. (h)
SKID — Same as *shoe*. (FR)
SKID, ICE — A device similiar to an ordinary skid, but with heavy teeth or cleats on the underside to dig into the ice. (FR)
SKY BOARD — The top panel of a circus wagon, usually cut in some decorative manner, and brightly painted. (O)
SLAB SIDE BODY — A deep side Prince Albert or other phaeton body, the quarters of which are made out of a thick plank and shaved down to give the required swell. (C)
SLAT — A thin piece of wood or metal used for various purposes in a carriage, such as roof-slat, blind-slat, bottom-slat, etc. (h)
SLAT BODY — A flat side body, the frame of which is made of a number of horizontal and cross slats, such as a grocery or express wagon. (C)
SLAT IRONS — Thin plates of iron, connected by a pivot at one end, to which the wooden bows of a top are attached, connecting them to the seat or body. The purpose is identical to that of bow sockets, except that slat irons are generally used on wagons rather than on carriages. (T)

Seamless thimble-skein, axle-box, and axle-nut. (S. D. Kimbark, Chicago, 1876.)

SLEIGH BELL — A small bell having a loose ball inside. Usually a number of these are attached to a strap, making a string of bells. (C)

SLICKER — (1.) A piece of wood, bone, or metal having the end shaped as a reverse bead, used for the purpose of creasing the edges of leather straps. (2.) A thick piece of glass, metal, or hardwood, the edge of which is rounded, used by trimmers for rubbing down leather or cloth upon the paste to make it adhere, or to burnish the edge of a piece of leather. (C)

SLIDE — A strip of bone, ivory, wood, or metal, attached to the garnish-rail of a carriage light, over which the glass string may slide easily when the window is raised or lowered. (h)

SLIDER BAR — Same as *sway bar*. This term was more common in England, while *sway bar* was more commonly used in America, though some United States manufacturers and the army did use the term *slider bar*. (O)

SLIDING DOOR — A door which slides, rather than being hung on hinges. Vehicles using this type of door are milk wagons, mail wagons, bakery wagons, some types of top wagons, etc. These doors slide on overhead rails which are mounted to the sides of the vehicle by means of short posts. The doors themselves are suspended from hangers which have small rollers to ride on the track. (Y)

SLIDING SEAT — An extra seat used in a phaeton or certain other carriages. It is so arranged that it can be hidden from view when not wanted, by sliding it under another seat or between the neck panels. (C)

SLIP LINING — Same as *false lining*.

SLIPPER — An iron slide or shoe used as a brake for a carriage wheel. (C)

SLIP TONGUE — A tongue which slips into a socket formed by the hounds and their bands. This arrangement keeps the tongue in a rigid position, even when it is not supported by the horses. It is also called a *stiff tongue*. (O)

SNIBILL — The most common spelling of the word *snipebill*. (O)

SNIPEBILL — (1.) A local New England term applied to a primitive substitute for a king-bolt, so named on account of its shape. This device was necessitated on certain vehicles wherein the neap or shafts were mortised rigidly into the front axle, so that raising or lowering the shafts caused the axle to rock slightly forward or backward. The action of this part is similiar to that of a universal joint. (2.) The term was also applied to a type of hinge used beneath the body of a dumping cart, which connected the body to the axle. The hinge consists of two interlinked eye bolts, and is generally made with a slot and key to hold it in position. (O & h)

SOCKET — A metal tube, open at one or both ends, used either to protect and ornament the end of some part, or to receive another part and hold it secure. See *whip socket, pole-socket, lamp socket, shaft-socket,* etc. (C,T, & h)

SOLDER — A metal alloy used for uniting other metals. Silver solder, or hard solder, fuses only at a red heat, while soft solder, made of lead and tin, fuses at a comparatively low heat. (C)

SPANNER — Same as *pole-hook*.

SPEAKING-TUBE — A rubber tube, covered with silk, passing from the inside of the coach to the driver's seat, by which the passengers may communicate with the driver. The ends are furnished with wooden mouth-pieces, the inside one being covered in imitation of a tassel. (T)

SPEAR — The pole and futchells of a wagon or cart, when made from a single piece, generally a sapling which has been split at the rear. Also called a *neap*. (O)

SPINDLE — One of the turned sticks used to support the seat-rail, and to fill the space between the seat-rail and bottom. (h)

SPINDLE-SEAT — See *stick seat*.

SPLASH-BOARD — Same as *dash*.

SPLICE-PIECE — That part of the framework of a coupe or clarence that defines the line of the neck on the underside, from the point of the coupe-pillar to the center of the arch. (C)

SPLINTER-BAR — A stationary front cross-bar, secured to the front gearing by futchells and draw-irons, to which the traces of a pair of horses are attached. Compare *evener*. (C & h)

SPLINTER-BAR SOCKET — An iron band, either square or round, that is fitted on each end of the splinter. (C)

SPLIT-KEY — A metal key, having two prongs which lie together, but are separated slightly after the key passes through the part which it secures. A cotter key. (O)

SPLIT SPOKE — A spoke made from timber that has been split out instead of being sawed, to insure that it is straight grained. (C)

SPOKE — One of the radiating wooden bars extending from the hub of a wheel to the rim. It has seven parts: *tenon,* entering the hub; *shoulder,* bearing against the hub; *face,* the area on the front side, adjacent to the hub; *throat,* the section changing from a rectangular to oval form; *back,* the rear side, adjacent to the hub; *body,* the oval portion, constituting the main portion of the spoke; and the *point,* which enters the felloe. (h)
SPOKE BAND — Same as *driving band.* (O)
SPOKE BOUND — Said of a wheel when the spoke bodies are too long, so that they prevent the ends of the felloes from drawing together. The term can also be used if the spoke points are too long, so that the tire cannot settle snugly on the rim. (O)
SPOKE DOG — A lever with a hook pivoted near the lower end, used in drawing spokes together when driving on the felloes. (O)
SPOKE POINTER — A tool used with a carpenter's brace to point the end of a spoke, so that the hollow auger can begin its cut easily and accurately. (O)
SPOKE-SET GAUGE — Same as *dish-staff.* (O)
SPRAG — (1.) A stiff timber used as a brake to a wagon wheel on steep grades. It is passed through between the spokes and locks against the bottom of the body. (2.) This term is also applied to a *drag-staff.* (C)
SPREAD-CHAIN — Same as *tie-chain.* (O)
SPREADER BAR — A wooden member which is attached by short spreader chains to the front end of a pole, to hold the singletrees when more than two horses are used. It is also attached to the fifth-chain when more than four horses are used. (O)
SPREADER CHAINS — A pair of short chains, joined together on the ring of a hook, and having hooks at the loose ends. They are used in conjunction with a spreader-bar.
SPRING — A term applied to a great variety of mechanical devices designed to give ease and elasticity to the movements of a carriage body, by intercepting the concussion caused by the unevenness of the surface over which the wheels pass. The most common variety of spring is composed of leaves, or plates of steel in graduated lengths, fastened together at the center. (h)
SPRING-BAND — The metal hoop, or band, used to bind the leaves of a spring together. (h)
SPRING-BAR — A bar of wood or iron placed on top of a spring, and parallel with it, to the ends of which the body loops are bolted. The term is also applied to the rear transom of a gearing, on which a whip, S, or C-spring is mounted. (C & O)
SPRING-BAR CLIP — One of the long clips used in place of bolts to attach a spring-bar to the spring. (C)
SPRING-BARREL — A tin tube, containing a spiral spring, around which the curtains of doors and quarter lights are rolled. It is operated by means of a cord attached to the lever of a ratchet wheel. Also called a *spring-curtain-roller.* (T)
SPRING-BLOCK — A small block attached to the axle or other part of the gearing to form a support for a spring. (C)
SPRING-BUFFER — A device such as a coil spring or a rubber block placed inside an elliptic spring, to serve as a shock absorber. (Y)
SPRING-CHAIR — Same as *spring block.* (I & Y)

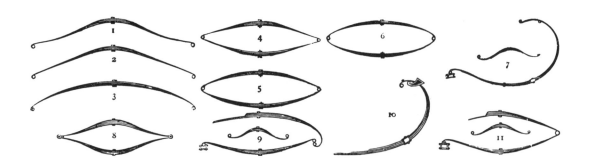

Carriage springs: 1, Double Sweep Concord; 2, Iver's Pattern Concord; 3, True Sweep Concord; 4, Philadelphia Shape Carriage; 5, Elliptic Bow Shape Carriage; 6, Round End Carriage; 7, Cradle Spring; 8, Double Sweep Elliptic; 9, French Scroll and Cross; 10, Full "C" Pattern; and 11, French Platform and Cross. (S. D. Kimbark, 1876.)

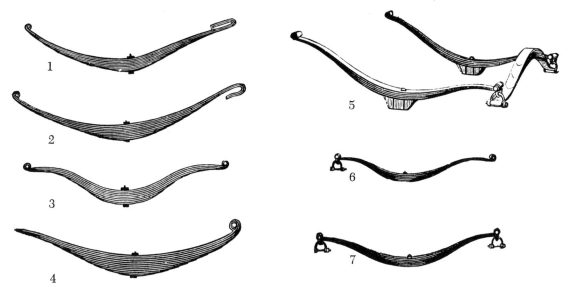

Springs for trucks and heavier carriages: 1. Truck Loop End; 2. Truck Hook End; 3. Truck Spring; 4. Truck Slide End; 5. Side and Cross Platform; 6. Side Platform; and 7. Cross Platform. (S.D. Kimbark, Chicago, 1876.)

SPRING-CLIP — A fastening resembling a U-bolt, used to secure a spring to another part of a vehicle. See *clip*. (K & I)

SPRING-CURTAIN-ROLLER — Same as a *spring-barrel*.

SPRING-LOCK — A door lock in which the bolt is moved not only by the handle, but also by a spring which always returns the bolt to the extended position. (K)

SPRING, MULHOLLAND — A torsion spring used with an ordinary half-spring of the type used on a side-bar gear. The eye is shackled to this half-spring, while the torsion spring passes through a journal secured underneath one of the sills, and the other end is bolted to the underside of the body, near the center. (Y)

SPRING-PERCH — A steel spring, used in place of the ordinary rigid perch, to connect the front and back gearings of a buggy or other light carriage. (C & h)

SPRING-SHACKLE — See *shackle*. (h)

SPRING, SHULER — A one-piece combination torsion and coil spring. The ends are attached to the underside of the sills, then take one complete turn as in a coil spring, then run to the support on the gear, which is equipped with roller bearings for the spring to turn in. (Y)

SPRING-SLIDE, TRUCK — A metal fitting in which the loose end of a truck spring can slide when the load presses the ends of the spring downward. (I)

SPRING, SOULE — A small horizontal spring attached to the underside of a side bar. The cross-spring is attached to the Soule spring, taking much of the jar and strain away from the side bar. (O)

SPRING-STAY — An iron brace used to support a spring. (h)

SPRING, TIMKEN — A type of cross-spring often used with a side-bar gear. The straight end is bolted underneath the floor, and the eye in the other end is shackled to the side bar. (O)

SQUAB — (1) Originally an extra cushion for a head rest. (2) Any quilted or stuffed lining, whether in the back or side of a carriage, and whether movable or stationary. (3) In some areas the term is applied only to that part of the trimming that is used for finishing the upper quarters of a closed carriage. (C)

SQUARE RULE — The application of practical geometry in the construction of carriage bodies. See *French rule*.

S-SPRING — A spring so named because of its resemblance to the letter S. It was the immediate successor of the whip spring, and the predecessor of the C-spring. Its mechanical construction is similiar to that of the C-spring. (h)

STABLE-SHUTTER — An ornamental paneled frame, used as a substitute for a carriage glass, to exclude observation, or to prevent light and dust from entering the carriage when it is in the stable or carriage house. (h)

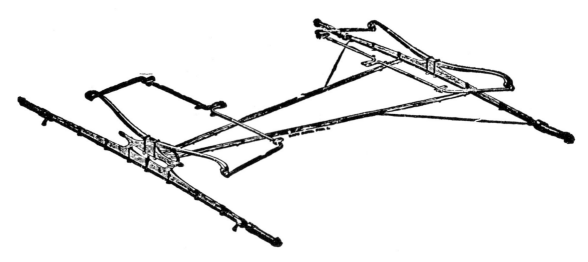

Mulholland springs. (Cray Bros., Cleveland, Ohio, 1910.)

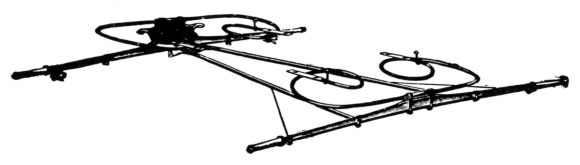

Shuler springs. (Cray Bros., Cleveland, Ohio, 1910.)

STAGGER — Same as *dodge*.

STAKE — (1) An upright framed into the end of a lumber wagon bolster. (2) Also the removable upright along the side or rear of a truck, dray, or other freight-carrying vehicle. (C & O)

STAKE-IRON — An iron fitting bolted to the side of a truck, dray, or wagon for the purpose of receiving a removable stake which assists in holding the load on the vehicle. It is sometimes called a *stake pocket*, especially when cast with four bearing surfaces rather than three. (I)

STAKE-POCKET — See *stake iron*.

STAKE-RING — One of the two rings attached to the side of a wagon stake or standard. They may be used to hold the extension stakes, or to secure ropes when tying down the load on a wagon. (Y & O)

STANDARD — (1.) A curved wooden support, generally ornamented by carving, used to support the hammer-cloth, and called a *hammercloth-standard;* or as a support connected with the page-board, in which case it is called the *footman's standard*. (h) (2.) Used synonymously with *stake*. (Y) (3.) One of the rods supporting a canopy top, as on a surrey or mountain wagon. (QMD) (4.) Generally, any upright member in a wooden framework. (O)

STANDING PILLAR — Any of the pillars that support the top or to which the doors are hinged or locked. The two latter pillars are called *hinge pillars* and *lock pillars*. (C & h)

STANDING TOP — A top that is permanently attached to the standing pillars. Also a top that cannot be lowered, such as a canopy top, as distinguished from the calash or folding top. (C)

STANHOPE PILLAR — A specific type of outside pillar extending from the top of a seat, at the front end, down to and finishing off with the moldings of the body, sometimes forming the duster as well. (C)

STAVE BOTTOM — A coach bottom made of narrow pieces, or staves, instead of being bent to the required shape. (C)

STAY — (1.) An iron brace or support to some part of a carriage, such as *spring-stay, perch-stay,* etc. Often called a *stay-brace.* (2.) A small piece of wood used to strengthen some part of a carriage body. (3.) Webbing used by the carriage trimmer to secure certain parts. (C & h)

STAY-CHAIN — One of a pair of short chains leading from the front axle of a vehicle to two points near the ends of the evener, for the purpose of preventing excessive movement of the evener. (K)

STAY-CHAIN HOOK — A hook made with an eye whereby it is attached to the end of a stay-chain, the hook attaching to the evener. Also a hook with a threaded shank, fastened to the axletree, which receives the ring of the stay-chain. (K)

STAY-IRON, WAGON BOX — An iron fitting secured to the side of a wagon box, just over the rear bolster, for the double purpose of holding the box firmly in position on the bolster, and receiving any wear caused by the stake. (K)

STAY-STRAPS — Leather straps which run from the front of a falling top, to some point forward on the vehicle body, for the purpose of holding a falling top in the raised position without the use of top joints. This feature was sometimes used late in the carriage era, and was popular on some early automobiles. (O)

STAY-WEB — A twilled web which will not stretch, used as a stay. See *web lace.* (T & C)

STEP — A metal support for the foot, to assist a passenger or driver in mounting a carriage. It is sometimes attached to the body, and sometimes to an axle or other part of the gearing. (h)

STEP-COVER — A protective cover, attached to a carriage door, and fitting closely over the step, to prevent mud from being thrown onto the step by the wheels. The cover removes by the opening of the door. (O)

STEP-PAD — A metal or wooden piece attached to the step-shank, forming the rest for the foot. Sometimes the surface is rubber covered, while in other cases a design may be worked into the metal, for antiskid purposes. (K, I, & C)

STEP-PIECE — The ogee sweep introduced in the curve of a bottomside, forming what is called an ogee quarter, to distinguish it from the Mexican-quarter, wherein two curves intersect. (h)

STEP-PLATE — (1.) A thin plate of iron placed against the body to prevent it from being injured by the feet when on the step. (2.) A plate that protects the body from wear by folding or unfolding steps. (C) (3.) Some manufacturers use the term to describe the step-pad. (I)

STEP-SHANK — An iron rod, extending from some part of a carriage or wagon, which supports the step-pad. (I)

STEP-STOP — A projection of iron, connected with either the step or the body, on which a folding step rests when let down. (h)

STICK SEAT — An old style of carriage seat having a number of perpendicular slats or spindles framed between the seat-rail and the seat bottom. (C)

STIFF TONGUE — A tongue fixed rigidly between a pair of hounds. It does not drop to the ground even when not supported by the horses. It is also called a *slip tongue.* (O)

STOCK — Same as *hub.*

STOP — A projection of iron, or a cord or strap of some kind, used to prevent some part of a vehicle from moving beyond a certain point, such as *step-stop, door-stop,* etc. (h)

STOPPER — A small plug used in the linch-pin hole of the point band to prevent the linch-pin from bouncing out of its seat in the axle-arm. In England these plugs were frequently of wood, but in America some vehicles, particularly Conestoga wagons, had the holes stopped with pieces of corn cob.

STORM HOOD — See *hood.*

STRAIK — A varient spelling of *strake.* (h)

STRAINER — An extra strip framed into the framework of quarters to act as a support to the panel, keeping it in shape. Also called a *batten* or *stretcher.* The term is also applied to the canvas glued to the inside of a panel. (C & h)

STRAKE — One of the short plates of iron formerly nailed onto the rim of a wheel, a number of which served as a tire, before the one-piece welded tire came into common use. The strakes were nailed on so that the ends fell between the felloe joints, thus allowing the strakes to serve as reinforcements to the felloe joints. Strakes were commonly used in the eighteenth century, and in many places into the nineteenth century, though welded tires were known many centuries before. In England strakes were used on some farm wagons even into the twentieth century. (O)

STRAP-BOLT — An iron strap bolted or riveted to the side of a wagon body. The lower end of the strap is formed into a bolt which passes through the cross-bar or transom, binding the side to the floor of the body. (K)

STRAP-HEAD — A bolt head that is made long and thin, so that it can be bent around the part which it passes through. (C)

STREAK — A term used in the trade during the earlier period, synonymous with *stripe*. (O)

STRETCHER — Same as *batten* or *strainer*. (h)

STRIPE — To draw fine lines of different colors, as ornamentation. (C)

STRIPING PENCIL — A small brush made of long hairs, used for striping. (C)

STUMP JOINT — The commercial name for a ready-made hinge used to make up a top joint. (C & h)

SUNKEN BOTTOM — A bottom hung below the outline of a body, in order to give more foot room. (C)

SUNSHADE, WAGON — A cheap three-bow top covered with duck, for use on a wagon seat. These tops are fitted to the seats by means of irons known as falling-top sunshade-irons, and they allow the tops to be fixed at any angle desired. (Y & I)

SUSPENSION — This term, in its technical application to carriage building, describes the means by which the body of the vehicle is removed from the concussion of the wheels. The vehicles of earlier times were without springs, and consequently had no true suspension. Suspension marks an era in carriage history, and to classify suspended and nonsuspended vehicles, as originally developed, is to draw the line between those of aristocratic and those of utilitarian origin,. This leaves us, on the one hand, the entire group of pleasure vehicles, and, on the other, those designed for practical uses, where *conveyance* is the only object. Suspension will thus be seen to have a vital relation to the development of modern pleasure carriages. The earliest known mode of suspension was the hanging of the body of the vehicle from the four corners, by means of leather straps, or braces, attached to four upright wooden posts built into the axles. The next step was to give these posts some elasticity, by making them lighter and curving them slightly; they were thus ultimately transformed into actual springs, first called *whip springs*, later *S-springs,* and finally *C-springs.* Various noteworthy experiments were made about a century ago (late eighteenth century), in suspension. In some instances long braces, called *thoroughbraces* (see also *brace*), were passed entirely under the body, as in the case of the Berlin. The improvements in springs, modifications of the perch, and changes in the construction of wheels, have almost entirely done away with the original forms of suspension; these changes are treated in detail under their proper heads.

The suspension of two-wheeled vehicles has also passed through a great number of phases, and here the mechanical problems are still more complicated than in four-wheel suspension. For the various forms of two-wheeled suspension see *Dennet spring, gig,* and *tilbury.* For other features of suspension also see *side bar, buckboard,* and *volante,* and the various kinds of springs, such as *mail, telegraph, nutcracker,* etc. (h)

SWAGE — A tool, variously shaped or grooved on the face, used by blacksmiths and other workers in metal for shaping their work. (C)

SWAN-NECK — (1.) A term sometimes applied to a *pole-hook.* (2.) See also *crane,* or *crane-neck.* (O)

SWAY BAR — A wooden bar, framed into the rear ends of the futchells, and bearing against the bottom of the perch, used to steady the front gearing. (h)

SWAY BAR PLATE — An iron plate screwed to the sway bar to strengthen it, and to prevent it from wearing. (C)

SWEDGE — The same as *swage.* (C)

SWEET'S CONCEALED-BAND HUB — A reinforced hub patented in 1886 by John M. Sweet of Batavia, New York, used to a somewhat limited extent. The iron reinforcing casting, with its spoke mortises cast in, is sunk in level with the surface of the hub. This was accomplished by turning down one end of the wood portion to a diameter that would permit pushing the iron fitting over it. The hub was then finished by pushing a second wooden part over the first, and against the iron fitting. Eventually this second part was also made of metal. (O)

SWING — The downward deflection of an axle-arm, which causes the wheels to have an outward cant, so that they are several inches farther apart at the tops than on the ground. This is necessary, because of the dish of the wheel, to give a plumb spoke (a spoke that stands at right angles to the ground). In England, the term, *dip,* is often used in place of *swing.* See *set* and *gather.* (O)

SWINGLE-BAR — A term sometimes applied to a variety of swingletree. In this case it is a bar with roller-bolts, pivoted to a splinter-bar by a vertical bolt. The purpose of this alteration is to prevent the collar from rubbing the shoulders of the horse, particularly when a breast collar is used. (FR)

SWINGLETREE — A short bar, attached to an evener or splinter-bar, to which the traces are attached for the purpose of draft. According to W.W. Skeat's *Etymological Dictionary of the English Language* (1882), the swingletree is often corruptly called a singletree, "Whence the term doubletree has arisen, to keep it company." The word *swingle* is said to mean swinger, or a thing that swings, and the part is so named from its swinging motion. The words swingletree, singletree, whiffletree and whippletree are often used synonymously by those in the trade, but preferably the first two terms should be applied to the swinging variety, while the last two terms should be applied to the type that is pivoted to the cross-bar of a pair of shafts. Mechanically the swingletree is a bar of wood about thirty to thirty-six inches long, with hooks at the ends for attachment to the traces, and a clip or staple in the center for attachment to the evener. In the later period some swingletrees were made of pressed steel. Whiffletrees were usually made of wood, pivoted to the cross-bar by a bolt, and in most instances a smoother action was provided by a set of couplings. The ends may be provided with metal fittings of various sorts, or they may be sword-end whiffletrees, wherein the wood tapers to a thin point, and the trace slips over the end and is held by a thong. Whiffletrees of metal were also known, such as those of the New England pleasure wagons. (O)

SWING POLE — A pole similiar to the regular pole, except that it is lighter in weight, to which the leaders are hitched when six horses are used. The pole has a hook at the rear end, for attachment to the regular pole, and a crab of the usual form at the front end. It is so named because it hangs between the swing team (middle team). (FR)

SWIVEL — A fitting inserted in a chain which allows the ends to turn in opposite directions without the chain becoming twisted. It consists of the shank of one eye turning in another eye. (I)

SWIVEL REACH-COUPLING — A swiveling coupling in the reach of a wagon, attached between the front ends of the rear hounds. It eases the passage of a wagon over uneven ground, by allowing either axle to move out of the plane of the other axle. (O)

SWORD CASE — A long narrow box permanently attached to the back panel of a carriage body. It opens under the top back squab, and was formerly used for carrying swords and firearms. Later this part became very small, and served only as an ornament, in many instances not even being accessible from the inside. It is a variety of *budget*. (C & h)

SYPHER— To overlap the chamfered edge of one plank upon the chamfered edge of another, in such a manner that the joint will be flush. (C)

SYSSING — An English term, used to describe varnish when it crawls together on a surface, leaving a series of cracks, ridges, and bare spaces. Synonymous with *alligator*. (O)

T

TACK — A small, sharp pointed cut nail, generally having a large flat head; the kind known as gimp tacks have small, round heads. They are put in papers supposed to contain a specified number, and designated by ounces, the different weights being due to the sizes of the tacks. (T)

TAIL BOARD — The back end of a wagon body, hinged or otherwise, attached to the bottom so that it can be dropped or removed for convenience when loading or unloading. Also called an *end board*, and *tail gate*. (C)

TAIL GATE — Same as *tail board*.

TALKING WAGON — A slang expression used to describe the dull, heavy, rumbling sound made by the wheels when they are playing free on the axle arms. (C)

TALMI GOLD — A beautiful gold colored alloy composed of eighty-six and four-tenths parts copper, twelve and two-tenths parts zinc, and one and one-tenth part tin. (C)

TANDEM — A team of two or more horses harnessed in line, one before the other. (h)

TANDEM-HORN — Same as *post-horn*. See *coach-horn*.

TAP BOLT — A bolt with a head on one end and a wood screw thread on the other, used for screwing into some fixed part instead of being passed through and fastened with a nut. Same as a *lag screw*. (C)

TARPAULIN — A waterproof cover used for the protection of goods; a wagon cover. The army used the abbreviated form, *paulin,* in referring to a wagon cover. (O)

TAR POT — A small bucket or pot carried with a wagon to hold the lubricant for the axles. The lubricant was frequently pine tar, thus the name, but lard and petroleum grease were also used. (O)

TASSEL — An ornamental pendant, having a closed top and loose-twisted pendants, the kind known as a coach tassel having a flat, braided head. (T)

TEE IRON — A T-shaped iron plate, screwed, bolted, or riveted over the junction of two wooden members, such as a shaft meeting the cross-bar. (K)

TELEGRAPH SPRING — Synonymous with *mail spring*. (h)

TEMPER — (1.) To harden steel to a certain degree, so as to make it suitable for use in tools, springs, etc. (2.) In painting, the mixing of the pigments with oil or turpentine in such proportions that the mixture is in good working consistency. (C)

TENON — The end of a piece of wood cut down, leaving a shoulder on two or more sides, to be set into an opening called a mortise so that the two pieces are joined firmly together. (C)

THILL — Synonymous with *shaft*. (C)

THIMBLE — A small roller on the bolt that supports one end of a thoroughbrace. (C)

THIMBLE-SKEIN — A skein for a wooden axle-arm, made in tubular form so as to receive all the wear from the axle-box. See *axle, thimble-skein*. (C)

THIMBLE-WASHER — A brass threaded ring, screwed into a carriage door to receive the handle and hold it more firmly in place, as a bushing. (h)

THIRD SEAT — A small portable folding seat, usually covered with carpet. It is not attached to the carriage in any way. The framework is usually of an X or *sawbuck* pattern. (O)

THOROUGHBRACE — When coach bodies were first suspended, it was by means of short straps, or braces, attached to the four corners of the body, and swung from upright pillars or posts. The introduction of the vehicle known as the Berlin seems to have marked an era in the mechanical development of the coach, especially with regard to suspension. Berlins, instead of being hung from four corners, had long straps called thoroughbraces which passed under the body, and were attached to and regulated by the windlasses or

A third seat such as this cost forty cents in 1910. (Cray Bros., Cleveland, Ohio.)

jacks at the axles. These thoroughbraces, made of a number of leather straps stitched together, served the combined purposes of spring and body loop. They continued in use into the twentieth century on such vehicles as the Concord coach, and other passenger wagons, in which cases they are generally not attached to windlasses, but have a turnbuckle in the center for adjustment, or they may be a long single strap, wound round and round the shackles so as to form a number of thicknesses of leather for the body to rest on.

By comparison, the ordinary brace, or *main brace,* is used wherever the C-spring is employed. (T & h)

THOROUGHSPRING — A patented steel spring offered by Doland and Scherb, of Sacramento, California, during the 1870s. Applied to light carriages, it was a continuous spring incorporating two C-springs with a connecting "thoroughbrace." It is unlikely that it was extensively used. (O)

THREE-QUARTER BODY — A buggy body made very narrow, being intended for but one person. There is no fixed width, but from seventeen to twenty inches is considered as a three-quarter seat. (C)

THREE-QUARTER SPRING — See *Dennett spring.*

THUMB NUT — A nut having projections or lugs so arranged that the nut can be turned on by the thumb and fingers, without the use of a wrench. It is also called a *wing nut.* (C)

THUMB SCREW — A workman's clamp, having a long screw bolt with projections on the upper end like those on a thumb nut. (C)

TIE-CHAIN EYE — The fitting by which a tie-chain is secured to the side of a wagon box. (I)

TIE-CHAIN, WAGON BOX — A chain attached to both sides of a wagon box, and spanning the inside of the box, to support the sides against the load. The chain may be unfastened in the center, and the length is generally adjustable. (I)

TILBURY SPRING — The spring combination which characterizes the Tilbury. In the rear is an iron framework supporting a transverse spring, from the ends of which run leather braces to two single elbow-springs extending from the body. In front, two more single elbow-springs extend from the body to the shafts, sometimes being connected to the shafts by short leather braces. In addition, double elbow-springs are sometimes interposed between the axle and shafts, in which case a total of seven springs are used. Bridges Adams, in his *English Pleasure Carriages,* says that "this mode of hanging is unscientific, inconvenient and unsightly, besides possessing considerable extra weight." Nonetheless, its use has continued to the present day (1882). (h)

TIP — A general name given to a small socket, such as a *shaft-tip, whiffletree-tip,* etc. (T)

TIRE — A band of iron or steel, surrounding the rim of a wheel, to which it is nailed, screwed, riveted, or bolted. Its purpose is to strengthen the wheel and protect the felloes from wear. In the latter part of the horse-drawn era, the term was also applied to rubber tires, either solid, cushion, or pneumatic, which were held on by wires or other devices. (C,O, & h)

TIRE-BENDER — A gear-driven machine having three small but stout rollers, used to bend flat stock to a uniform radius for use as a tire. (Y)

TIRE BOLT — A small bolt, having a countersunk head, used to hold a tire on a wheel. The head may have a slot like a screw head, or it may be unslotted, depending on friction to hold it in place while the nut is being tightened. (C)

TIRE RIVET, CALIFORNIA — A rivet run horizontally through the felloe of a wheel, just inside the tire, with the head and washer extending over the edge of the tire, to hold the tire on the wheel if it should become loose. (S1887 & O)

TIRE-SHRINKER (or UPSETTER) — A machine for shortening a tire. The tire is grasped in two places by strong jaws, and the pairs of jaws are moved closer together by a powerful leverage system. Only the short section between the two sets of jaws needs to be brought to a red heat. (Y)

TOE BOARD — The board attached to the brackets of heavy coaches, used as a foot rest for the driver, and as a dashboard. Also a similiar board used on certain types of passenger and delivery wagons. (C & Y)

TOE BOARD HANDLE — A metal handle attached to the toe board to assist the driver in mounting. (C)

TOE RAIL — A raised rail placed on the floor of a light carriage, just behind the dash, to serve as a foot rest. (Y & B)

TOILET BOX — A small box made of rosewood or ebony, highly polished, or of other wood covered with leather. It is attached inside a coupe, or other closed carriage, and carries a small looking glass, brush, comb, and other small toilet articles. (T)

TONGUE — Same as *pole*. (C)

TONGUE BOLT — The bolt that passes through the front hounds, securing the tongue to the gearing. (C)

TONGUE CHAIN — Same as *pole-chain*.

TONGUE HOUNDS — Short hounds attached to the sides of the rear end of a *drop tongue*. (O)

TONGUE PLATE — An iron plate fastened to the hounds or tongue, around the hole for the doubletree pin. This plate receives the wear caused by the swiveling action of the doubletree. (K & Y)

TONGUE SUPPORT — Any of the patented devices to hold the weight of the tongue from the horses' shoulders, or to hold the tongue up when it is not in use. It may be a spring at the rear of the tongue, or some sort of post at the front end, though the latter is only for supporting the tongue when it is not in use. (Y)

TOOL BOX — A box attached to a wagon body for carrying tools. On Conestoga wagons they are generally on the side, while on other wagons they are frequently on the front end gate. (O)

TOP — All that portion of a carriage body, whether standing or folding, above the belt-rail. (C)

TOP BOARDS — Thin pine boards used for covering a top; roof boards. (C)

TOP END-RAIL — The top end piece that holds the ends of the top rails and the tops of the standing pillars together. (C)

TOP JOINTS — The hinged props or braces that support a bow top in the raised position. (C)

TOP JOINT RIVET — A rivet having a broad head, used to connect the hinge of a top joint. (T)

TOP LEVER — A lever inside the top of a falling-top carriage, used to break the top joint and lower the top without the necessity of reaching outside the carriage. Also called an *outside joint breaker*. (I & Y)

TOP PROP— A bolt having a long, flat base, by which it is secured to the bows of a top, used as a support and connecting bolt for the top joints. (T)

TOP PROP NUT — A metal nut, threaded and capped, used on the end of the bolt of the top prop. (T)

TOP RAIL — The rail running parallel with the body, into which the standing and corner pillars are framed, and which acts as a support for the curves and boards forming the roof. (C)

TRACE-POST — Same as *roller-bolt*. (O)

TRACK — The width of a vehicle, measured from the center of one tire to the center of the other, on the ground. This distance varies according to the locality, but fifty-six inches and sixty-two inches are popular tracks. (O)

TRACTION — The act of drawing, or the state of being drawn. (C)

TRAIL — The rear portion of a gun carriage, which rests on the ground when the gun is unlimbered, and attaches to the limber when the gun is to be moved. (W)

TRAIL TONGUE — A short variety of drop tongue, used to couple one wagon to another. This was a feature common on western freight wagons, where two or more wagons were coupled together and drawn by a

Tire benders, such as this Boynton & Plummer model, had one adjustable roller, the adjustment of which determined the radius of the tire. The benders could be operated either by power, or by a hand crank. (*The Hub,* October 1900.)

Tire shrinkers, such as this Western Chief model, upset both tires and axles to guarantee accuracy. In the type shown, the iron was heated red, and then compressed by means of powerful leverage. (Cray Bros., Cleveland, Ohio, 1910.)

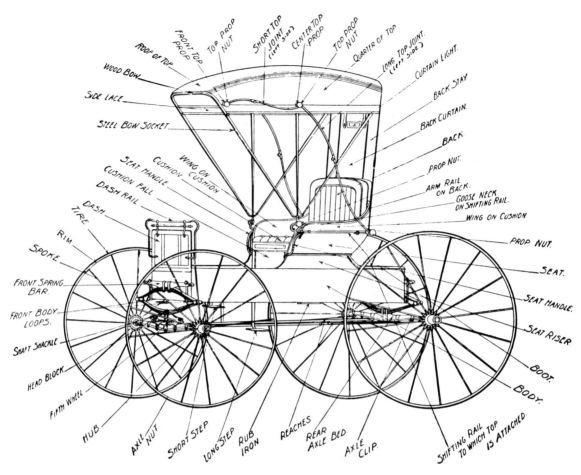

The parts names of this PIANO-BOX BUGGY are applicable to many other vehicles as well. (Cash Buyers' Union, Chicago, Catalog 3m, 1903.)

large number of animals harnessed to the lead wagon. A short heavy chain was attached to the underside of the trail tongue, and this chain was in turn attached to the long chain running under the wagons, from the tongue of the lead wagon. (S226)

TRANSFER ORNAMENT — See decalomania, definition 2.

TRANSOM — The bar of wood to which the front of a perch is attached. It rests on the front axle, to which it is attached by the king-bolt. The transom occupies a position in a carriage similiar to the position of a bolster in a wagon, and in some areas, the term *transom* is applied to an ordinary bolster. The term is also applied to the hindmost cross-bar of a gearing, to which a *whip-*, *S-*, or *C-spring* is attached. This hind transsom is also called a *spring-bar*. The terms *transom* and *bed* are often used synonymously. (C & FR)

TRANSOM PLATES — Iron wearing plates attached to the top of the front axle, and to the bottom of a transom, at the king-bolt. (FR)

TRAVELER — A small wheel affixed to a handle, used for measuring the circumference of a tire after it has been welded. (C)

TREAD — That portion of the step upon which the foot rests. Technically, the width of a tire or a sleigh runner is called the tread. (C)

TREE — A term frequently applied to a piece of timber used for some specific purpose, such as *whiffletree*, *doubletree*, *axletree*, etc.

TRIMMER — A workman employed at the trade of carriage trimming. (C)

TRIMMING — The upholstering of a carriage with leather, cloth, or other material. (T)

TRIPOLI — Same as *rotten stone*.

TRUCK — A small wheel, especially the small strong wheel of certain types of gun carriage. (O)

TRUCK SPRING — A heavy, half-elliptic spring used on trucks, similiar to a platform spring, but without a cross spring. A truck spring is shackled to the body of the truck at its forward end, while the rear end is unshackled, but slides in a box. (O)

TRUSS ROD — An iron or steel rod bowed away from other parts, and then tightened against them by means of end nuts or a turnbuckle in order to form a more rigid framework. (O)

TRUSS ROD STRUT — A fulcrum against which the truss rod bears. (I)

TRUSS SKEIN — A type of thimble skein having a hole in the outer end, through which a long rod passes. The rod exits underneath the axletree, near the back end of the skein, passes under a fulcrum placed between the rod and the tree, and does the same thing at the other end. Nuts on the outer ends of the rod are drawn up tightly, thus forming a truss for the axletree. (Y)

TUFT — A knob of silk or worsted, used in trimming, to prevent the tie cord from drawing through the cloth; also used as an ornament. (T)

TUFTING BUTTON — A button used in place of a tuft. See *button*. (T)

TUFTING CORD — Fine strong cord, used to secure the tufts and bind cushions and squabs. (T)

TUFTING NAIL — A nail with a large head, covered with cloth or other material, used in place of a tuft or tufting button where these types cannot be used. See *nail*. (T,K, & Y)

Top joint. (Cray Bros., Cleveland, Ohio, 1910.)

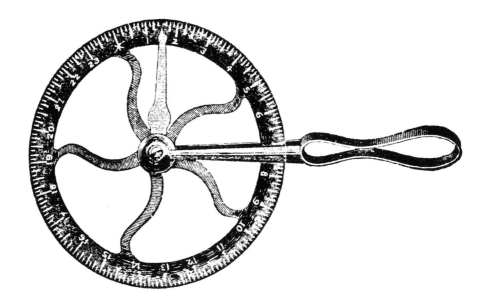

Traveler, or tire measuring wheel, which is run around the wooden rim of a wheel, and then around the inside of an iron tire, to determine how much a tire needs to be shrunk before it is set on the wheel. (Cray Bros., Cleveland, Ohio, 1910.)

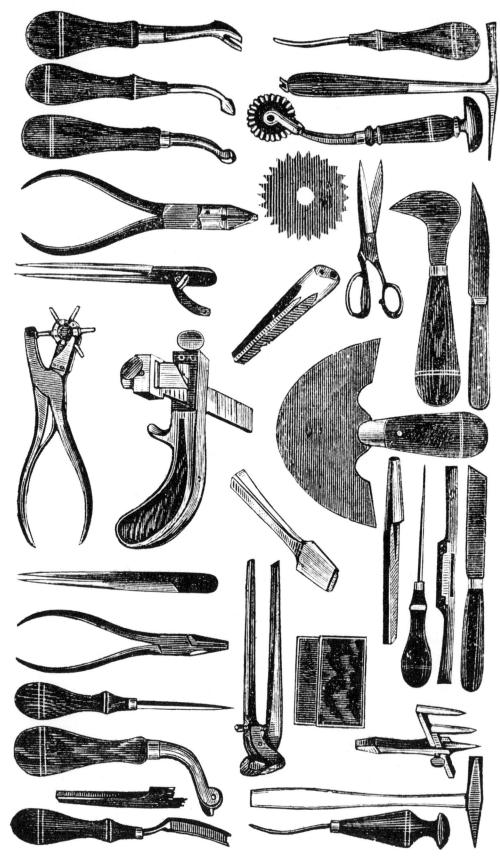

Carriage Trimmers' Tools. (From William N. Fitzgerald: *The Carriage Trimmers' Manual and Guide Book*, 1881.)

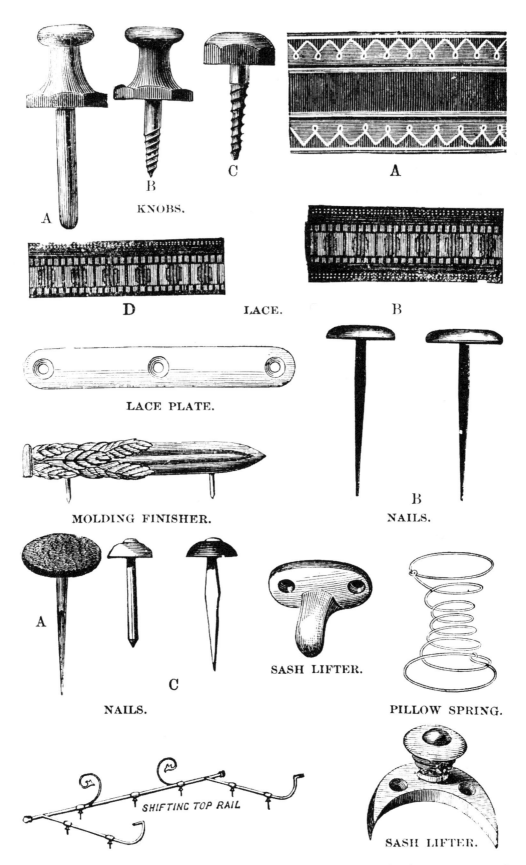

Trimmings and hardware used in the trimming of carriages. (From William N. Fitzgerald: *The Carriage Trimmers' Manual and Guide Book*, 1881.)

TUG IRON — An iron eye attached to the shaft of a wagon, to which the breeching strap is attached, by which the carriage is held back when going down hill. It is most frequently called a *hold back*. (C)

TURN-OUT SEAT — A seat which folds into the rear of a vehicle such as a buggy. When in use, it folds out and extends beyond the rear of the vehicle body. (O)

TURPENTINE — A resinous substance exuded naturally or on incision from several species of trees, such as pine and fir, from which spirits of turpentine is distilled. (C)

TURPS — Spirits of turpentine.

Trimmings and hardware used in the trimming of carriages. (From William N. Fitzgerald: *The Carriage Trimmers' Manual and Guide Book*, 1881.)

TURRETED BELT — A broad belt around a body, painted to imitate the top of a turret. (C)

TURTLE BOOT — A detached boot for heavy coaches hung off upon iron loops. It is made of wood and shaped on the top and back end like the back of a turtle. (C)

TUYERE — (Pronounced *tweer.*) The nozzle of a bellows in a blacksmith's forge or smelting furnace. The term was applied not only to the nozzle of the bellows, but also to the air pipe into which this nozzle was inserted. As forges and blowers became more sophisticated, the term was collectively applied not only to the air passage, but to any and all parts and openings that assisted in introducing the air to the fire-pot.

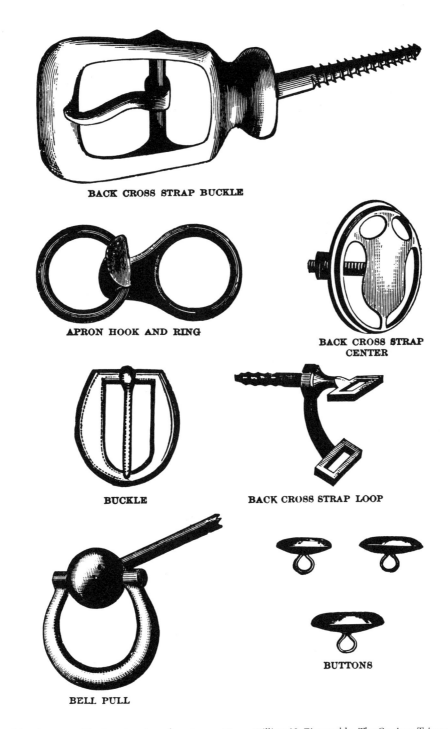

Trimmings and hardware used in the trimming of carriages. (From William N. Fitzgerald: *The Carriage Trimmers' Manual and Guide Book,* 1881.)

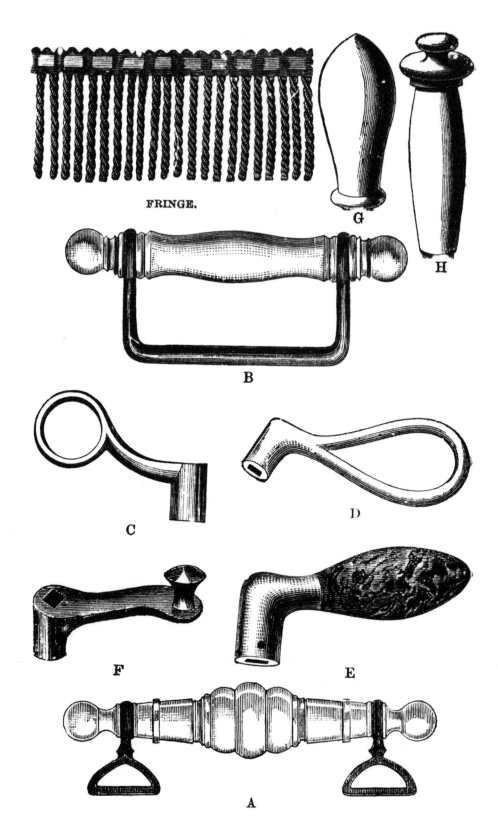

Trimmings and hardware used in the trimming of carriages: A, ivory pull handle; B, lace handle; C, D, E, and F, inside handles; G and H, French lock handles. (From William N. Fitzgerald: *The Carriage Trimmers' Manual and Guide Book*.)

U

ULTRAMARINE — The richest and most beautiful of all blues. It is prepared from lapis lazuli, a mineral of great beauty. The mineral is heated and thrown into water to cause it to pulverize easily. Then it is reduced to a fine powder, made into a paste with varnish, and afterwards put into a linen cloth and kneaded in hot water. The first washing is of no value, the second water gives the purest pigment, and the third is less valuable. Its extravagant price has caused it to be counterfeited, but the prime article can be detected by nitric acid, which instantly destroys the color but has little effect upon the adulterated article. (C)

UMBER — An impure native oxide of iron and manganese, brought from Umbra, Italy, from which its name. It is of a blackish brown color, and is the only simple brown in use. It is used to a great extent in compounding colors. (C)

UMBRELLA HOLDER — An iron support used for attaching an umbrella top to a carriage. (T)

UMBRELLA TOP — An adjustable carriage top, constructed like an umbrella. (T)

UNDER-CARRIAGE — Same as *gearing.* (C)

UNDER TOP — The underside of a squabbed top. (T)

UNHUNG — A carriage body removed from the gearing. (C)

UPHOLSTERER — A carriage trimmer. (C)

UPPER BACK PANEL — The back panel between the belt-rail and the top end rail. (C)

UPPER QUARTER — The panel of a coach body that extends from the belt-rail up to the top rail. (C)

UPSET — To thicken and shorten a piece of iron or steel stock by heating the area that is to be upset to a red heat, and striking the stock on the end. Also see *tire-shrinker.* (C)

VALANCE — The strip of leather placed against the front edge of the front bow; also the covered strip on the back edge of the back bow. The term is also applied to the loose fall attached to the inside of a dash, and in some areas, to a *seat fall*. (T)

VAN DYCK BROWN — A valuable brown, but very hard to grind. It is a good dryer, possesses a good body, and is particularly useful for making fancy shades. (C)

VARNISH — A thick, viscid liquid, consisting of a solution of resinous matter in volatile liquid. That known as coach varnish is made of gum copal, linseed oil, and turpentine. (C)

VARNISH BRUSH — A flat brush made of badger's, or other fine hair. (C)

VEHICLE — In painting, a liquid, such as a solvent, carrier, or binder, with which various pigments are applied. (C)

VELVET — A cut pile fabric, the pile of which originally was made of silk, and is now also made of wool or mohair. (O)

VELVETEEN — A kind of cloth made of cotton in imitation of velvet, now used to a considerable extent for trimming light carriages, sleighs, etc. (C)

VENETIAN BLIND — A blind made of thin slats, set into a frame, the edges overlapping each other, so as to exclude the sun's rays without preventing a free circulation of air. (T)

VENETIAN EMERALD — A beautiful green lake made from coffee and river water precipitated by pure soda. It is said to retain its color under the influence of sunlight, and to resist the action of acids. (C)

VERDIGRIS — One of the best simple greens used by carriage painters. It has a bluish tint, but when mixed with a little yellow produces a rich green. (C)

VERMEIL — Vermilion, or the color of vermilion. Also gilt silver, copper, or bronze. (C)

VERMILION — The most brilliant of all light reds. It bears a good body, works smoothly, and grinds fine. (C)

VERONA GREEN — A durable green, but one that requires much care in mixing, as it does not readily incorporate with oil. (C)

VISE-MAN — A workman who works at the vise in a blacksmith shop. A finisher, filer, or whitesmith. (C)

VOIDER — A piece of horn, three to four inches long, used by painters in place of a palette knife. (C)

W

WABBLING WHEEL — A wheel in which the box is set off-center, giving it an uneven motion when revolving. (C)

WADDING — Prepared sheets of raw cotton. (T)

WAGON STANDARD — A short upright post framed into and near the end of a bolster, used to hold the wagon box in place on the gear. Also called a *bolster stake*. (K)

WAGON STANDARD BRACE — A metal fitting used to render the wagon standard more rigid in the bolster. They are sometimes formed like a corner brace, and sometimes like a rod, flattened at both ends and set in between the two wooden parts at approximately a 45-degree angle. (K)

WAGON WRENCH — A wrench for removing axle nuts, the handle of which serves as the doubletree pin, in the same manner as the hammer pin of a linch pin running gear. Since there are no linch pins to be removed, the handle of a wagon wrench is not chisel-shaped as is the end of the hammer pin. (Y)

WAIST-RAIL — Same as *arm-rail*.

WARNER WHEEL — One of the best and most popular patent wheels, patented in 1867 by A. Warner. It employs a wooden hub, over which a malleable iron fitting is pressed. The fitting has mortises for the spokes cast in, and the spokes are driven through these mortises into the hub, the shoulders of the spokes resting on the wooden part of the hub.

WARP — (1.) Said of a board that is twisted edgewise. (2.) The threads that extend lengthwise of a woven fabric, parallel with the selvage. (C)

WASHER — A flat ring of metal, leather, or other material, used to relieve friction; when used with a bolt, to secure a tight joint without marring the timber. (C)

WASH LEATHER — Same as *shammy*. (C)

WATER GILDING — The gilding of metal surfaces by covering them with a thin coating of amalgam of gold, and then volatilizing the mercury by heat. (C)

WEAR IRON — An iron plate or shield affixed to the side of a wagon body to protect it from injury from the wheel when turning. (C)

WEATHER STRIP — A beveled strip of wood, attached to the outside of a carriage body over a door, to prevent rain from blowing in the crack above the door. (O)

WEB LACE — A strong hempen lace woven in such a manner as to be difficult to stretch. It is used for stays to the inside of calash tops. Also called *stay web*. (C & T)

WELD — To unite two pieces of iron by heating almost to fusion, and afterward beating or pressing them together to form one part. Today two additional methods of welding may be used in carriage work: oxyacetylene welding and electric welding. (C)

WELT — A cord covered with cloth or leather, used as an ornament, and for the purpose of strengthening seams on borders. It is also used in place of seaming lace. (C)

WHEEL — A circular framework that is capable of turning on an axle. It may be solid, as a wheel cut from end of a log; or it may be constructed with a center hub, from which the spokes radiate, the outer ends of which are bound by a rim or felloe. (C)

WHEEL, SUNBURST — A type of wheel used on circus wagons. These are ordinary wheels with wedge-shaped pieces of wood inserted between the spokes, held by grooves cut in the sides of the spokes. This presents a continuous surface for decoration. (O)

WHEEL-DOG — A *wear iron* having sharp prongs, which are driven into the wood in the place of screws or bolts, to hold the piece in place. (C)

407

WHEEL-HORSE — A bench used to hold a hub in position while the spokes are being driven, and the felloes are being fitted. (C)

WHEEL-HOUSE — The arch cut into the underside of a carriage or wagon body, to allow the wheels to turn sharply under the body when the vehicle is being turned. (O)

WHEEL-IRON — Same as *draw-iron*.

WHEEL-PIT — A pit used for the same purpose as the wheel-horse, except that the hub rests on a level with the floor with its axis in a horizontal position. This arrangement is used for driving spokes of the heaviest kind. (C)

WHEEL-PLATE — A metal washer used on linch pin axles to prevent the end of the box from wearing against the linch pin. See *axle washer*. (C)

WHEEL-SHOE — See *shoe*.

WHEELWRIGHT — A workman whose business is building wheels. (C)

WHIFFLETREE — The bar to which the traces are hitched for the purpose of draft. See *swingletree*. (C)

WHIFFLETREE BOLT — A carriage bolt with a special head, used to secure the whiffletree to the cross-bar. The head is usually an elongated oval, but may be diamond-shaped, a four-pointed star, or some other ornamental design. (K)

WHIFFLETREE BRACE — The D-shaped iron fitting which fits around the whiffletree and cross-bar. The bolt passes through both ends of the brace. (K)

WHIFFLETREE CLEVIS — Same as *whiffletree brace*.

WHIFFLETREE COUPLINGS — A pair of plates, usually of malleable iron, on which the whiffletree swivels, the upper one being mounted on the whiffletree and the lower one on the cross bar. Generally the surfaces of the plates are formed so as to fit into each other, increasing the bearing surface and lessening the strain on the whiffletree bolt. Sometimes they are called whiffletree plates, though those designated as **plates** usually do not have the last-named feature. (K, I & C)

WHIFFLETREE HOOK — The hook attached to the end of a whiffletree, to which a trace is attached. (C)

WHIFFLETREE KEY — A small leather strap which passes through the *cock-eye* of a whiffletree, holding the end of the trace in place. This is also used on a sword-end whiffletree. (K)

WHIFFLETREE PLATES — See *whiffletree couplings*.

WHIFFLETREE SOCKET — A metal fitting on the end of a whiffletree, such as a *cock-eye, hook,* etc., to which the trace is attached. It differs from a *whiffletree tongue* in that it is made integral with a ferrule, and slips over the end of the tree, while the tongue is inserted into the end of the tree and bound with a separate ferrule. (K)

WHIFFLETREE STAPLE — See *singletree staple*.

WHIFFLETREE TIP — A metal fitting for the end of a whiffletree, on which a trace is fastened. Besides the cock-eye, there are various other types such as L-end, fan tail, T-end, button-end, and numerous patent types with mechanical latches which prevent the traces from slipping off. This is a general term that may be applied either to a whiffletree socket or tongue. (K)

WHIFFLETREE-TONGUE — Any type of metal fitting on the end of a whiffletree, such as *cock-eye, hook,* etc., to which a trace is attached. It differs from a *whiffletree socket* in that it is inserted into the wood, and used with a ferrule, while the socket has an integral ferrule, and slips over the end of the tree. (K)

WHIPPANCE — A variation of the word whiffletree, but referring in reality to a type of swingletree. This is an English term. (O)

WHIPPLETREE — Same as *whiffletree*. See *swingletree*.

WHIP SOCKET — A short tube of wood, leather, metal, rubber or other material, attached to the forepart of a vehicle for the purpose of holding a whip. (C)

WHIP SOCKET FASTENER — A device for securing the whip socket to the dash or front part of the body. (T)

WHIP SPRING — The predecessor of the S-spring and the C-spring, consisting of a series of overlapping metal plates or leaves, bent in a slight curve (suggesting the name), to which are attached the braces which suspend the body. (h)

WHITE LEAD — Lead calcined by means of vinegar and a low heat. That known as keg lead is ground in linseed oil. (C)

WHITESMITH — A workman who files and polishes iron or steel, working in conjunction with a blacksmith. (C)

WHITEWOOD — Any of numerous trees having white or light colored wood; also the wood of any of these trees. The wood of the tulip, cottonwood, and linden or basswood trees is called whitewood. Yellow poplar, or tulip, is much used for panels of carriages on account of its ability to retain its shape and because of its smooth grain which is easily covered with paint. (C & O)

WILLIAMSON'S BLUE — Same as *Prussian blue*. (O)

WINDOW — An opening in the body or curtain of a carriage, intended for the admission of light or air. It is usually closed by some transparent material, such as glass, so that only light is admitted. (O)

WINDOW FASTENINGS — Metal devices for securing the window frames in any desired position. (T)

WINDOW FRAME — The frame enclosing the glass of a window or door. Same as *sash*. (O)

WING — (1.) An iron frame covered with leather or canvas, and placed in a position to protect the passengers from the dust or mud thrown by the wheels when the carriage is in motion. Wood is sometimes used in place of iron and leather. (C) (2.) The term is also sometimes applied to the flare boards of a wagon. (S226) (3.) Occasionally hounds are called wings. (FR) (4.) Any winglike projection, such as a cushion wing. (O)

WING NUT — A nut having one or two projections, or wings, so that it can be put on with the fingers, without the use of a wrench. Also called a *thumb nut*. (O)

WOOF — The threads that run crosswise in a woven fabric, at right angles to the selvage. (O)

WROUGHT IRON — A pure form of iron containing only about five-tenths percent carbon. It can be made directly from the ore, as in a Catalan forge, or by purifying cast iron in a reverberatory furnace. It is very tough and malleable, and was the principal metal used in carriage construction during the earlier period. Even during the latter part of the horse-drawn era, it was still extensively used on many of the better vehicles.

X Y Z

XC PLATE — Iron or steel hardware such as bits, buckles, and other harness mountings that have been tinned in order to prevent rusting. (O)

YANDELL TOP — A name sometimes applied to the peculiar style of top used on a doctor's buggy or phaeton. It is named after the designer, a Louisville, Kentucky, physician. (M)
YARD-OF-TIN — See *coach-horn*.
YOKE — See *neck yoke* and *ox yoke*. (C)
YOKE TIP — A socket-type metal fitting used on the end of a neck yoke. The fitting has a loop running parallel to the yoke, through which the pole strap passes. (K)

ZINC — A bluish white metallic element used especially as a protective coating for iron and steel parts; and also, mixed with copper to form various alloys. (O)
ZINC-WHITE — An oxide of zinc, not much used by carriage painters. (C)

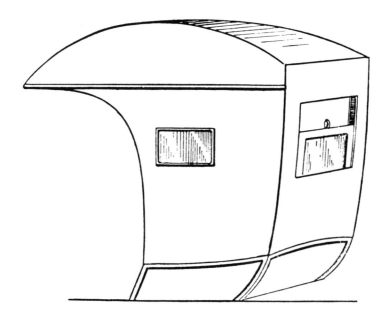

The Yandell top as shown in *The Hub*, June 1887.

HARNESS NOMENCLATURE

A

ALUMINUM BRONZE — A fine yellow alloy, closely resembling gold in color, and susceptible of a high polish. It is used to a limited extent for harness mountings. The component parts are, by weight: copper, ninety parts, aluminum, ten parts. (C)

ANKLE BOOT — A covering for the ankle of a horse to protect it from injury when struck by the other feet. It is made of leather or heavy felt, with a small piece called a shield placed over the part to be protected. (C)

ANNOTTO — A species of yellowish red dyeing material, used in connection with other dyes for staining leather brown. (C)

ANTIMONY — A brittle white metal much used in compounding alloys. (C)

AWL — A pointed tool for piercing leather in preparation for the stitches. (O)

B

BACK BAND – (1.) That portion of the harness attached to the gig saddle under the jockey, and used as a support for the shaft tugs. It is sometimes called a *tug bearer*. (2.) Also the band used instead of a pad on common harness. (3.) The English use the term synonymous with *ridge-tie*. (C & O)

BACK STRAP – (1.) In many areas this term is synonymous with turnback, either the single or double variety. (2.) Also the band used instead of a pad on common harness, in which case the term is synonymous with *back band*. (C & O)

BALL MOUNTING – A harness mounting having a ball at the point where the ring is attached to the base. (C)

BAND MOUNTING – A harness mounting, the ring of which is broad and flat, with square edges. (C)

BASE – The portion of a terret or other harness mounting which rests upon the leather, and to which the bolt and ring are attached. (C)

BEAD – The small roll on each side of a gig saddle housing. (C)

BEARER LOOP – A metal loop attached to a gig tree for attachment of a tug bearer. (O)

BEARER, TUG – That portion of the harness attached to the gig saddle under the jockey, and used as a support for the shaft tug. It is also called a *back band*, and may be one continuous strap, hanging down on both sides, or two separate straps attached to the gig tree just below the jockeys.

BEARING CHAIN – A term, used largely by the military, synonymous with *fifth-chain carrier*.

BEARING REIN – Same as *check rein*. (FR)

BELL STRAP – A leather strap to which small bells are attached, used in connection with sleigh harness. (C)

BELL THROAT BAND – A band attached to the bridle in the same manner as the throat latch, to which are fastened a number of small bells. It was used to a considerable extent throughout Europe, but there is no evidence that it was extensively used in this country. (C)

BELLY BAND – The strap that encompasses the belly of a horse, to which the billets of a pad or gig saddle are attached. In some areas the term is applied both to the band securing the gig saddle, and to the band which attaches to the shaft tugs. In other areas, belly band is applied only to the latter, while the term girth is applied to the former. (C & O)

BELLY BAND BILLET – A billet attached to a pad, or to a trace if no pad is used, to which a belly band is secured. (O)

BELLY MARTINGALE – Same as *choke strap*.

BEVELER – A tool resembling a creaser, but forming a deeper, wider groove for ornamentation. (O)

BILLET – A short strap punctured with holes, attached to various parts of a harness. It fastens into a buckle to connect different straps and portions of the harness. (C)

BIT – The part of the bridle that is inserted in the horse's mouth, whereby the rider or driver may, through the reins, guide and control the horse. It consists of a mouthpiece, to the ends of which are attached the bit cheeks. On some types there are no cheeks, except for the side rings. (C)
While the following types of bits are only a portion of the many variations used, these are some of the better known.

BALDWIN – A patent bit having two mouthpieces, the inner one working on levers. It is believed to possess superior advantages over other bits for driving horses that are hard to control, or have bad or vicious habits. (C)

BAR – A general term for a solid mouthpiece bit not having lever action.

Bits (continued)

BARREL HEAD — A bit having a mouthpiece with loose rings. The mouthpiece, however, instead of being attached to the rings with simple eyes, has long barrellike eyes, with the rings running through from top to bottom.

BEN LANE — A half cheek snaffle, having a loop upon the upper side of the ring for receiving the cheek strap of the bridle, by which the bit is held in position. The ring is slightly curved upon the upper and lower sides, and straight at the point where the reins are attached; the cheekpieces have a small loop upon each, the same as those on the well-known *Dan Mace* bit. (C)

BENTINCK — A severe riding bit having a large movable port which, when in the horse's mouth, stands above where the port is ordinarily placed. This bit is but little used. (C)

BRADOON — A small bit having loose rings, and either a stiff or jointed mouth. It is often used as a second bit, for the bearing reins in a curb bridle. Gag bradoons have loops on the rings, through which a strap is passed, the ends of which are connected with the reins; when made in this way they are very

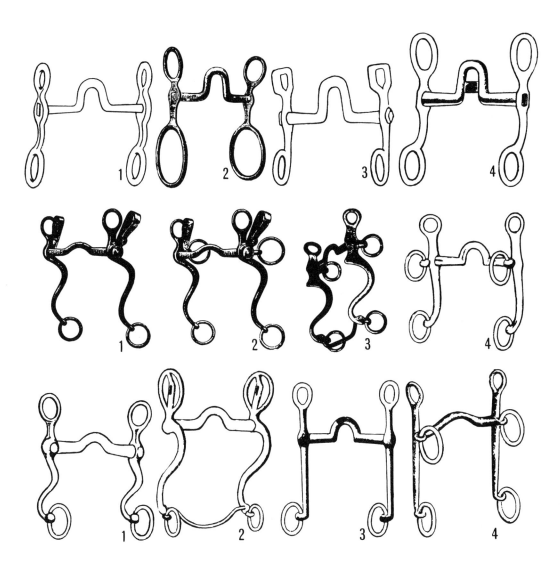

Bits. (Smith, Worthington & Co., New York, 1901.) Left to right, *top row*: Junker, port; globe cheek, port; Indianola; port, with roller. *Middle row*: Whitman, 2-ring; Whitman, 4-ring; military, 4-ring; Pelham, sabre cheek. *Bottom row*: U. S. Standard Cavalry, 2-ring; U. S. Shoemaker; port, 2-ring; port, 4-ring, with lip strap loop.

Bits (continued)

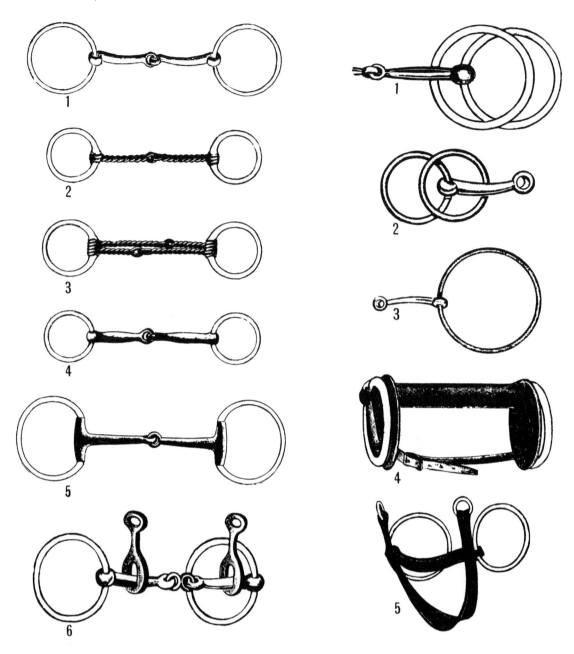

Bits. (Smith, Worthington & Co., New York, 1901.) Top to bottom, *left*: Bradoon; Bradoon, single twisted wire mouth; Bradoon, double twisted wire mouth; Bradoon, Weymouth; barrel head, race; Rockwell. *Right*: Wilson; Wilson; Bradoon, watering or race; stalker or charter oak, leather mouth; humane, leather mouth.

severe, and are used only for riding bridles. Link and T bradoons are bits having links or Ts attached to the rings, by which the strap known as the head collar is attached; they are used only for military purposes. (C)

BREAKING SNAFFLE — A very heavy jointed bit, the sections of the mouthpiece being joined by a ring on which are hung three or more players. (C)

BUXTON — A popular English style of driving bit, the peculiar feature being a straight section on the cheek piece, upon which the mouthpiece slides, and by which its severity is increased or diminished at the will of the driver. The mouthpiece is stiff, with or without a port, and the cheeks are made in a variety of fanciful patterns. There is also a bit of the same name without the sliding mouth. (C)

Bits (continued)

CAVALRY — A bit having an extended S cheek piece, with a loose ring at the lower end, and a loop above the mouthpiece for receiving the bridle cheek. The cheeks and mouth are solid, the latter being made with or without a port. (C)

CHAIN — A bit having a chain mouthpiece. (C)

CHECK — A slightly curved bar bit, with a small ring in each end.

CHIFNEY — A severe curb bit having a short, movable arm attached to the cheek just above the mouthpiece (in addition to the stationary arm) for receiving the cheek strap of the bridle; the gag chain or strap is attached to the stationary upper arm of the cheekpiece. Also see *Hanoverian Chifney*. (C)

COACH — A bit having a large scroll or straight cheek made stationary on the mouthpiece, loops for the driving reins being placed at different distances from the mouthpiece, by which the leverage is increased or diminished. The mouthpiece is always stiff, with or without a port, and the lower ends of the cheeks are connected by a small bar. (C)

CRESCENDO — An American style of bit having the rings hinged to the mouthpiece. The mouthpiece is made of two or three bars, and may be stiff or jointed. If jointed, the joint is made like a hinge.

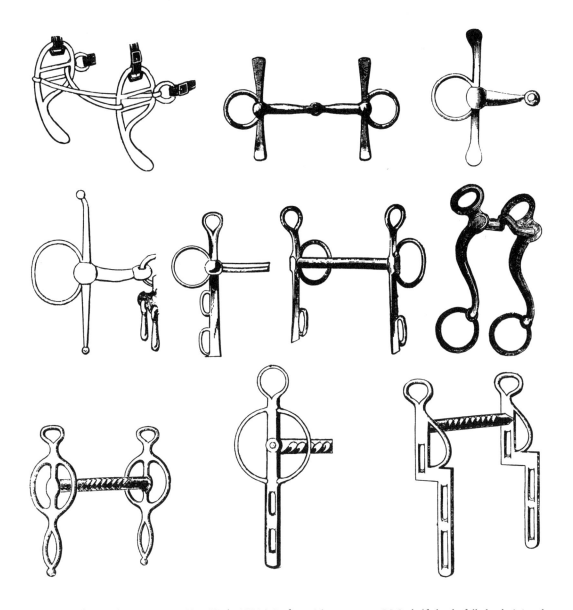

Bits. (Smith, Worthington & Co., New York, 1901.) Left to right, *top row*: J.I.C., half cheek; full cheek, jointed; full cheek, jointed. *Middle row*: full cheek, colt, with drabs; Hanoverian, 2-loop, loose cheek; Hanoverian, 1-loop, stiff cheek; port, racking. *Bottom row*: Liverpool, 1-loop; Liverpool, 2-loop; elbow.

Bits (continued)

CURB — A general name applied to all bits having cheeks, either straight or curved, extending above and below the mouthpiece, with rings or loops upon the lower arms for receiving the driving reins. The upper arms have loops by which the bit is attached to the bridle, and to which the gag chain or strap is attached. (C)

DAN MACE — A popular driving half cheek snaffle. The straight cheeks have small loops which receive a strap running from one cheek to the other, which when taut serves to cause them to act together. (C)

DEXTER — A bit having a mouthpiece that is very broad at the cheeks and thin in the center; the mouthpiece may be stiff or jointed. The most popular pattern is the half cheek snaffle, but they are also made with full snaffle cheeks, or with rings. (C)

DOUBLE JOINT — A bit having two mouthpieces, both of which are jointed; the sections of each are of unequal length, so that a short and long section is attached to each cheek. When the joints are sprung the mouthpieces present the appearance of the letter W, by which term the bit is also known. It is one of the most severe bits in use. (C)

DRENCHING — A bit inserted into a horse's mouth for the purpose of giving the animal medicine. The mouthpiece is hollow, with a hole toward the rear, and one of the outer ends is fitted with a funnel, into which the bottle of medicine is placed. The horse's head is then drawn up by means of a rope and pulley, the funnel is inverted, and the medicine runs out of the hole in the mouthpiece, and down the horse's throat.

DUMPY — A curb bit having the length of the lower arm of the cheek reduced so that it is but a trifle longer than the upper arm. (C)

EGG BUTT — A bit similiar to the *barrel head bit,* except that the eyes are generally shorter, and more egg shaped. The two terms are sometimes used synonymously.

ELBOW — A bit with a flat bar cheek which runs downward a short distance, turns back at a right angle, and then turns a second time to proceed downward again. It has loops above for attachment of the bridle cheeks, and slots below for the reins.

FULL CHEEK — A bit with a straight cheek attached to the ring, half above and half below the ring.

GAG — Same as a *curb bit*.

GAG BRADOON — A bit like a *gag snaffle,* except that there are only rings, and no cheeks. (C)

GAG SNAFFLE — A bit having small loops on the top and bottom of the side-ring, through which a round leather strap passes that is attached to the rein, producing a pulley action, and one that is very severe if desired. (C)

GIG — A bit the same type as the Hanoverian, except that there is only one loop on the cheek. It is sometimes called a short Hanoverian, and also a stiff Buxton. (C)

HALF CHEEK — That portion of a bit like the full cheek, except that it has only the lower portion of the cheek.

HANOVERIAN — A straight or curved cheek bit with two or more loops on the lower arm, a ring at the mouthpiece, and a loop at the end of the short arm for attachment to the bridle cheek. (C)

HANOVERIAN CHIFNEY — This bit differs from the regular Hanoverian by having short movable arms attached, in the same manner as the Chifney bit; the curb chain is attached to the stationary cheeks. (C)

LIVERPOOL — A popular English style of driving bit, characterized by a flat bar cheek with a loop at the top for the bridle cheek, and an integral ring around the end of the straight mouthpiece to which the rein and line is attached. Slots in the lower branch of the cheek accommodate the line if a curb chain is used.

LOLLING — Same as *tongue bit*.

Bits (continued)

MARTIN — A stiff bar bit, having a spoon-shaped port, from four and three-fourths inches to six inches long and one-half inch wide; the top of the port is convexed and polished. When in use, this long port rests against the roof of the horse's mouth, and when necessary it becomes unusually severe, because of the powerful leverage; yet it does not injure the mouth of the animal as much as other port bits, owing to its large surface and smoothness. This port may be used on any bit with a lever cheek. (C)

MEDICINE — Same as *drenching bit*.

MEXICAN — A stiff cheek bit, having a high port, to which is attached a large ring which, when the bit is in the horse's mouth, encircles the jaw. The cheeks are long, and have rein rings at the lower ends; they are also wide and very ornamental, and chains and small drops are attached to various parts for the purpose of ornamentation. It is the most severe bit in use. There are many styles of Mexican bits, but they all possess the general peculiarities of the one described above, although in some cases the ring is omitted. (C)

MOUTHING — Same as *breaking snaffle*.

PELHAM — A hybrid bit, used either for driving or riding, having cheeks like a curb bit, and either a jointed or stiff mouthpiece, with or without port. There are many varieties of this bit, but the most common style is the *half-moon,* having a slightly curved mouthpiece, while the cheeks have loops above for attachment to the bridle, rings at the mouthpiece, and loose rings at the lower ends. Generally there are small eyes near the centers of the lower branch, for use with a lip strap. The upper reins are attached to the rings at the mouthpiece, and the lower reins to the loose rings of the lower bars, serving as a curb. This bit is often used with a double cheek bridle, a bradoon bit being attached to the other pair of cheeks. (C & O)

PORT — A general name for all bits having a port mouthpiece. (C)

RACING — A jointed ring bit, made very light, the loose rings varying in size from three to six inches. (C)

REARING — A bit having a curved mouthpiece which forms the flattened side of a ring, to each side of which are attached side rings, while on the lower side is another ring of the same size, into which the martingale strap is buckled to prevent the horse from lifting his head when rearing. (C)

REVERSIBLE MOUTH — A patent bit having a rule joint (like the hinged rule of a carpenter); when in one position it works the same as a jointed mouth bit, but when reversed, it becomes a stiff mouth bit. (C)

RING — A name given to all bits having a ring cheek, whether loose or otherwise. (C)

RUBBER MOUTH — A bit having an ordinary iron mouthpiece that is covered with hard or vulcanized rubber to protect the animal's mouth. (C)

SHERIDAN — A patented bit having a tubular mouthpiece. The eyes are secured by one end being put into the tube and fastened by a rivet; between the ends of the mouthpiece and the cheeks there are patent leather washers, serving the double purpose of ornaments, and guards to prevent the cheekpieces from bearing against the mouth. (C)

SNAFFLE — This name is given by American makers to all bits having cheek and side ring in one piece, the straight cheeks serving to prevent the rings from being drawn into the horse's mouth. They are designated as jointed, stiff, twisted, or double-mouth snaffles, according to construction. In some areas any bit having a jointed mouth is called a snaffle, regardless of the form of the cheek. (C)

TONGUE — A bit having a stiff mouth, to which is attached a plate or shield, so placed as to prevent the horse from getting his tongue over the mouthpiece. (C)

TWISTED — A bit having a mouthpiece made of square stock, and afterwards twisted. (C)

TWISTED WIRE — A jointed mouth bit having either a single or double mouthpiece made of twisted wire.

WALKER — A milder form of Mexican bit, but retaining the characteristics of the Mexican type. (C)

WILSON SNAFFLE — A jointed mouth ring bit, the rings of which are loose and made in the usual manner. Upon the mouthpiece, however, there are two loose rings, into which the cheek straps of the bridle are buckled. (C)

BIT STRAP — A short strap used to attach the bit to a short cheek bridle, or to a halter. (C)
BITTING HARNESS — A harness used for breaking colts, also known as *breaking harness*. The most common variety consists of a plain halter, a girth with wooden uprights set at an angle in the center of the pad, and leather reins with rubber web ends. Rings are attached to the uprights and to the sides of the girth, into which the ends of the reins are buckled after being passed through the rings of the bit. (C)
BLACKING — A liquid or paste used to give black color to leather. (C)
BLANKET — A covering of wool or other fabric. (C)
BLIND — A leather, or metal and leather, shield attached to the cheekpiece of a bridle to prevent the horse from seeing objects at his side. It is frequently made in an ornamental form, or decorated with ornaments. Also called a *blinker* or *winker*. (C)
BLIND BRIDLE — A bridle having blinds attached to the cheekpieces. (C)
BLIND STITCH — An ornamental stitch placed upon the outside cover of a blind or strap, the underside of which is covered by the lining. Also a stitch that is shown on one side of the leather only. (C)
BLINKER — Same as *blind*. (C)
BODY — The main strap of a breast collar or breeching, consisting of a heavy fold, or a single strap with a narrow layer strap stitched to it. Also the section of a turnback at the point where the hip straps cross. (C)
BOLSTER — The cushion or padded part of a saddle. (C)
BOLT HOOK — A check rein hook. The base by which it is attached sits flat upon, or passes under, the plates of the saddletree, and it is secured by a bolt. Also called a *check hook*. (C)
BOLT PIECE — The short piece used to connect the throat and chin straps of a halter. It is so called because of a metal bolt which passes through the lower end, and supports the halter strap ring. (C)
BOOT — A covering for any portion of a horse's leg, made of leather, or of felt and leather. (C)
BOSAL — The nose loop of a hackamore.
BOSS — A metal ornament placed upon the cheek of a bit. (C)
BOW — Two pieces of wood joined, or an iron, shaped as a bow to give form to the front of a saddletree. (C)
BRACE — A strap used to give additional support to straps that cross one another. (C)
BREAKING HARNESS — Same as *bitting harness*. (C)
BREAST APRON (or CLOTH) — The small horse blanket used to cover the breast of a horse.
BREAST CHAIN — A short chain, the ends of which are attached to the hame breast rings. It supports either a pole chain or strap, or the end of a neck yoke, and serves to both guide and hold back a vehicle. (C)
BREAST COLLAR — A leather strap that encircles the horse's breast, supported by neck straps. The traces are attached to the ends of the collar. It is the oldest style of horse collar in use. (C)
BREAST COLLAR IRON — A metal fitting extending across the front of a breast collar, standing several inches in front of the collar. The choke straps and pole straps are attached to this iron when double harness is used. It is sometimes called a *yoke iron*.
BREAST GIRTH — A strap or piece of web attached to the front ends of a blanket, or to the roller, to keep the ends in position across the breast. (C)
BREAST PLATE — A leather or web stay used to prevent a riding saddle from sliding backward; it may be attached to the martingale by rings, or be made in one piece and used without the martingale. (C)
BREAST STRAP — A heavy strap used in the same manner as the breast chain. (C)
BREAST STRAP ROLLER — A fitting with a roller on one end and a hook on the other, for the purpose of attaching the breast strap to the neck yoke ring or the pole chain. The roller protects the breast strap from wear. (Y)
BREAST STRAP SLIDE — A fitting which slides on the breast strap, to protect it from wear at the point where the neck yoke ring or pole chain attaches. (C & Y)
BREAST YOKE — A part resembling a short singletree, which attaches to the neck yoke. The side backers attach to the breast yoke, and serve to back, or hold back a vehicle. (Y)
BREECHING — That part of the harness which encircles the breech of a horse, consisting of a breeching body and hip straps. Its purpose is to enable the animal to back, or hold back the vehicle. (C)
BREECHING BODY — The main portion of a breeching, fitting around the breech of a horse. It is a broad, heavy strap, folded over, or sometimes with a second narrower strap stitched on over it. It is sometimes called the breeching seat. (C)
BREECHING DEE — A D-shaped ring, one of which is attached to each end of the breeching body. (C)
BREECHING, JERSEY — A variety of breeching employing a sort of back band, to which the hip straps are attached. Double turnbacks are used with this breeching.

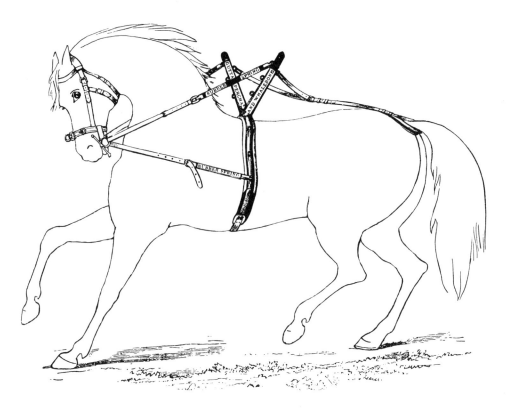

Blackwell's bitting harness cost $70. (Smith, Worthington & Co., New York, 1901.)

Plain breast-collar harness (with housing); commonly used for COACHES and CHARIOTS, especially if the draught is light. Not so good for heavy draught, but is lighter, and easy to put on and take off. (From William Felton's *Treatise on Carriages,* 1796, vol. 2, plate 50)

BREECHING RING — A full ring, used for the same purpose as the *dee*. (C)

BREECHING SEAT — Same as *breeching body*.

BREECHING STAYS — The pair of curved leather pieces attached to the lower end of a hip strap, for supporting the breeching body at two points from a single hip strap.

BREECHING STRAP — A strap attached to the breeching ring, and extending forward to the trace buckle, or to the hold back on a shaft. Also called a *holdback strap*. (C)

BREECHING TRACE BEARER — A small ornamental drop attached to the breeching body, acting as a support to the trace. (C)

BREECHING TUG — A short tug with buckle, attached to the body of the breeching, by which the hip strap is attached. (C)

BRIDGE — A metal loop or arch attached to a harness, generally to a pad, through which a strap passes for support. Double turnbacks are often so supported.

BRIDLE — That part of the harness which is used on the horse's head in connection with a bit, by which the animal is controlled; they are properly divided into two classes, namely, *harness* and *riding*, the latter being without blinds. Those for harness are generally designated by the width of the cheekpiece, as one-half inch, five-eighths inch, three-fourths inch, seven-eighths inch, one inch, etc. The basic parts of a bridle are: the *crownpiece*, the part passing over the top of a horse's head; the *bridle front* or *browband*, the strap passing over the horse's forehead, connecting at both sides with the crownpiece; *cheekpieces*, the side straps, attaching above to (and sometimes integral with) the crownpiece, and below to the bit; *throat latch*, the strap extending from the ends of the crownpiece, and encircling the horse's throat; *nose band*, the strap over the top of a horse's nose, extending from one cheek to another just above the bit; and the *reins*, being the straps extending from the side rings of a bit, for controlling the horse. (C & O)

DOUBLE CHEEK — A riding bridle having two pairs of cheeks, to which are attached a gag and bradoon, each being controlled by a separate pair of reins. (C)

PORT — A bridle using a port bit; either single or double, according to the number of reins used. (C)

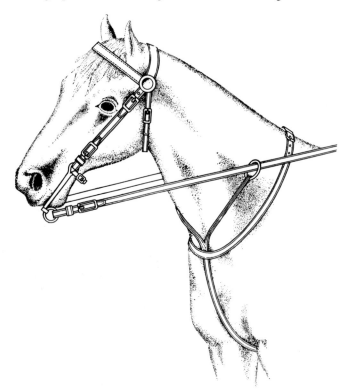

English bridle with martingale. (The Strecker Bros. Co., Marietta, Ohio, 1921.)

Bridles (continued)

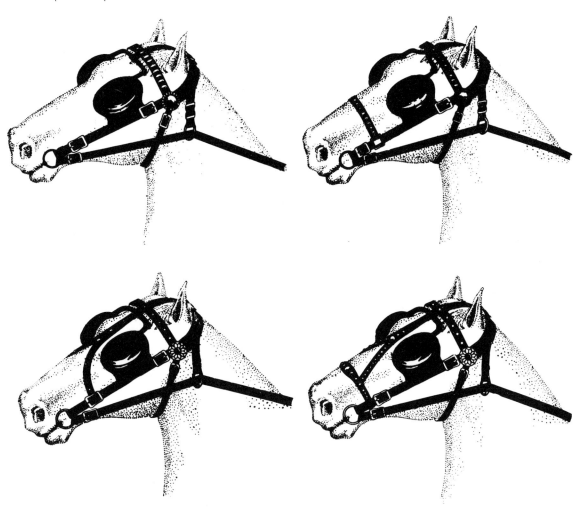

Team bridles. (The Strecker Bros. Co., Marietta, Ohio, 1921.)

Riding bridles, the one on the right being the Texas pattern. (The Strecker Bros. Co., Marietta, Ohio, 1921.)

Bridles (continued)

> ROUND — A bridle made with round straps throughout. Under this heading may be classed a great variety of bridles, full or partially rounded. (C)
>
> SHORT CHEEK — A blind bridle having short cheekpieces with rings at the lower ends, into which the bit strap is buckled. (C)
>
> SNAFFLE — A riding bridle with a snaffle bit, and generally, single reins. (C)

BUCKLE — A metal rim having a straight bar upon which is placed a loose piece called a tongue, which is passed through a hole in a strap to hold it in a fixed position. These, like other harness mountings, are divided into numerous classes, of which only the general names are given below:

> CENTER BAR BUCKLE — A buckle having the bar to which the tongue is attached placed at or near the center, both ends being finished alike. (C)
>
> CONWAY BUCKLE — A type of center bar buckle made without a tongue, but having an upright post on the center bar, made integral with the buckle. This buckle is not sewn to a strap, but both portions of the strap are held by slipping them over the center post.
>
> HALF BUCKLE — A buckle without a center bar, the tongue being attached to one of the side bars.
>
> HARNESS BUCKLE — A general name for a buckle designed to be used with various parts of a harness. (C)
>
> LOOP BUCKLE — (1.) A pattern of buckle having a metal loop back of the tongue bar, to hold the loose end of a strap after it passes through the buckle. (2.) A buckle having a metal loop on the side bar by which a strap may be attached. (C)
>
> ROLLER BUCKLE — One having a loose roller on the top bar, to ease the passage of a strap through the buckle. (C)
>
> SCREW BAR BUCKLE — A buckle having a machine screw for the bar on which the tongue is mounted. This type of buckle is used for repairs, as it can be put on a strap without the necessity of putting in new stitching.
>
> TRACE BUCKLE — A heavy buckle designed to be used on traces only. They frequently have loops on both sides, for attachment of other straps. (C)
>
> TRACE LOOP BUCKLE — A heavy center bar buckle having small loops upon the side bars. The belly band and market strap attach to these additional loops. (C)
>
> WEDGE BUCKLE — A patented buckle which permits the strap to lay straight, it being held by a wedge with a spur, instead of by a tongue. (C)

BUFF LEATHER — A light colored, oil-tanned leather, soft and tough, made originally from the hides of buffaloes, but now often made of cattle hides. Driving reins are sometimes made of this leather. (C)

BURNISHED MOUNTING — One to which the luster is imparted by a steel burnisher instead of being polished in other ways. (C)

BURNISHER — A tool with a hard, smooth, rounded surface, used for burnishing or polishing metal mountings.

BUTT CHAIN — Same as *heel chain*.

BUTT STRAP — See *spider, mine*.

C

CACOLET — A pack saddle that is similiar to a horse-litter, used to transport sick, wounded, or feeble persons. The passengers generally sit in folding chairs attached to the sides of an animal, though in some instances a litter might replace one of the chairs. The cacolet, reportedly developed by the French during the Crimean War, was used to a very limited extent during the American Civil War. A few were built by the famed carriage builders, Lawrence, Bradley and Pardee, of New Haven, Connecticut. (O)

CANTLE — The hind bow or raised rear portion of a saddle tree. (C)

CAP — A small crescent-shaped piece of hard leather placed on the top of a collar. (C)

CARRYING STRAP — A short strap running from the hame to a side backer, to carry the latter. Also called *yoke strap,* and *jockey strap* or *chain.*

CART HARNESS — A heavy single harness used for drawing carts. It consists of an ordinary bridle, hames, collar, breeching, and a specially constructed cart saddle which bears the weight of the load on the horse's back. (C & O)

CART SADDLE — The heavy saddle used with cart harness. It is made of wood; running across the top, and in the center of the saddle, is a channel, in which rests the chain that supports the shafts. (C)

CAVESON — A portion of the apparatus used in breaking a colt. The main feature is the nose piece, which is buckled around the nostrils, having a long rein attached by which the colt is controlled from the ground until he becomes accustomed to the bit. (C)

CHAFE — A flap of leather, generally cut in an ornamental fashion, attached to and behind a harness at any point where buckles or other metal fittings are found. The purpose is to protect the body of a horse from the metal fitting. Also called a *safe.*

CHAIN PIPE — A leather tube, two and a-half to three feet long, which covers a trace chain to prevent the chain from chafing the body of the animal.

CHAPE — The short piece of strap by which a buckle is attached to any part of a harness. The term is also applied to the loop close to a buckle, through which the end of a strap is passed after it goes through the buckle. (C)

CHECKER — A tool resembling a small double creaser, used to ornament a piece of leather by checkering. The one point of the tool is run in the last line made, thus serving as a guide to keep the lines parallel and equally spaced. (O)

CHECK HOOK — Same as *bolt hook.*

CHECK LINE — Same as *lead rein.*

CHECK REIN — A short rein attached to each cheek of the bit, and extending back to the saddle or pad, where it is held by a hook. Its purpose is to hold the horse's head high. (C)

CHECK STRAP — Same as *rein strap.*

CHEEK — That portion of a bit outside of the horse's mouth. (C)

CHEEKPIECE (or BRIDLE CHEEK) — The piece of the bridle extending from the crownpiece, down the sides of the horse's head to the cheek of the bit. (C)

CHIN CHAIN — A short chain attached to the rings of a bradoon bit, to which the coupling strap is attached. The term is used especially by the military.

CHIN PIECE — The strap attached to a halter or bridle, passing from cheek to cheek under the horse's jaw. Its position is opposite that of a nose band. Also called a *chin strap.* (C)

CHOKE STRAP — The strap running from the belly band, between the front legs, to the collar of an animal. In the United States it is sometimes called a martingale. This strap may also run directly to a neck yoke, in which case an additional strap, known as the collar strap, runs from the collar down to the choke strap, supporting it by means of a ring or loop.

CINCH (or CINCHA) Same as *girth*. It is often made of web or braided horsehair.

CLIP — See *hame clip*.

CLOSE PLATING — Silver plating done with thin sheets of silver. The sheet is carefully formed around the object to be plated, then sweated, or soldered in place. Harness mountings are sometimes plated in this manner. (C)

COACH HARNESS — The heaviest kind of double harness, used for pleasure driving. It consists of bridles, round collars, hames, traces, pads, and breechings. (C)

COCKEYE — A metal attachment having an eye at one end, and arms spreading to a loose or stationary bolt, by which it is attached to the end of a strap, particularly to a trace. It is used as a connecting link where a short chain is attached to a strap. (C)

COCK HORSE HARNESS — A simplified harness used with a riding saddle on the extra horse that is sometimes necessary to assist the regular team on a steep hill. Generally it consists of a bridle, collar and hames, traces, turnback, crupper, hip straps, and butt straps to support the singletree. (FR)

COLLAR — That portion of the harness against which an animal exerts his weight to move a load. Collars are of two types: breast collars, which pass around the chest and have traces directly attached; and round collars, resting against the neck, and supporting the hames to which the traces are attached. In the trade, however, the term collar refers to the round, or ordinary variety, unless otherwise specified. (C) The term is also applied to the neck strap used with a martingale.

COLLAR FOREWALE — An English term for the front roll of a horse collar.

COLLAR, KAY — An ordinary collar wherein the lining on the back side of the collar is attached in such a way that the front roll does not show from the rear. It is named after the inventor. (FR)

COLLAR LOOP — A round ornament used as an assistant stay to the back strap or the gig saddle of a single harness. (C)

COLLAR PAD — A pad of leather, felt, metal, or other material, placed underneath the top of a collar to provide greater comfort for the horse. The term is sometimes applied to a *sweat pad*.

COLLAR, RIM — An ordinary collar wherein the roll in front is visible from the rear. It differs from the Kay collar in this respect, for the latter is formed smoothly on the rear side, so that the roll does not show. (FR)

COLLAR STRAP — A strap fastened around the lower part of a collar, having a ring at the lower end to support the choke strap when the latter runs directly to the neck yoke. Also the neck strap used with a martingale.

CONCORD HARNESS — A trade name given to a class of heavy harness made in Concord, New Hampshire, by James R. Hill & Company; it is used with Concord coaches, and other types of vehicles. By 1871 the term was used as a trademark. (C)

COUPE HARNESS — A heavy single harness used for pleasure driving; it is one of the most showy styles used. (C)

COUPLING STRAP — This term is especially used by the military, and means the same as *tie strap*.

CREASER — A piece of wood, bone, or metal having the end shaped as a reverse bead, used for the purpose of creasing (ornamenting the edge with a line) the edges of leather straps. It is sometimes called a *slicker*. (C)

CREASER, ADJUSTABLE (or SCREW) — A double pointed creaser having a screw adjustment to vary the distance between the points, the one point serving as a guide along the edge of a strap being creased. (O)

CREST — A harness ornament made of metal, on which is stamped or cast a coat of arms or other heraldic device, monogram, or initial letter. A crest is commonly used on the center of a blinker. (C)

CREW PUNCH — A harnessmaker's punch that cuts an elongated hole for a buckle tongue. (O)

CRIB MUZZLE — A muzzle used on a horse to correct the habit of cribbing, or wood gnawing. (C)

CROSS REIN RING — A plain or ornamental ring used on the inside driving reins of a double harness, at the point where the reins cross each other. (C)

CROTCH BREECHING — A breeching employing crotch straps, or double turnbacks.

CROTCH STRAPS — Same as *double turnbacks*.

CROWNPIECE — That portion of a halter or bridle that rests upon the horse's head, to which the cheekpieces are attached. (C)

Fine COACH harness. (The Strecker Bros. Co., Marietta, Ohio, 1921.)

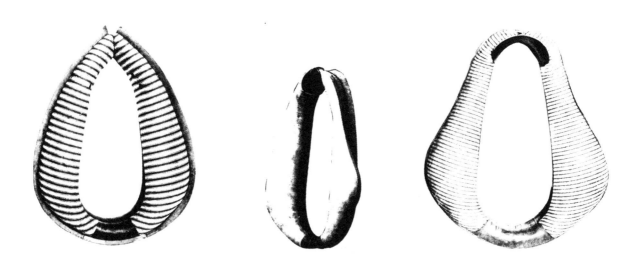

Irish collar, Concord style; half Sweeney collar; full Sweeney collar. (The Perkins-Campbell Co., Cincinnati, Ohio, 1925.)

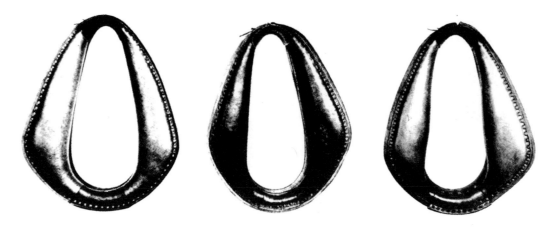

Imitation Scotch collars. (The Perkins-Campbell Co., Cincinnati, Ohio, 1925.)

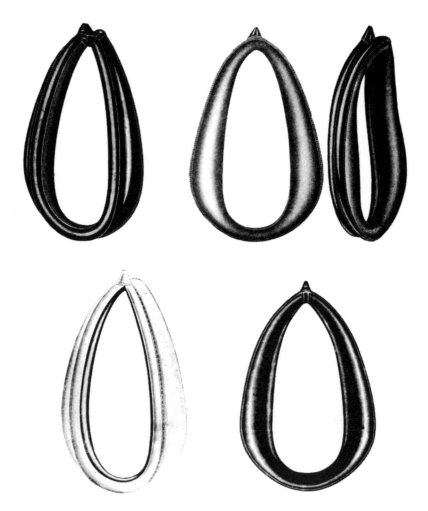

Upper row: Imitation case collars; *lower row:* coach collar and Kay collar. (The Strecker Bros. Co., Marietta, Ohio, 1921.)

English COUPE harness. (The Strecker Bros. Co., Marietta, Ohio, 1921.)

CROWNPIECE LAYER — A short layer used upon a crownpiece of a coach bridle. The gag runners are often attached to the ends of this part. (C)

CROWN SOAP — Soft soap, made of whale or cod oil and lye of potassa, with a small amount of tallow added after the lye has cut the grease. (C)

CRUPPER — A strap that is attached to the back of the saddle, and to the dock in the rear. Sometimes the term applied to the entire assembly — the turnback, fork, and dock; at other times it is applied to the dock. (C)

CRUPPER FORK — A strap with two branches which buckle into the crupper dock.

CRUPPER LOOP — A metal loop attached to the back of a riding or gig saddle, to which the crupper is attached. (C)

CURB CHAIN — A small chain extending from cheek to cheek of a bridle bit, by which the horse's jaw is severely cramped when the driving reins are pulled. (C)

CURB HOOK — A small metal hook attached to the upper arm of a bit cheek to hold one end of a curb chain.

CURB STRAP — A leather strap used for the same purpose as the *curb chain*. (C)

CURRICLE BAR — A bar running from the pad of one horse to the pad of his mate, to support the pole of a curricle. The two arms were constructed in a manner that allowed them to move freely in and out of the main body of the bar, so as to permit the movements of the horses without these movements being transmitted to the pole. (O)

D E

DOCK — That part of the crupper which encircles the horse's tail. (C)
DRAB — Same as *player*.
DRAFT EYE — That portion of a hame to which the trace is attached. (C)
DRAW GAUGE — An adjustable tool with a small blade used to cut leather straps to a uniform width. (Y)
DROP — An ornamental leather pendant hanging from some part of a harness, such as the face drop which rests on the forehead between the eyes, the kidney drop, or the martingale drop. It is sometimes called a *frog*.
DUTCH COLLAR — Same as *breast collar*. (FR)

EAR DROP — A leather ornament which is also used as a gag swivel. (C)
EDGE TRIMMER — A small cutting tool used to cut away the sharp edge of a leather strap; also a machine made for the same purpose. (C & O)
ELECTROPLATE — (1.)*(v.t.)* To cover iron or other metal with a film of silver, gold, nickel, etc., by electrolysis. (2.)*(n.)* Harness mountings plated by this process. (C)
EMBOSS — To cover the skirts of saddles or other articles with raised figures, produced by tools having the desired device cut in them. (C)
ENAMEL LEATHER — See same in Carriage Nomenclature section. (C)

FACE PIECE — That portion of a bridle which is attached to the crownpiece, and extends down the horse's face, finished with an ornamental end; or, branching off in two parts which are attached to either the bridle cheeks or the bit cheeks. (C)

FAIR STITCH — The stitching which forms an exposed seam, as distinguished from that in which the stitches are sunk into a channel and hidden from view. (C)

FALSE STITCH — An imitation stitch formed by a prick wheel in channels produced by a creaser. (C)

FELT — Cloth made of wool or fur, fulled into a compact mass. (C)

FELT BOOT — A horse boot having a body made of felt. (C)

FELT PLATE — A heavy felt blanket of small size, used as a saddle cloth. (C)

FENDER — A large flap on a stirrup strap, extending toward the rear.

FIFTH-CHAIN — A chain running from the front end of a pole, to carry the spreader and singletrees of the leaders, when more than four animals are employed.

FIFTH-CHAIN CARRIER — A chain running from the inside hame of one swing horse, to the inside hame of his mate, for the purpose of supporting the fifth-chain.

FITTER — A workman who confines himself to fitting up and laying out the ornamental work on harness before it is stitched. (C)

FITTING — That portion of the work on a harness performed by the workman who prepares the various straps for the stitchers. (C)

FLAP — The broad, heavy piece of leather that is attached to the saddle tree, and extends down the sides. It is also called the *skirt*. (C)

FLAP SQUAB — A small squab attached to the saddletree under the flap. (C)

FLY — A small metal disc or other figure hanging from a pivot at the top, used as an ornament in a terret or check rein hook. (C & O)

FLY CAP — A woven cover, or net, worn over the ears of a horse to prevent flies from getting into the ears.

FLY HOOK — A check rein hook having an ornamental top, with a metal figure suspended on a pivot in the top of the ring as an ornament. (C)

FLY NET — See *net*.

FLY TERRET — A harness mounting shaped like the terret, but having a fly suspended in the ring. It is used only as an ornament to the crown piece of a bridle. (C)

FOLD — Various parts of the harness, such as the body to the crown piece, breeching, breast collar, martingale, etc., formed of thin leather, folded with the edges in the center of the strap. (C)

FOLDED HAND PART — The hand part of a driving rein made as a fold, the center seam being covered with a narrow strip of leather stitched on, or sometimes the edges are overlapped and stitched together. (C)

FOREHEAD BAND — Same as *bridle front*.

FRENCH PAD — A pad so shaped that when on the horse's back the sides stand in a horizontal position; this is done by folding it so that the line of the fold will be at a slight angle with the center line of the pad top or plate. (C)

FROG — Same as *drop*.

FRONT — That portion of a bridle which encircles the forehead of a horse; also called the *forehead band*, or *brow band*. (C)

FURNITURE — Generally, the harness of a horse. In England the term is sometimes applied only to the harness mountings. (C)

Leather fly net. (R. S. Luqueer & Co., New York, 1890.)

G

GAG CENTER — The center part of the gag runner, made of leather or metal. (C)

GAG CHAIN — Same as *curb chain*.

GAG HOOK AND EYE — That part of a gag runner by which it is attached to the crownpiece of the bridle. (C)

GAG REIN — The rounded portion of the check rein, which is attached to the cheek of the bit, and slides in the gag runner. (C)

GAG RUNNER — The part of a bridle attached to the crownpiece near the rosette, through which the check rein passes; it is made of leather or metal. (C)

GAG STRAP — Same as *curb strap*. (C)

GAG SWIVEL — A swiveling metal loop used as a *gag runner*.

GALLUS STRAP — See *spider, mine*.

GIG SADDLE — The small saddle used on single harness to support the shafts. There are strap saddles and pad saddles. The former are made with jockeys, housings, and skirts, while the latter have no housings, but have skirts which run all the way up, under the jockeys, the skirts being padded for their entire length. (C)

GIG TREE — The wood or iron frame which forms the bearing portion of a gig saddle. (C)

GILT MOUNTING — A harness mounting covered with a coating of pure gold, deposited by electrolysis. (C)

GIRTH — The strap or web that passes under a horse's belly, by which a saddle or pad is secured on the horse's back. Also called a *belly band*. (C)

GIRTH WEB — A heavy cotton or linen web, woven in such a manner that it will not stretch, used in place of leather for girths. (C)

Strap gig saddle. 1, Saddle; 2, bottom; 3, jockey; 4, housing; 5, bead; 6, skirt; 7, tug bearer, or back band; 8, point strap. (The Strecker Bros. Co., Marietta, Ohio, 1921.)

H

HACKAMORE — A simplified bridle resembling a halter. It is made of rope, leather, rawhide, or braided horsehair, and in place of a bit has a loop or bosal which can be tightened around the animal's nose. It is sometimes used in breaking or training horses.

HALTER — A head stall, usually with a nose band, throat strap and chin strap, made of rope, web, or leather, and used for confining or leading horses. A style known as the bridle halter is made with a front like a riding bridle and is provided with a loose ring at the lower end of each cheekpiece into which the bit strap is buckled. A style known as the five-ring halter has been adopted by the government for military use. (C)

HALTER, NECK — A single strap, secured around the horse's neck, used in place of a regular halter for leading or confining the animal.

HALTER SQUARE — The square ring used in a halter, where the cheek meets the nose band.

HALTER STEM — The long strap, rope, or chain used with the halter. (C)

HAME — A curved piece of wood or metal, two of which are fitted to the collar. The hames are provided with draft eyes, to which the traces are attached. The collar, hames, and traces constitute the main portion of the harness for draft purposes. (C)

HAME BACK STRAP RING (or TURNBACK RING) — The ring just above the draft eye to which one of the double turnbacks or back straps is attached.

HAME BALL — An ornamental metal ball on the top of a hame. These are often of brass, but may be of malleable iron, japanned, tinned, or plated.

HAME, BOLT — A hame having for a draft eye a bolt which runs through a pair of eyes. The bolt serves as the center for a roller, around which the tug is secured.

HAME BREAST RING — A ring near the bottom of a hame, to which the breast strap or chain is secured. Also called a *hame holdback*. (Y)

HAME CHAIN — A short chain used to secure the lower ends of the hames when on the collar, the center having a ring for attaching the pole chain or strap. (C)

HAME CLIP — A metal link attached to the draft eye, and having two thin forged straps by which the hame tug is attached. (C)

HAME DOG — Same as *hame link*.

HAME FASTENER — A mechanical device having some sort of latch for securing the lower ends of the hames. It is used in place of a hame link or chain.

HAME HOLDBACK — The ring to which a breast strap or chain is attached.

HAME HOLDBACK PLATE — A metal plate to which the holdback ring is attached. This plate generally has two holes through which the branches of the hame staple pass.

HAME, JUG HANDLE — A hame having a draft eye and terret resembling a jug handle, rigidly attached to the hame.

HAME LINE RING — A ring near the top of a hame, through which a line passes. Same as *hame terret*. (Y)

HAME LINK — A metal attachment for securing the lower ends of the hames when they are in position on the collar. The ends are provided with hooks or eyes, and the center has a loop or ring for the attachment of the pole chain or strap. Also called a *hame dog*. (C)

HAME, LONE STAR — A style of hame having a ratchet with a series of holes near the bottom, into which a double hook is fastened as a draft eye, enabling the point of draft to be adjusted.

HAME LOOP — (1.) A single metal loop at the top or bottom of a hame, for receiving a hame strap. This loop

Five-ring halter. (The Strecker Bros. Co., Marietta, Ohio, 1921.)

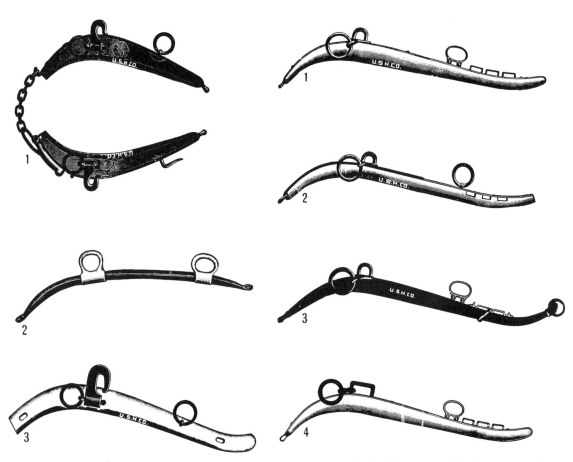

Hames. (The Strecker Bros. Co., Marietta, Ohio, 1921.) *Left*: 1, mine hook; 2, jug terret and draft eye; iron with japanned body; 3, red Richmond lace. *Right*: 1, red clip; 2, Concord varnished clip; 3, tubular steel; 4, black square staple.

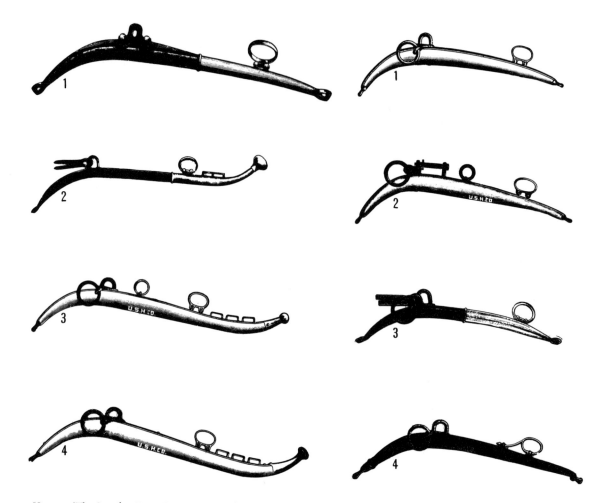

Hames. (The Strecker Bros. Co., Marietta, Ohio, 1921.) *Left*: 1, hollow iron, low top; 2, tubular iron, ball top; 3, red clip; 4, black clip, brass ball top. *Right*: 1, red or black clip; 2, black bolt; 3, tubular iron; 4, tubular clip, low top.

is sometimes called an eye . (2.) Also a metal fitting near the top of a hame, for receiving the upper hame strap. Generally these have two or three openings, so that the strap may be inserted at different levels. In some instances wood hames have notches which take the place of these loops, with a metal plate over the tops of the notches to keep the strap in place.

HAME RING – (1.) A ring made with or without a loop on the side, and attached to the hame by a narrow strap. It is used in place of a hame terret on the inside hame of coach harness. (2.) Generally, any of the various rings attached to a hame, for the purpose of supporting a line or breast chain, for attachment of a turnback, etc. (C)

HAME STAPLE – A heavy iron staple on a hame, serving as a draft eye, and to which a hame clip is attached.

HAME START – One of the eyes through which the bolt of a *bolt hame* passes.

HAME STRAP – A strap used to connect the upper and lower ends of hames, holding the hames in position on the collar. (C)

HAME STRING – A stout thong, of leather or rawhide, used for the same purpose as a hame strap.

HAME TERRET – The rein ring, which is attached to the hame near the top. (C)

HAME TUG – The forward end of the trace, which is attached to the hame by the hame link, having a buckle or loop at the back end for securing the trace. (C)

HAND – A hand's breadth, or four inches, being a system used to measure the height of horses. (C)

HAND-IRON (or PALM-IRON) – A metal device to fit the palm of the hand, and serving the same purpose as a thimble, to push a stout needle through heavy leather. (O)

HAND KNIFE (or SHOE KNIFE) – A leather worker's knife having a square point. (O)

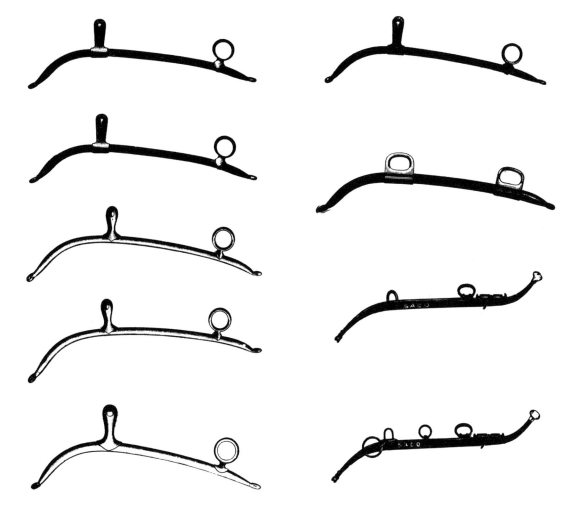

Iron hames, either plated or japanned. The Saco hames are of hollow iron and are japanned. (The Perkins-Campbell Co., New York, 1925.)

HAND PART — That portion of a driving rein handled by the driver. (C)

HARNESS — The complete equipment by which the horse is attached to a vehicle of any kind, consisting of the bridle, collar, hames, traces, pad or gig saddle complete, and breeching. They are classified as single and double; of the single there is the track, buggy, coupe, express or grocer, and cart; of the double there is the light road, light coach, coach, team, farm, stage, and express. (C)

HARNESS BUCKLE — See *buckle*.

HARNESS LEATHER — See listing for same in Carriage Nomenclature section.

HARNESS RING — Any of the small rings used on the various parts of a harness, the large rings being designated by specific names. (C)

HARNESS SNAP — A metal hook having a spring catch, by which the opening in the hook is closed, thus preventing the hook from detaching itself. (C)

HEAD COLLAR — The name by which stable halters are known in the English market. (C)

HEAD COLLAR REIN — The same as *halter stem*. (C)

HEAD KNIFE — A knife with a beak-shaped blade, used by harnessmakers to cut curves, and to cut holes for buckle tongues. (O)

HEAD PIECE — Same as *crownpiece*.

HEAD STALL — The portion of a bridle or halter which encompasses the head. (C)

HEAD TERRET — A terret placed in the center of the crownpiece as an ornament. These are sometimes fitted with a fly, or swinger. (C)

HEEL CHAIN — A short stout chain attached to the rear end of a leather trace.

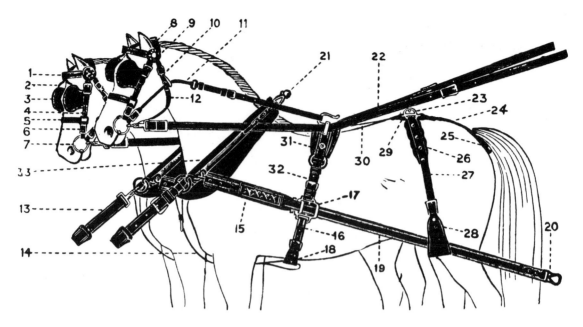

Nomenclature of double team harness. (The Strecker Bros. Co., Marietta, Ohio, 1921.) 1, Bridle front; 2, winker brace; 3, blind; 4, bridle cheek; 5, nose band; 6, bit strap; 7, bit; 8, crown piece; 9, rosette; 10, swivel; 11, round part to side rein; 12, throat latch; 13, breast strap; 14, choke strap; 15, hame tug; 16, belly band billet; 17, three-loop trace buckle; 18, belly band; 19, trace; 20, cockeye; 21, hame; 22, lines; 23, trace carrier; 24, crupper piece and crupper; 25, crupper; 26, safe on hip strap; 27, hip strap; 28, hip strap end; 29, safe under hip strap; 30, turnback; 31, team pad; 32, market strap.

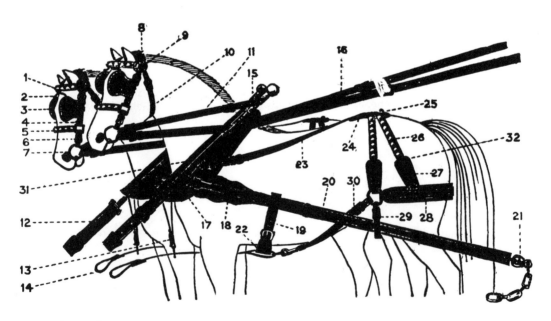

Nomenclature of team harness. (The Strecker Bros. Co., Marietta, Ohio, 1921.) 1, Bridle front; 2, winker brace; 3, blind; 4, bridle cheek; 5, nose band; 6, bit strap; 7, bit; 8, crown piece; 9, rosette; 10, throat latch; 11, side rein (or check rein); 12, breast strap; 13, collar strap; 14, choke strap; 15, hame; 16, line; 17, collar; 18, trace safe; 19, belly band billet; 20, trace; 21, chain; 22, belly band; 23, turnback; 24, safe under hip straps; 25, trace carrier; 26, hip strap; 27, lead-ups on breeching; 28, breeching body; 29, lazy strap; 30, side strap; 31, back strap ring on hame; 32, safe on lead-up.

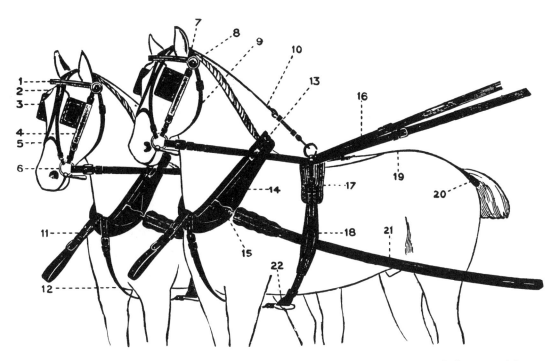

Nomenclature of double buggy harness. (The Strecker Bros. Co., Marietta, Ohio, 1921.) 1, Bridle front; 2, winker brace; 3, blind; 4, bridle cheek (box loop); 5, nose band in overcheck; 6, bit; 7, crown piece; 8, rosette; 9, throat latch; 10, overcheck; 11, pole strap; 12, choke strap; 13, hame; 14, collar; 15, hame clip; 16, line; 17, coach pad; 18, bearer and pad skirt; 19, turnback; 20, crupper; 21, trace (attached to hame); 22, belly band.

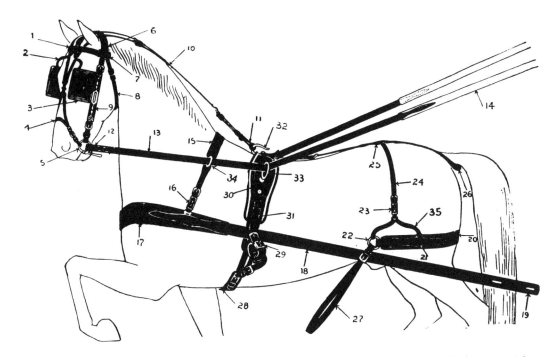

Nomenclature of single buggy harness. (The Strecker Bros. Co., Marietta, Ohio, 1921.) 1, Bridle front; 2, winker brace; 3, blinds; 4, noseband in overcheck; 5, bit; 6, crown piece; 7, rosette; 8, throat latch; 9, bridle cheek (box loop); 10, overcheck; 11, overcheck end loop; 12, spring billet on line; 13, line (front part); 14, hand part on line; 15, neck strap; 16, breast collar lead-up; 17, breast collar body; 18, trace; 19, heel of trace (showing dart hole); 20, breeching body; 21, breeching layer; 22, breeching ring; 23, breeching tugs; 24, hip strap; 25, turnback; 26, crupper; 27, breeching strap; 28, belly band; 29, shaft tug; 30, gig saddle; 31, gig saddle bearer; 32, bolt hook; 33, terret; 34, neck strap terret; 35, breeching stay.

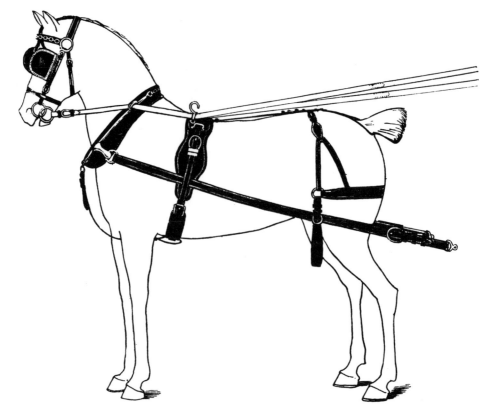

This HANSOM CAB harness cost $130. (Smith, Worthington & Co., New York, 1901.)

Double harness for use with a TRAP or ROCKAWAY—priced at $60. (Smith, Worthington & Co., New York, 1901.)

This four-in-hand (English) road harness sold for $425. (Smith, Worthington & Co., New York, 1901.)

Single DRAY harness, for two-wheel DRAY. (The Strecker Bros. Co., Marietta, Ohio, 1921.)

Single EXPRESS harness. (The Strecker Bros. Co., Marietta, Ohio, 1921.)

Double New York EXPRESS harness—$84. (Smith, Worthington & Co., New York, 1901.)

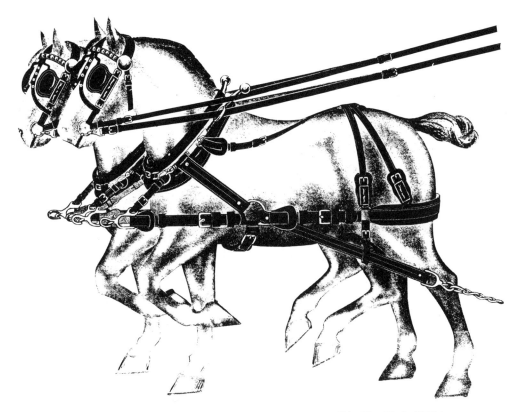

New England TRUCK harness was priced at $147. (Dorries & Co., Inc., Buffalo, New York, 1917.)

City TRUCK harness. (The Strecker Bros. Co., Marietta, Ohio, 1921.)

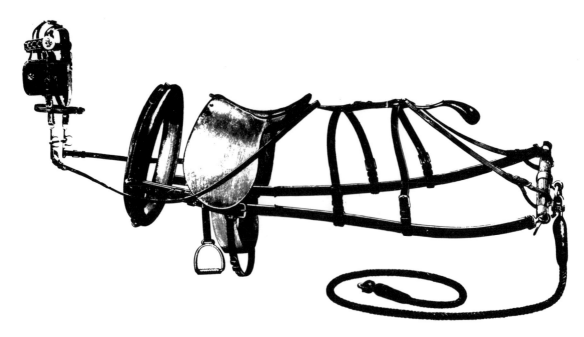

Cock Horse Harness. (From Fairman Rogers: *A Manual of Coaching,* 1901.)

HIP STRAP — That portion of the breeching which passes over the rump of the horse, and acts as a support to the breeching body. Also, where no breeching is used, the hip strap supports the lazy strap, which in turn supports the trace. (C)

HIP STRAP SLIDE — A loose ornament of metal or leather placed upon the hip strap. (C)

HIP STRAP TUG — The short tug attached to the breeching body, into which the hip strap is buckled. (C)

HITCHING WEIGHT — A weight of iron, or sometimes of other metal, carried in a carriage, and set on the ground when there is no post or rail to tie a horse to. The weights vary from ten to fifty pounds, averaging around twenty pounds, and they have a ring on top for securing the horse. They come in a variety of shapes, but are most often a truncated cone, or a hemisphere. They are sometimes called *horse weights*.

HOBBLE — A fetter for horses or other animals when they are turned out to graze. Same as *hopple*. (C)

HOLDBACK, BELLY — A term sometimes applied to the unit consisting of a pair of side straps and a choke strap.

HOLDBACK CHAIN — A chain used for the same purpose as a holdback strap. These chains are frequently used with carts.

HOLDBACK STRAP — A strap attached to the breeching ring, and secured to a shaft to hold back the vehicle. The term is also applied to a choke strap, or to a side backer on double harness, the purpose being the same as above. (C)

HOLSTER — A leather case attached to the fore part of a riding saddle, for the purpose of carrying a pistol. (C)

HOOD — That part of a horse blanket which covers the horse's head and neck. (C)

HOOF BOOT — A leather boot made to fit the hoof of a horse, with an iron shoe attached, used as a protection to the hoof in case a shoe is lost or when the hoof is in a condition that will not permit nails to be driven into it. (C)

HOPPLE — A fetter for horses or other animals when they are turned out to graze. Same as *hobble*. (C)

HORN — The raised piece on the pommel of a saddle, used principally on Texan or Mexican saddles. (C)

HORSE BLANKET — See *horse clothing*.

HORSE CLOTH — A cloth for covering horses. (C)

HORSE CLOTHING — A general term for all kinds of blankets used on horses. A full set consists of the main blanket, the hood, breast cloth, pad cloth, and roller or girth. (C)

HORSE MUZZLE — A wire or leather covering used to cover a horse's mouth, to prevent injury from biting, and as a cure for cribbing. (C)

HORSE NAIL — One of the nails used to secure a horseshoe to the hoof of an animal.

HORSESHOE — A U-shape iron bar fitted to a horse's hoof as a wearing surface, and to prevent injury to the hoof.

HORSE WEIGHT — Same as *hitching weight*.

HORSE WHIP — A heavy thong whip used by teamsters. (C)

HOUSING — (1.) A saddle pad, placed under a riding saddle to prevent chafing. They are often of felt or sheepskin and, unlike the numnah, are much larger than the saddle, so that a fairly large border, frequently ornamented, is exposed. (2.) An ornamental piece of leather or other material placed under a team pad, or under the jockeys of a gig saddle. (C & O)

HOUSING, HAME — A heavy leather cover put on over the hames and collar.

I J K

INTERFERING BOOT — A boot for preventing a horse from cutting the opposite foot while traveling. It is generally made of some soft or elastic material, such as long bristles, India rubber balls, etc. (C)

JACK STRAP — See *side backers*.
JAPANNED AND LINED — A mounting that has been japanned, having the inner side of the rings lined with a lustrous metal. (C)
JAPANNED MOUNTING — A harness mounting that has been coated with japan, as a finish, and to protect the metal from rusting. (C)
JERK LINE — A single line used for driving four or more horses. It is attached to a ring near the center of the lead rein on the near leader, and the horse is turned to the left by a steady pull, and to the right by a number of quick jerks. Also called a *lead line*.
JOCKEY — A short skirt, generally of patent leather, stitched and attached to a gig saddle at the lower edge of the seat. It is sometimes made of iron, integral with the seat. The short skirt of a riding saddle is also called a jockey. (C)
JOCKEY CHAIN (or STRAP) — See *carrying strap*.
JOCKEY SIDE STRAP — See *side backer*.
JOCKEY STICK — A bar of wood or iron connecting with the left hame of the near leader and the right bit cheek of the off leader, for the purpose of transmitting movements of the near leader to his mate.
JOINTED BIT — A bit having the mouthpiece made in two or more pieces, and jointed in the center or elsewhere. (C)

KEEPER — Same as *loop*, meaning the leather variety. (FR)
KEMBLE JACKSON CHECK — A style of overcheck used to compel the horse to hold his head up, and to prevent him from lagging on the bit. It is made of two narrow face pieces, which join at the crown of the the bridle to a single strap, which being carried to the saddle hook, acts as a check. The face pieces branch off and the ends are attached to a small ring bit; the action of the check causes this bit to bear heavily against the upper side of the mouth. It is a most severe and useless contrivance. (C)
KICKING STRAP — A strap used to prevent horses from kicking. It is generally attached at the center of the turnback body, and the ends are secured to the shafts. (C)
KIDNEY DROP — An ornamental leather pendant, often with brass spots, which hangs down from the turn-back about midway between the saddle and hip strap. In cases where no breeching is used, it might also support a hip strap, which in turn supports the lazy strap and the trace.
KIDNEY LINK — A kidney-shaped link, with the concave side open, used to join together the lower ends of a pair of coach hames. On the kidney link hangs the ring through which the pole chain passes. A type of *hame link*. (FR)
KIDNEY STRAP — A term sometimes applied to the forward hip strap, when the two hip straps are nearly parallel, rather than originating at the same point as is generally the case. It is more commonly called a hip strap.
KNEE CAP — A covering for a horse's knee, made of felt or soft leather, having a hard leather shield directly over the knee, to protect the horse from injury when jumping or training. (C)

L

LAYER — A strap, sometimes ornamental, stitched over the seam of a fold, or to a wide single strap body; also the extra piece attached to the crownpiece of a bridle, to which is fastened the gag runner ring. (C)

LAYER LOOP — A metal loop, sewn under a layer, such as on a breeching body. The loop serves the same purpose as a ring, for the attachment of a strap.

LAZY STRAP — A short strap attached to a hip strap, or to a breeching ring, or elsewhere on the breeching body, to support the trace. The term is sometimes used to describe any strap that carries another, such as the strap which carries a side backer.

LEAD LINE — Same as *jerk line*.

LEAD REIN — The short rein attached to the rings of a near leader's bit, to which the jerk line is attached.

LEAD-UP — Same as *tug, breast collar*, or *tug, breeching*.

LEATHER COVERED MOUNTING — A mounting having the rings or other parts covered with leather, stitched on with either a hidden or exposed seam. (C)

LEG BOOT — A horse boot which extends from the hoof to the knee, made of soft leather and fitted closely to the leg. It is intended to protect the horse's leg from injury by ice or mud. (C)

LEISURE STRAP — Same as *lazy strap*.

LIGHT HARNESS — A term generally used to designate any harness, double or single, having a trace one inch or less in width. (C)

LINE — A term frequently applied to a driving rein.

LINE GUARD — A metal device consisting of a bar with a ring at each end and one above center, to prevent lines from dropping and fouling on the end of a shaft. The throat latch passes through the center ring and the lines through the end rings. (O)

LINE RING — A ring placed high on a hame, through which the line passes for support. Same purpose as a *hame terret*.

LINK — See *hame link*.

LIP STRAP (or STRING) — A strap which is passed through a ring in the center of the curb chain, with the ends attached to the lower branches of the bit cheeks. It is intended to prevent the horse from putting his lip outside either of the lower branches of the cheeks, and from turning over the curb bit. (O)

LOIN CLOTH — A *quarter blanket* used to throw over the horse's quarters to prevent him from taking cold.

LOIN STRAP — A term used by the military to designate the strap used to carry the trace loop (lazy strap), and in turn, the trace. When used without breeching, it approximates a hip strap, but it may also be used in addition to the breeching and hip straps. (QMD)

LONG BREECHING — A breeching in which the kicking strap answers in place of the regular hip strap. (C)

LONG TUG HARNESS — Harness having the trace attached to the hame by a clip instead of being buckled to a short tug. The back end of the trace is furnished with a cockeye, or a ring and chain. (C)

LOOP — (1.) A piece of leather stamped or worked up so as to form a square channel; it is attached to the harness straps below the buckles, the channel forming a recess into which the loose end of a strap passes after leaving the buckle, and by which the strap end is held in place. It is sometimes called a *chape*, or *keeper*. (2.) Also a small metal bar on the side of, and made integral with, a buckle, ring, or other fitting, for the purpose of attaching a strap. (3.) A metal ring attached to a part of the harness for receiving a strap, such as a *layer loop*. (C & O)

LOOP RING — A ring having a loop on one side, used either as an inside hame ring or a breeching ring. (C)

LOOP STICK — A metal or hardwood form on which leather loops are shaped.

M

MANE COMB — A long toothed comb used to comb a horse's mane. (C)

MANE SHEET — A covering for the mane; that portion of a blanket connected with the hood; the hood. (C)

MARKET STRAP — A strap for supporting a trace, and holding a pad in place on double harness. The upper end passes through a ring on the pad while the lower end passes through a loop on the trace, or on a three-loop trace buckle. In some instances no pad is used, and the market strap then fastens to a ring at the juncture of the shoulder strap and turnback (double variety).

MARKET TUG — A tug terminating at the upper end in a ring, backed by a safe; from the ring a turnback runs toward the rear, and a shoulder strap toward the front. This tug is a continuation of the belly band billet, or in some cases it might terminate at a loop on a trace buckle, while the belly band billet extends from the loop on the opposite side of the buckle.

MARTINGALE — A strap attached to the belly band of a harness, passing between the horse's legs and up to the collar or collar strap, from which point all types but the standing martingale divide into two branches having rings at the ends through which the riding or driving reins pass. Its purpose is to keep a horse's head down. The term is sometimes applied to a choke strap. (C)

MARTINGALE, BELLY — A term sometimes appled to a *choke strap*.

MARTINGALE BOTTOM — That part of a martingale forming the loop through which the belly band passes. (C)

MARTINGALE CHAIN — An ornamental chain occupying the same position as a martingale, but used more as an ornament than as a functional piece. (C)

MARTINGALE FOLD — See *fold*.

MARTINGALE RING — A large ring of metal or ivory, attached to a branch end of a martingale. (C)

MINE CAP — A leather shield on a mine bridle, running from the crownpiece to the winker brace, to protect the forehead of an animal should he strike his head on some low object inside the mine.

MINE HARNESS — A variety of harness used on animals inside mines. See *spider, mine,* for description.

MOGUL SPRING — A short, stout coil spring attached to the rear end of a trace, to reduce the shocks incident to drawing exceptionally heavy loads. These springs are used extensively in the artillery for drawing pieces of ordnance. (QMD)

MOUNTINGS — The general term given to the metal trimmings of a harness. In England they are often referred to as furniture. (C)

MOUTH — An abbreviation of the word *mouthpiece*.

MOUTH GAUGE — A device for measuring the horse's mouth, consisting of a gauge which answers as the mouthpiece, a stationary cheek on one end, and a sliding cheek on the other, with a set screw to hold it in any desired position. On the lower arm of the sliding cheek there is another slide, held to its place by a set screw. The first slide is used to obtain the exact width of the horse's mouth. The slide on the lower arm of the sliding cheek is used to measure the height of the bars of the mouth. Each of the slides is marked off in inches. (O)

MOUTHPIECE — That part of a bit which is placed in a horse's mouth. (C)

MUZZLE — A covering for the horse's mouth to prevent him from biting, or from cribbing. It may be made of heavy leather, galvanized wire, or block tin covered with leather. (C)

MUZZLE STRAP — A broad strap buckled around a horse's mouth, in place of a muzzle, to keep him from biting. (C)

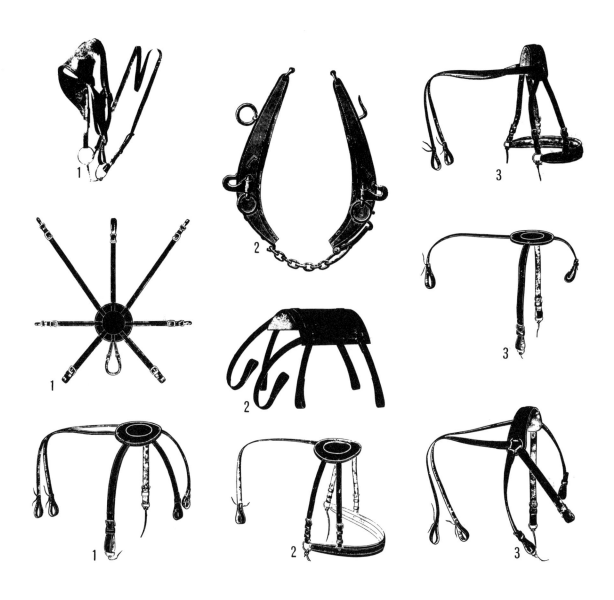

Mine harness. (The Perkins-Campbell Co., New York, 1925.) *Top row*: 1, mine blind bridle, equipped with mine cap; 2, Beagle mine hames; 3, West Virginia mine breeching. *Middle row*: 1, Buckeye mine spider; 2, consolidation mine spider; 3, Pennsylvania mine hip strap. *Bottom row*: 1, Pennsylvania mine spider; 2, Pennsylvania mine breeching; 3, West Virginia mine spider.

NO

NECK CHAIN – The chain attached to a neck strap, or neck halter, for leading an animal. The term is used particularly by the military.

NECK PIECE – Same as *neck strap,* for supporting a breast collar.

NECK STRAP – (1.) A strap that goes over the back of a horse's neck for the purpose of supporting a breast collar. It is sometimes called a *neck piece*. (2.) A *neck halter* is sometimes called a neck strap, particularly by the military.

NECK YOKE – A wooden bar which bears against the holdback on a pole, to the ends of which the breast straps or chains, or the pole straps are attached. Through this yoke the horses both guide and hold back the vehicle.

NET – An open-work covering for a horse, made of cord, light leather thongs, etc. It is intended to keep flies off of a horse by the movement of the net on the horse's body. (C)

NICKEL PLATE MOUNTING – A harness mounting having the surface of the metal covered with a coat of nickel, deposited by electrolysis. (C)

NOSE BAG – A bag, made of leather, duck, or other material, used for feeding horses on the street. (C)

NOSE BAND – That part of a bridle that encircles the horse's nose a few inches above the nostrils. It may be attached to the bridle cheeks, or to the branches of an overcheck. It is also called a *nosepiece*. (C)

NOSEPIECE – Same as *nose band*. (C)

NUMNAH – A saddle pad, placed between a riding saddle and the animal's back, to prevent chafing. They are often made of felt or sheepskin, and are shaped to fit the saddle.

OUTSIDE BELLY BAND – An extra belly band used on single harness, to prevent the shafts from rising too high. (C)

OVAL MOUNTING – A harness mounting having an oval cross section. (C)

OVERCHECK – A strap having two branches, running from the bit cheeks, over the top of the horse's head, to the hook on the pad or saddle of the harness. Its purpose is to hold the horse's head up.

OVERCHECK END – A small fitting, usually of leather, attached to the rear end of an overcheck, for attachment to the bolt hook.

OVERDRAW – Same as *overcheck*.

P

PACK SADDLE — A heavy saddle of special design, or more often a saddletree only, fitted with hooks, rings, straps, etc., and used to carry baggage or freight in areas where vehicles cannot go. (C)

PAD — That part of a double harness placed across a horse's back, and used in place of the gig saddle that is used on single harness. (C)

PAD CLOTH — Same as *housing*. (C)

PAD HOOK — A check rein hook, made in the same style as a terret, but having an opening in the back of the ring to receive the rein. (C)

PAD PLATE — An iron bow, either malleable or wrought, upon which the pad is made. It gives stiffness to the pad and provides a means for attaching the mountings. (C)

PAD SCREW — A screw bolt having an ornamental head, used for securing the pad sides to the pad plate. (C)

PAD SIDE (or SKIRT) — The strip of leather attached to the end of the pad, to which a point strap is attached. (C)

PAD TOP — The ornamental leather that forms the top or finish to the pad. (C)

PAD TRACE BEARER — A strap attached to the outside of the pad skirt, under the end of the pad top, extending down to the lower edge of the pad skirt where it is again attached. It is three-fourths inch to one inch longer than the distance between the points where it is attached, giving it a fullness that causes it to stand far enough outside the skirt to permit the passage of the trace between it and the skirt. It is used only in double harness. (C)

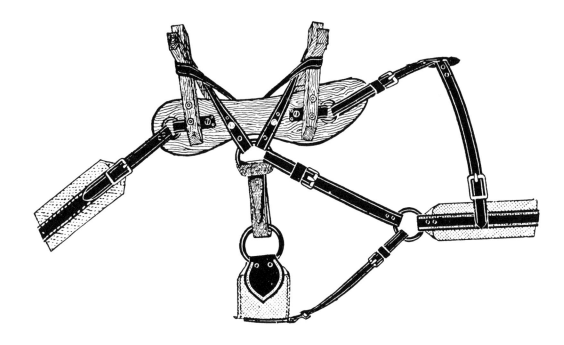

Nevada pack saddle. (Davis Saddlery Co., Sacramento, California, ca. 1915-1925.)

PAIR HORSE HARNESS — The general name given to double harness in England. (C)

PANNEL — A pad used to prevent a saddle from rubbing, or a crude pad used as a saddle.

PATENT LOOP — A hard pressed leather loop, extensively used about 1871. (C)

PHILADELPHIA FACE PIECE — A cross face piece to a bridle, having an ornament at the point where the straps diverge. (C)

PIPE LOOP — The name given to a long narrow loop. (C)

PIPING, TRACE — A leather tubing, with stitched seam, used to house a chain trace.

PLAYER — A small pendant hanging from the center of the mouthpiece of a bit. Generally several are used. Also called a *drab*.

PLUME — An ornament, of feathers or other materials, used on horses. (T)

POINT STRAP — A short strap stitched to a wide one for the purpose of attaching the latter to another strap by a buckle. The end of any strap that is provided with holes for a buckle tongue is designated a point strap. (C)

POLE STRAP — A heavy leather strap by which a horse is harnessed to the front end of a pole. It extends from the pole crab to the breast strap or chain, or from the neck yoke to the collar. It is also called a *pole piece*, and when used with a neck yoke, a *yoke strap*. (C & T)

POMMEL — The front bow, or raised front portion of a saddletree.

PORT — The curve in the mouthpiece of a bridle bit. (C)

POST HOOK — A check rein hook with an ornamental post that extends above the opening for the rein. (C)

POULTICE BOOT — A large boot used for applying a poultice to a horse's leg. It is made of soft leather, with a heavy sole leather bottom. (C)

PRESSED LOOP — A leather loop made in the harness shop, the top being pressed up with an ornamental stamp. (C)

PRICKING IRON — A tool that is struck by a hammer to mark the points where stitches are to be put in, serving the same purpose as a *prick wheel*. (O)

PRICK WHEEL — A tool with a toothed wheel, used to prick off the work for the harness stitcher, the wheel being made to produce a given number of marks per inch. (C)

QUARTER BLANKET — A small blanket generally used under the harness, covering the horse's back from the shoulders to the hips, though in some cases it extends no further forward than the front of the pad. They are cut so as to fit well, and are frequently ornamented with monograms, etc. Also called a *loin cloth*. (C)

QUARTER BOOT — A leather boot designed to protect the heels of the horse's fore feet from injury by overreaching with the hind feet. (C)

QUARTER STRAP — Same as *shoulder strap*.

QUILOR — A term sometimes applied to the breeching of certain wagon harness, particularly the type used by the army during the last half of the nineteenth century.

R

RACE CLOTH — The cloth used in connection with race saddles; they are provided with pockets to hold the weights needed to meet the requirements of the race course. (C)

RACE SADDLE — A very small, light saddle used only for racing purposes. (C)

REIN — The strap attached directly to the cheek of the bit, by which the horse is guided by the driver, or the head held in a certain position, as by a check rein. Of driving reins there are lead reins, and wheel reins. Both types are further divided into draught reins, which run straight through from the hand to the bit, and coupling reins, which are shorter and run from the bit to the point where they buckle onto the draught rein. The draught reins are on the outside of the team, and the coupling reins are on the inside. Driving reins are often called lines. (C & FR)

REIN HOLDER — A spring clasp attached to the dash of a vehicle, for the purpose of holding the reins when they are not in the hands of the driver. (C)

REIN, SIDE — Same as *check rein*.

REIN STRAP — A strap extending from the rump ring or trace carrier to the check rein, where it hooks into a ring sliding on that rein. This strap is often used on harness having double turnbacks, but it is sometimes used with a single turnback and, in some instances, takes the place of a single turnback. It is also called a *check strap*.

RIDGE-TIE — A term used in England, meaning the chain which crosses the horse's back to hold up a pair of shafts. It was sometimes called a *back band*. (O)

ROLLER — The broad padded surcingle used as a girth to hold a heavy blanket in position, generally made of twilled web, with leather billets and chapes. (C)

ROLLER, BOLT HAME — The steel roller around the bolt of a bolt hame.

ROSETTE — A leather or metal ornament, placed on a bridle or halter at the point where the front joins the crownpiece. (C)

ROUND COLLAR — See *collar*.

ROUND KNIFE — A knife with a crescent-shape blade, used especially by saddlers. (O)

ROWEL — The small toothed wheel of a spur.

RUBBER — A coarse towel used for drying horses. (C)

RUBBER MOUNTING — A harness mounting made by a patented process, by which the metal is covered with vulcanized India rubber, in imitation of leather-covered work. They are very durable, of a fine color, and take a beautiful polish. (C)

RUGGING — A coarse cloth used for the body of horse boots. (C)

RUMP PROTECTOR — See *spider, mine*.

RUMP RING — The large metal ring on the rump of the horse, at the junction of the turnback and hip straps.

Plain neck, or round-collar harness (with saddle); commonly used for CURRICLES, CHAISES, HACKNEY and STAGE COACHES, and POST-CHAISES. Horses work with more ease in them than in breast-collars, "but their advantage is disregarded from the prejudice of custom, and the absurdity prevails of using breast-collars to heavy four-wheeled carriages." (From William Felton: *Treatise on Carriages,* 1796, vol. 2, plate 51.)

S

SADDLE — A seat made upon a frame, fitted to a horse's back for the purpose of carrying a rider; also the small metal centerpiece used on a gig saddle. (C)

SADDLE BAGS — Bags made of leather, and attached to the saddle to carry small articles on horseback. (C)

SADDLE BOW — The raised front or rear portion of a saddletree. (C)

SADDLE CHAIN — A chain which goes across the saddle of a horse that is harnessed to a cart. It attaches to the upper cart shaft hooks, and supports the shafts of the cart. (I)

SADDLE CLOTH — A cloth placed under a saddle, extending back of the seat. (C)

SADDLE GIRTH — The girth by which the saddle is held in position. (C)

SADDLE NAIL — A short nail having a large head, used for making saddles. (C)

SADDLER — A workman who makes riding saddles. (C)

SADDLE REED — A small reed used in place of cord to form the edge of a gig saddle side. (C)

SADDLERY HARDWARE — A general term for all metal parts used in making saddles and harness. (C)

SADDLETREE — A frame of wood or iron, forming the support of a saddle. (C)

SAFE — Same as *chafe*.

SAFETY REIN — An extra rein buckled to the mouthpiece of a curb bit, for use in extreme cases. (C)

SAFETY STRAP — An extra back band passing over the seat of a gig saddle, having holes through which the terrets pass to keep it in position, the ends being buckled to the shaft tugs. It is used as a safeguard on light trotting harness. (C)

SCALLOPING IRON — A leather working tool having a scalloped cutting blade, used to cut ornamental designs in leather. (O)

SCRAPER — A thin piece of wood shaped like a knife blade and provided with a handle, used to scrape sweat from horses. (C)

SCREW-RACE — An adjustable tool for cutting a shallow channel in leather, where it is desired to sink stitches below the surface for protection. (O)

SEAT — The broad part of a saddle upon which the rider sits; also the small saddle-shaped piece on top of a gig saddle. (C)

SERGE — A coarse woolen stuff used for lining gig saddles, etc. (C)

SEWING PALM — A device made of heavy leather, and fitting into the palm of a leather worker's hand, used for the same purpose as the *hand-iron*. (QMD)

SHABRACK — The name given to the cloth of a military saddle. (C)

SHAFT FENDER — A large, heavy leather flap attached to the forward end of a trace along the lower edge, to prevent the shaft of a vehicle from chafing the shoulder of the horse. (O)

SHIN BOOT — A horse boot having a long leather shield to protect the shin of a horse from being injured by the opposite foot. It is used on trotting horses. (C)

SHOE POCKET — A small leather pocket attached to a saddle for the purpose of carrying one or more extra hoseshoes when on a journey on horseback. (C)

SHOE STIRRUP — A stirrup having a foot rest shaped like a shoe. (C)

SHOULDER STRAP — A strap used with one of the double turnbacks, the two being joined by a ring, to the underside of which a market tug or strap is attached. The forward end of the shoulder strap runs to a ring on the hame. Sometimes called a *quarter strap*. (O)

SIDE BACKER — One of the heavy straps running from the breeching rings to the breast yoke, for holding back or backing a wagon. Manufacturers use a variety of terms for this part, four different ones giving

Whitman military saddle. (The Perkins-Campbell Co., New York, 1925.)

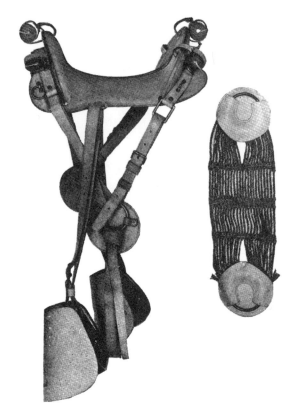

Regulation McClellan saddle. (The Perkins-Campbell Co., New York, 1925.)

457

Morgan saddle. (The Perkins-Campbell Co., New York, 1925.)

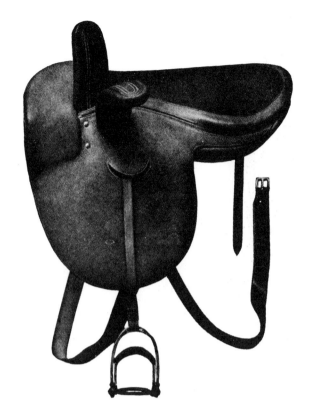

Ladies' English side saddle, on French cutback tree. (The Perkins-Campbell Co., New York, 1925.)

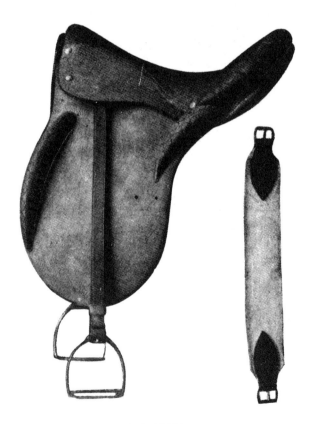

Somerset saddle. (The Perkins-Campbell Co., New York, 1925.)

Wagon saddle, used on the left wheel-horse of a wagon team. (The Perkins-Campbell Co., New York, 1925.)

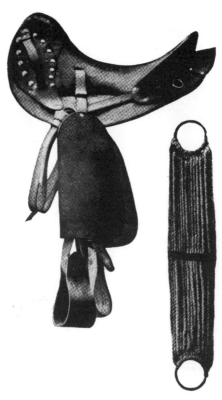

Mosby saddle. (The Perkins-Campbell Co., New York, 1925.)

Texas or stock saddle, believed to have been offered by the G. H. Schoellkopf Saddlery, Dallas, Texas. (C. E. Arnold photo.)

    as many terms: jack strap, side backer, jockey side strap, and yoke strap. Likewise, the supporting strap which runs from the side backer to the hame is also given four names: carrying strap (most common), breast strap, yoke strap (confusing it with the other usage of yoke strap), and jockey strap or chain. (O)

SIDE BACKER, BOSTON — A side backer which buckles into a trace ring. The term is applied also to an assembly consisting of a neck yoke and a pair of breast yokes, to which the side backers are attached. The variety of neck yoke used may differ from the ususal type, resembling an evener, and attached to the pole by pole chains. (O)

SIDE RING — A ring at either extremity of the mouthpiece of a bit. It may be made integral with the bit cheek or it may be separate.

SIDE STRAP — A strap running from the breeching ring to the ring at the rear end of a choke strap or martingale; also a strap from the breeching ring to the shaft of a vehicle. The variety used with single harness is often called a *holdback strap*.

SKIRT — That part of a gig saddle or pad lying under the jockey or the pad top, but on top of the housing. The point straps are attached to the skirt. The flaps hanging below the seat of a riding saddle are also called skirts.

SLICKER — (1.) A polished tool of hardwood, metal, glass, or other hard material, used to burnish edges of leather, smooth stitching, etc. It is also called a *rubber*. (2.) The term is sometimes applied to a *creaser*, but it is best reserved for the first usage.

SLITTER (or SLITTING MACHINE) — An adjustable machine with a wide blade, used for slitting leather straps or belting to a uniform thickness. (Y)

SNAP — See *harness snap*.

SOMERSET— The name given to a saddle padded before the knee and behind the thigh. It was originally made for a person who had lost his leg below the knee, from whom it takes its name. (C)

SPEEDY CUT BOOT — A peculiar, shaped, knee boot, designed to protect the knee from injury from the opposite foot of a high-stepping trotting horse. (C)

SPIDER, MINE — The rear portion of a special harness used on animals in mines. Spread on a flat surface, it resembles a spider. It generally consists of six straps, though there are numerous variations. Two straps, known as gallus straps, run forward to the hames; two known as trace straps, are approximately like hip or kidney straps, and support the traces; and two butt straps hang down in the rear to support the singletree. A large piece of leather known as the rump protector is sometimes located at the junction of the straps to protect the rump of the animal from low hanging mine timbers. (O)

SPOT — A small metal ornament, usually round or oval, attached to some part of the harness. These are often put in rows.

SPREAD STRAP — A short strap attached to the crownpiece of a wheel-horse, having a ring at the lower end through which a line from the lead-horse passes, to support the line and keep it from fouling on some other part of the harness. These straps are often decorated with colored celluloid rings.

SPRING BAR — See *stirrup bar*.

SPUR — An instrument having a rowel, or small wheel with sharp points, worn on a horseman's heels to prick the horse to hasten his speed. (C)

STAGE CHAIN — A variety of heel chain having a hook at the rear end of the chain, which fastens into a ring about midway in the chain.

STIRRUP — A piece of metal, wood, or leather, having a flat section for a footrest, with a bow-shaped section above into which a stirrup strap is buckled to attach it to a saddle. It is used as a support for the foot while riding on horseback. (C)

STIRRUP BAR — The part of a saddle to which the stirrup strap is attached. Some of the patented types are called spring bars. (C)

STIRRUP LEATHER (or STRAP) — The strap by which the stirrup is attached to the saddle. (C)

SURCINGLE — A strip of leather or web which passes over a saddle, blanket, or other article on a horse's back, to hold it fast. (C)

SWEAT PAD — A pad of felt or other fabric placed under a collar to provide greater comfort for the shoulders of the horse.

SWING (or SWINGER) — Same as *fly*.

SWIVEL — A fitting having two eyes, turning freely in the center, for use in such things as trace chains, to prevent the chains from becoming twisted. The term is also applied to a *gag runner*.

T

TANDEM HARNESS — That designed for use on two or more horses, harnessed in line, one before the other. (O)

TAPADERO — A hoodlike stirrup cover used with a western saddle. They are sometimes highly ornamented, and are often cut to a point at the bottom.

TERRET — A ring attached to the pad or saddle, and the hames of a harness, through which the driving reins pass. (C)

TERRET, NECK STRAP — A terret placed on the neck strap of a single harness, through which a line is passed to prevent fouling. (O)

THROAT LASH — Same as *throat latch*.

THROAT LATCH — The strap of a bridle or halter that encircles the horse's throat and assists in holding the bridle or halter in position. (C)

THROAT STRAP — The part of a halter which corresponds with the *throat latch* of a bridle. The army uses the two terms sysnonymously with either bridle or halter. (C & O)

TIE STRAP — A long strap having a buckle and chape on one end, used as an estra strap to a bridle for tying an animal. (C)

TRACE — A strap, chain, or rope attached to the hames or collar of a harness, by which the vehicle is drawn. (C)

TRACE BEARER (or SUPPORT) — A leather strap attached at two points to a saddle or pad skirt, forming a loop which the trace passes through. In some instances it continues downward to form the belly band billet.

TRACE BUCKLE — A long heavy buckle designed for use on traces only. (C)

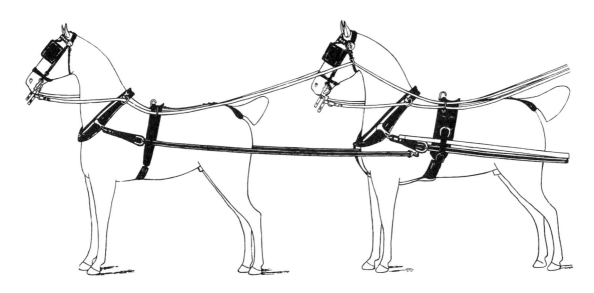

Tandem harness. In an area having steep hills, breeching was often added to the harness of the wheel-horse. (Smith, Worthington, & Co., New York, 1901.)

TRACE CARRIER — The metal fitting at the rear end of the turnback, having loops at the sides for the attachment of the hip straps, and a loop at the rear for the attachment of the crupper, if the latter is used. The term is sometimes applied to the *lazy strap*.

TRACE CHAIN — A chain designed to be used as a trace. (C)

TRACE LOOP — A square metal loop used to attach a coach trace to the roller-bolt or whiffletree. The army uses the term to designate the strap which carries the trace, or the *lazy strap*. (C & O)

TRACE PIPE — See *piping, trace*.

TRACE RING — A ring in a two-piece trace, into which the belly band and side backers also buckle.

TRACE STRAP — See *spider, mine*.

TRACK HARNESS — A breast collar single harness, made up in the lightest and plainest manner. (C)

TRAINING HALTER — A halter made in the same manner as a riding bridle, except that it has short instead of long cheeks, which are provided with rings into which the bit straps may be buckled. (C)

TRIAL BIT — A skeleton bit used by some practical bit makers to determine the exact width of a horse's mouth, as well as the breadth and height of the port. Same as *mouth gauge*. (C)

TRIMMINGS — A general name for the ornaments hardware, etc., used on a harness. (C)

TUG, BREAST COLLAR — A short strap and buckle secured to the breast collar for the attachment of the neck strap. Also called a *lead-up*, or *up-tug*.

TUG, BREECHING — A short strap and buckle secured to the breeching body for the attachment of a hip strap. Also called a *lead-up*, or *up-tug*.

TUG, HAME — A short strap and buckle secured to the hame, for the attachment of a trace.

TUG, SHAFT — A loop of leather or metal, with buckle, secured to a back band for the purpose of supporting a shaft.

TUG BEARER — Same as *back band*.

TURNBACK — Sometimes called a *back strap*, the turnback is a strap extending from the trace carrier on the rump of an animal to the hames, the pad, or the gig saddle. It may also extend as far as the crupper, particularly if no breeching or hip straps are used. Double turnbacks run from the trace carrier to the hames, one dropping downward on each side of the horse. These double turnbacks may be interrupted at the market strap or tug, in which case a shoulder strap continues on to the hame. (O)

TWISTED MOUTH — The square mouthpiece of a bit that has been twisted to make it more severe than it would otherwise be. (C)

Single-strap, or track harness—$57. (Smith, Worthington & Co., New York, 1901.).

U W X Y

UP-TUG — See *tug, breast collar,* or *breeching.*

WHIP — (1.) An instrument of leather or wood, or a combination of the two, used to correct or urge an animal. Whips come in a variety of styles, and may be from five to eight feet long. Some have braided rawhide centers, covered with leather, such as a teamster's *black snake whip;* others may have an elastic wooden stock, such as a *Loudon whip;* and still others may be a plain wooden switch, with or without a lash. (2.) One who handles a whip; a coachman or driver of horses.

WINKER — Same as *blind.* (C)

WINKER LEATHER — A heavy, glazed leather used as the outside piece of a winker. (C)

WINKER PLATE — A metal plate, slightly dished, used to give form and firmness to the winker. (C)

WINKER STRAP — A strap attached to the crownpiece of a bridle, extending down the forehead a few inches, then branching off to each side, where the lower ends are attached to the winkers to hold them in position. Also called a *winker stay* or *winker brace.* (C)

XC PLATE — Iron or steel hardware, such as bits, buckles, or other harness mountings, that have been tinned in order to prevent rusting.

YOKE — See *neck yoke* and *ox yoke* in the Carriage Nomenclature section.

YOKE IRON — Same as *breast collar iron.*

YOKE STRAP — A variety of *pole strap,* running from the collar to the neck yoke. Also see *side backer,* and *carrying strap.*

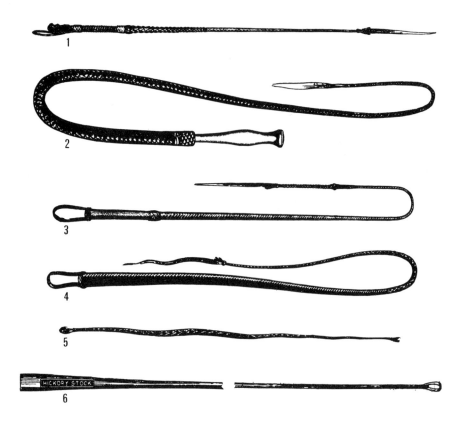

Whips. (Smith, Worthington & Co., New York, 1901.) 1, Dog whip; 2, drover's whip; 3, drover's whip; 4, blacksnake whip; 5, white horsehide lash; 6, white hickory handmade stock.

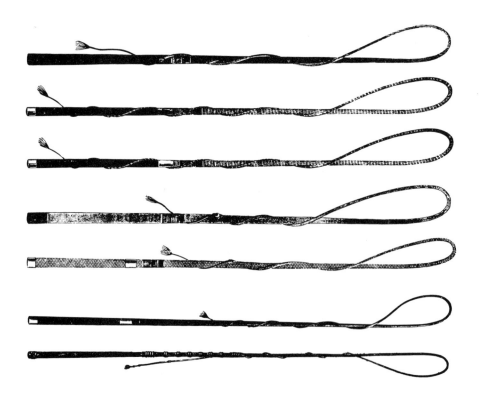

Express whips ranging in length from 3½ feet to 7½ feet. (Smith, Worthington & Co., New York, 1901.)

WILLIAM FELTON'S 1796 GLOSSARY

Since the terms shown in the Carriage Nomenclature section of this dictionary are largely those of the latter part of the nineteenth century, this glossary of eighteenth-century terms is given from William Felton's *A Treatise on Carriages*, published in London in 1796. The original glossary, as printed in the 1796 edition, has been broadened by additions from Felton's text. Felton's quaint eighteenth-century spelling has been retained.

A

APRON — Same as *knee boot*.

ARMS — The distinction of families, which are mostly painted on the pannels.

AXLETREE — A piece of wrought iron work, fixed to the under part of the carriage, on which the wheels are placed.

AXLETREE ARM — That part of the axletree which passes through the center of the wheel, and on which it turns.

AXLETREE BED — The timber, in which the axletree is let or bedded.

AXLETREE BOXES — Iron tubes fitted to the arms of the axletree, fixed firm in the wheel's stock, and which contains the grease or oil.

AXLETREE HOOP — An iron hoop, which fixes the axletree to the timber or bed on which it rests.

AXLETREE NUT — An iron screw, with a large surface, fixed to the fore or hind end of the axletree, for the purpose of keeping on the wheels.

AXLETREE WASHER — An iron collar or shoulder, fitted to the body or large end of the axletree, against which the back of the wheel wears, for the purpose of keeping in the grease.

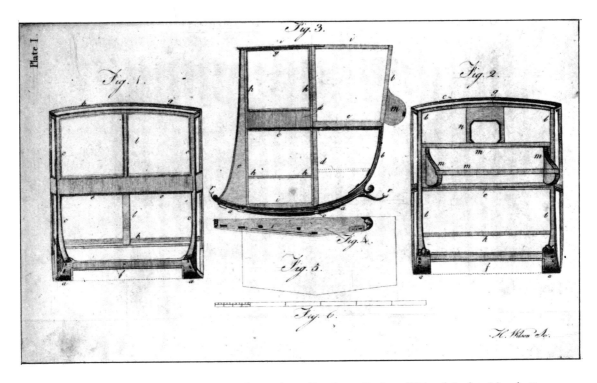

CHARIOT or POST-CHAISE body. (From William Felton: *Treatise on Carriages*, 1796, vol. 1, plate 1.) a, bottom side; b, corner pillar; c, fore pillar; d, standing, or perpendicular pillar; e, middle rails; f, bottom bars; g, roof rails; h, door pillars; i, door top and bottom rails; k, fore and back seat-rails; l, front or middle pillar; m, sword-case; n, back-light piece; o, rockers; p, compass rails, or hoop sticks; q, rest-piece for the glasses; r, body loops.

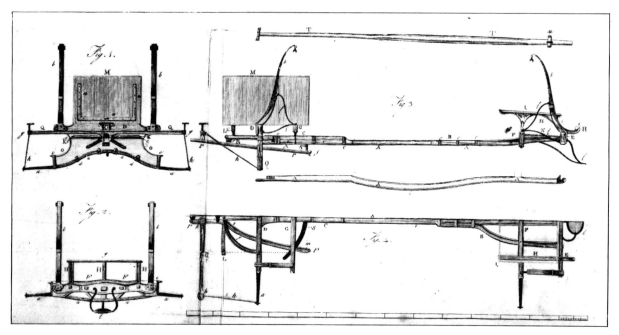

A perch carriage for a CHARIOT. By a slight modification of length, it will also serve for a COACH or PHAETON. (From William Felton: *Treatise on Carriages,* 1796, vol. 1, plate 6.) A, Perch; B, hind hooping-wings; C, fore hooping-piece; D, fore transom, or fore spring-bar; E, hind transom, or hind spring-bar; F, Hind axletree-bed; G, budget bar, or horn bar; H, hind blocks; I, footboard, or platform; K, wheel-piece; L, fore block; M, boot; N, nunters; O, fore axletree-bed; P, futchels; Q, splinter-bar; R, front felly-piece; S, sway bar; T, pole; U, pole-gib. a, hind and fore axletrees; b, hind and fore springs; c, perch- and axle-hoops; d, axletree clips; e, transom and wheel-plates; f, spring-stays; g, splinter-bar sockets; h, wheel-irons; i, side perch-plates; k, splinter-bar rolls; l, footman's step; m, sway bar plates; n, budget plate.

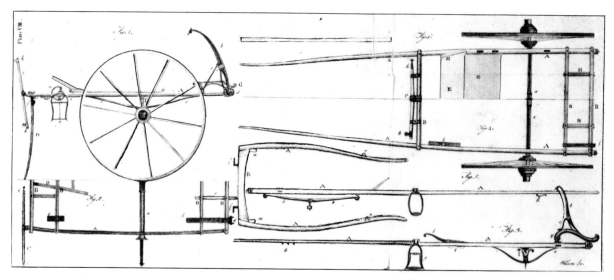

Views of carriages for two-wheel vehicles. (From William Felton: *Treatise on Carriages,* 1796, vol. 1, plate 8.) Figures 1 & 2, CURRICLE (side and half top view); figures 3 & 4, GIG (side and half top view); figures 5 & 6, WHISKEY (side and half top view); figure 7, Shafts for a one-horse, four-wheel carriage. A, Shafts; B, cross-framings; C, splinters, or splinter-bars; D, ladder-prop for a CURRICLE; E, brackets, footboard, and bottom of a WHISKEY; F, hind and fore blocks; G, small blocks (for supporting a platform); H, cross-framings, or nunters; I, raisers. a, axletree; b, springs; c, spring-stay; d, spring jacks; e, main or bottom stays; f, ladder joints; g, steps; h, splinter sockets; i, curricle sockets, for shafts; k, tug plates or stops; l, hooks, by which these shafts hang on the splinter-bar; m, hooks the traces are fixed to; n, breeching staples.

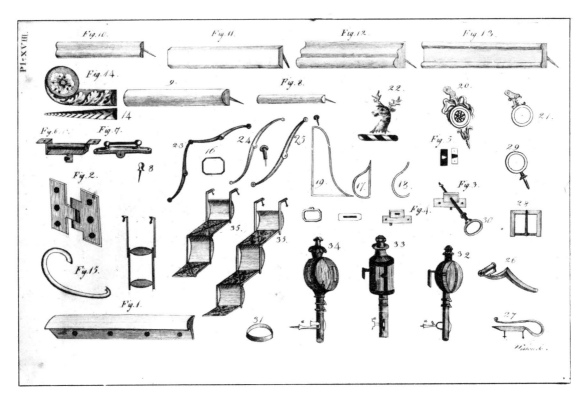

Eighteenth-century carriage parts. (From William Felton: *Treatise on Carriages,* 1796, vol. 1, plate 18.) Figure 1, Door plates; 2, door hinges; 3, door box lock, with handle; 4, door box lock, requiring key; 5, dove-tailed catch; 6, glass roller; 7, glass roller; 8, (Original error) 2nd figure 8 is a small quill bead; 9, common moulding; 10, moulding; 11, flat moulding; 12, fancy moulding, seldom used but for HANDSOME CARRIAGES; 13, common moulding; 14, scroll and tip ornament; 15, plated wing frame; 16, octagon frame for back light; 17, sword case frame, whole; 18, sword case frame, half; 19, sham or real door frame; 20, fancy-worked head plate; 21, bead-rim head plate; 22, crest; 23, real and sham joints; 24, real and sham joints; 25, real and sham joints; 26, body loop; 27, pole hook; 28, plated buckles; 29, check-brace ring; 30; door handle; 31, wheel hoops; 32, globe lamp; 33, Italian lamp; 34, oval lamp; 35, inside folding steps, double and treble; 36, outside CHAISE steps.

B

BACK BAND — Part of a one-horse chaise harness, which crosses the saddle, and supports the shafts.
BACK STRAP — A part of the harness looped on the crupper, and buckled to a loop or tug to keep up the traces.
BARS — Timbers of various sorts, particularly described in alphabetical order.
BATTENS — Strips of wood, which are fixed on the outside of the pannels to form the framing, and are then moulded; but when fixed on the inside of the pannels, are to mend or strengthen them.
BEADS — The mouldings which ornament the carriage.
BEARING REIN — The rein which holds up the horse's head.
BELLY BAND — A leather which buckles round the horse's belly, and fixes on the pad or housing.
BIT — An iron instrument, which is put into the mouth of the horse, by which he is governed.
BLINDS — Such as Venetian and spring blinds; see each in their order.
BLOCKS — Wooden raisers to the springs of phaetons; foot-boards, budgets, shafts, &c., mostly ornamented by carving, and are described by what is raised upon them, such as budget blocks, etc.
BODY — That part of the carriage which contains the passengers.
BODY LOOPS — Strong iron loops, screwed or bolted to the bottom corners of the body, and by which it hangs.
BOLTS — Iron pins of various lengths, headed at the one end, and screwed at the other, and are in general about

half an inch thick.

BOODGE or SWORD-CASE — A prominence from the back of the body to carry parcels in.

BOOTS and BUDGETS — Large leathered boxes, fixed on the fore part of the carriage, and distinguished by the various names of Salisbury, platform, or trunk boots and budgets.

BOTTOM or PANNEL BARS — The bottom end framings of the body, on which the end pannels rest.

BOTTOM BOARDS — Boards which form the bottom of the body.

BOXES — See axletree box, seat box, coach box, driving box, cap box, &c.

BOX LOCKS — Are the locks used for the doors of the body.

BRACES — The leathers by which the bodies are hung, or checked.

BRACKETS — Parts of the framing of the body, which support the foot-board, and also the carved ornaments, fixed on each side the top of the coach box foot-board.

BRASS BEAD EDGINGS — Brass plates, which are screwed to the side of doors for them to shut on.

BREAST COLLAR — A part of the harness which is placed round the horse's breast, by which he draws.

BREECHING — That part of the harness which goes round the breech of the horse.

BRIDGE — Part of the furniture of the harness, mostly made in the shape of the buckle, but has no tongue, only two cross-bars or bridges, round which the strapping is looped.

BRIDLE — That part of the harness which is put on the head of the horse, by which he is managed.

BRIDOON — An additional temporary bridle, made similiar to a riding or watering bridle.

BRIDOON BIT — The bit which is used to the bridoon bridle.

BRIDOON CHAIN, or LINKS — Small ornaments, through which the bridoon reins run.

BUDGET — See *boot*.

BUDGET, BOOT or HORN BAR — The inner cross bar to the front of the carriage, on which the fore spring stay and budget rest.

BUGGY — A small phaeton or chaise, made only to carry one person.

BUTTONS — Nails or screws with large brass heads, for the purpose of hitching on the straps, mostly silvered, but sometimes plated.

BUTTON HANGERS — Small ornamented tassels, which are placed on the fringe.

C

CABRIOLE — A two-wheel carriage, with the body somewhat like a chariot, built and used mostly in France.

CAP — A small piece of leather, used to confine a temporary pin or bolt, such as a pole pin cap, &c. Also an iron ferrule on the end of a wooden part, either for strength, or for the attachment of some other part.

CAP BOX — A long leather case, used for the purpose of carrying ladies head dresses.

CARPETING — Covering the bottom of the body or step treads with carpet.

CARRIAGE — That part, on which the body is placed, and to which the wheels are united.

CHAIN BELT — A thin wire chain, covered with leather, made in the form, and to answer the use of a strap, for the purpose of securing trunks, &c. behind a carriage.

CHAIR — A light chaise without pannels, for the use of parks, gardens, &c. a name commonly applied to all light chaises.

CHECK BRACE — A single strip of leather, which is looped through a ring at the corners of the body, to check it from swinging too much endways.

CHECK RING — An iron ring screwed into the corner pillars of the body for the check braces.

CHECK STRING — A worsted line, by which the coachman has notice to stop.

CLIP — An iron fitting used to bind parts together, secured by nuts and a plate.

COACH BOX — The fixture on which the driver sits.

COLLAR — That part of a harness, by which the horse draws; it is of two sorts, the breast and the heam (round). Also a shoulder or middle to an iron stay or bolt.

COLLAR BOLT — A bolt with two nuts, and a collar in the middle.

COLLAR BRACE — A strong leather strap, fixed under the body, to check it from swinging sideways.

COLLAR BRACE RING — An iron ring, through which the collar brace is looped.

COMPASS IRONS — Irons which support the foot-board and stays.

COMPASS RAILS — See *hoop sticks*.

CORK LEDGE – A long strip of cork, nailed on the coachman's foot-board, against which his feet are placed.
CORNER PILLARS – The corner framings of bodies.
CORNICE RAILS – The top framing of the body of a coach or chariot, called roof rails.
COUNTER SUNK BOLT – A bolt, the head of which is let in level with the surface of the plate it fixes.
COUPLING REINS – The reins which couple the horses together.
CRADLE – A leather convenience fixed to opposite bearings, for any thing to be carried safe, and the coachman to ride easy upon.
CRANES – Strong iron bars, which form the sides of the upper carriage, and unite the back and fore timbers, shaped like a crane's neck, for the purpose of the fore wheels to pass under.
CRANE NECK CARRIAGE – A carriage that is made with cranes.
CRANE SHAFT – Wood instead of iron, for the same purpose.
CROWN PIECE – That part of the bridle which lies on the horse's head.
CURB – The small chain which goes round the horse's jaw, and hooks to the bit.
CURB HOOK – A hook which the curb is hitched to.
CURRICLE – A two-wheel carriage, drawn by two horses abreast.
CURRICLE-BAR – An iron contrivance supported by the backs of two horses, to carry the pole of a curricle. The two ends slide freely in the main body to allow a freer movement of the horses.
CURTUERS or CUTTOS – Projections left at the ends of the axletree bed, which lie over the back part of the wheels to shelter the axletree from gravel or other dirt.

DASHING or SPLASHING LEATHER – A large iron frame, covered with leather, preventing the dirt from splashing against the passengers or pannels.
DEE – A ring in the shape of a D, for a strap to pass through.
DOOR PILLARS – The side framings of the doors.
DOOR STYLES or MIDDLE RAILS – The middle framing of the doors.
DOVETAIL KETCH – A small iron ketch, fixed on the side of the door, to prevent it settling.
DRAG CHAIN – A strong chain, with a large hook to hitch on the hind wheel, and keep it from turning when descending a hill.
DRAG STAFF – A short pole, which is fixed under the hind part of the carriage, and to be let down when ascending a hill, to give the horse more ease, by occasionally resting.
DRIVING BOX – A portable box, on which a cushion is placed, to raise the driver.
DRIVING CUSHION – A deep cushion, made purposely for the driver to sit on.
DROP BOTTOM – The bottom of a coach, chariot, or chaise body, when sunk deeper than the surface of the framing, to give more room.
DROP SEAT BOX – A box which is made to hang between the seat rails, to carry luggage.
DUKE'S BIT – A bit of peculiar form on the outside.

EAR BOWS – Leathers bent across the horse's ears, lapped with tape, the same as the fronts and roses.
ELBOW CASE – A cavity in the inside of the body, at the elbow part, for bottles, &c., seldom used but to travelling carriages.
ELBOW RAILS – The middle part of the framing to a coach or chariot, and the upper part to a chaise or phaeton body, on which the elbow rests. Inside elbows are projections within the body for the elbow to rest on.
ELBOW SPRINGS – Are those that rise in an oblique direction from their bearings, mostly used to one-horse, or phaeton carriages.
EMBOSSING – A method of raising the crests, etc., in silver or plated metals, &c., the same as in relievo.
ENGLISH POLE PIECES – The pole pieces that are fixed to the pole-end.

F

FOOT BOARDS — Are what the feet of the servant or driver rest on.
FOOT BOARD LEDGE — A small piece of timber fixed on the footboard, against which the coachman's feet are placed.
FOOTMAN CUSHION — A wooden frame stuffed, and covered with stout leather, to ease and elevate the servant behind the carriage.
FOOTMAN HOLDERS — Lace, with tassels, hung to the back of the body, by which the footman holds.
FOOTMAN STEP — An iron step, fixed to the hind part of the carriage, for the servant to mount by.
FORE BAR or BLOCK — A bar framed in the front of a carriage.
FORE CARRIAGE — The under part, or conductor of a four-wheel carriage, to which the fore wheels are placed.
FOREHEAD PIECE — An ornament, which hangs loosely on the forehead of the horse.
FORE PILLARS — That part of the framing of a chariot, on which the doors hang, and which forms the front sweep.
FORE RAILS — The cross framing rails to the fore end of a body.
FORE TRANSOM — The timber which crosses the perch, on which the springs are placed, and through which the center pin, or perch bolt, passes to the fore carriage.
FRAME HEADS — The head of a chaise or phaeton, made on an iron frame, for the purpose of taking off occasionally.

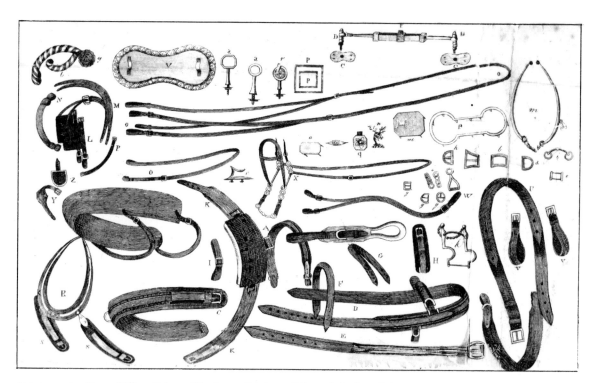

Harness parts. (From William Felton: *Treatise on Carriages,* 1796, vol. 2, plate 49.) A, Housing or pad; B, crupper; C, breast collar; D, breeching; E, traces; F, back strap; G, hip strap; H, tugs; I, Newmarket strap; K, belly band; L, winker; M, head stall, or crown piece; N, front, or forehead piece; O, reins; P, throat band; Q, false belly band; R, heam, with round or neck collar; S, heam tugs; T, false collar; U, back band; V, shaft tugs; W, martingale; X, bridoon head, or rein; Y, nose band; Z, forehead piece. a, Territs; b, trace rings; c, watering, or bearing hook; d, collar dee; e, bridge; f, collar buckles; g, buckles; h, throat-band dee; i, swivel; j, chain-links; k, bit; l, bridoon bit; m, heams; n, housing and winker plates, or pieces; o, studs; p, frames; q, forehead piece, nose piece, breast piece or side pieces (ornaments); r, fly head ring; s, roses; t, earbows; v, pad cloth. Illustrated at top, near tear in plate —curricle bar. The square center portion of the bar holds two sliding iron bars, compensating for the movement of the horses, the ends of the bars being fixed in the stands, D, which in turn are fastened to the plates, C, which are mounted to the saddles. F is a staple from which the brace runs that supports the pole.

FRENCH POLE PIECES – Pole pieces which are made double, so as to be taken off occasionally.
FRENCH REINS – Long coupling reins, which buckle at the upper part of the long hand reins.
FRONT – A broad stripe to the front of the bridle, mostly covered with taping to match the roses.
FUTCHELLS – The timbers of the under carriage, in which the pole is fixed.
FALLS – That part of the lining, which hangs loose from the seat rails.
FALSE BELLY BAND – A leather strap, which buckles on each side of the collar to keep it down, so as to save the use of a breeching.
FALSE COLLARS – Those that are occasionally added under the others, to prevent the horse from being galled by friction.
FALSE LINING – A linen cover, to preserve the cloth lining clean.
FELLY – A divided part of the rim of a wheel; also a small part of a circle which is fixed on the futchells, and forms a bearing for the whole or half wheel front.
FENCE – A rabbet round the edges of the lights, to prevent the weather getting between it, and the glass or shutter frame.
FESTOON CURTAIN – A silk curtain trimmed with silk fringe. Intended as ornament only, these curtains are fixed over windows and doors of a carriage, and are frequently made to hang in drapery form.
FILLET – A narrow painted border, not exceeding one inch broad.

G

GALLING LEATHER – A broad strip of leather, sewed under that part of the harness, where there is a buckle to prevent it from galling the horse, or placed under the coachman's seat.
GIB – A small half-round wedge, which keeps the pole from rising.
GIB STRAPE – Two straps nailed to the gib, to confine it in its place.
GIG – A one-horse chaise built in a fanciful style.
GLASS ROLLERS – A brass machine, which eases the weight of the glass when drawing up.
GLASS STRINGS or HOLDERS – The lace which is nailed to the frames, to draw the glasses up by.
GLOBE LAMP – A lamp, the body of which is of globular form.
GRASSHOPPER SPRING – A peculiar formed spring, which fixes under the shaft of a one-horse chaise to the axletree.

H

HAMMER CLOTH – An ornamented covering to the coachman's seat.
HAND REINS – The reins which the driver holds, and by which the horses are guided.
HANGING and UNHANGING – Is taking the body from the carriage for any material repair, and re-fixing it when done.
HAT BOX – A leather case used to carry gentlemen's hats.
HEAD – The top or cover of a phaeton, chaise, &c., or the top of a bridle.
HEAD PLATES – Metal ornaments, placed at the upper parts of bodies.
HEAD PLATE PINS – Small nails, with plated heads, to fasten the head plates with.
HEAD RING, or HEAD TERRIT – A ring, placed on top of the bridle of the wheel harness, through which the leading reins pass, when four horses are drove in hand, and sometimes used for ornament only.
HEAD STALL – The bridle without the bit or reins, and sometimes means the crown piece only.
HEAM COLLAR – A padded or stuffed collar, which goes round the horse's neck, and by which he draws.
HEAM LINKS – The links, which unite the heams at the bottom.
HEAMS – Two compassed irons, with links at one end, and loops to buckle at the other, fitted to the neck collar, by which the draught is taken.
HEAM STRAP – A small strap, which confines the heams at the top.
HEAM TUGS – A part of the harness rivetted to the heams, to which the traces are fastened or buckled.

HEDGE HOG — A leather stuck full of nails, to buckle on the pole with the points upward, to prevent the horses gnawing it.
HEEL BOARDS, or HEEL LEATHERS — Boards or leathers nailed under the coachman's seat, to shelter the legs from the cold.
HIND STANDARDS — An ornamented platform, on which the footman stands behind the carriage.
HIP STRAPS — A part of the harness, which lies on the hips of the horse, and buckles to the breeching tugs, which it supports.
HOLDERS — Broad lace with tassels, by which the person in the carriage holds, or draws the glasses up by.
HOOPED WHEEL — The wheel whereof the iron rim is one entire piece.
HOOPING PIECE — A strong timber, which unites the perch to the fore end of the carriage.
HOOPING WINGS — Two extending timbers, which unite the perch to the fore end of the carriage.
HOOPS — Iron rims, which are tightly drove on, to strengthen or unite two things together.
HOOP STICKS — Thin compassed rails, which form the roof.
HORN BAR — Same as *budget* or *boot bar*.
HOUSING — A small square pad, which lies on the horse's back, to which most of the harness is fixed.
HOUSING CUSHION — The soft stuffed under part of the housing.

I

IMPERIAL — A leathered case, which is placed occasionally on the roof of the body, for the purpose of carrying cloaths.
ITALIAN LAMPS — A lamp of an oblong or cylindrical round form.

J

JACK — A small machine, in which the brace is fixed, to be let out or taken in by.
JAPANNING — Painting, with a black glossy preparation, the leathered part of the body and carriage.
JEW'S HARP STAPLE — An iron staple, in the shape of a Jew's harp, and a connected part of the grasshopper spring, which it raises from the axletree.
JOINTING — The cleaning of the mouldings, and levelling the joints of the framing, previous to new painting.
JOINT PROPS — What the joints are placed on.
JOINTS — The irons, by which the heads of chaises or landaus are let up and down.

K

KNEE BOOT or KNEE FLAP — The leather which covers the knees, when sitting in an open carriage.
KNEE BOOT CHECKS — The flaps on the sides of the knee boots.
KNEE BOOT FALL — The strip of cloth, which covers the top of the knee boot, made of the same materials as the lining is.
KNEE BOOT STRAP — What fastens the knee boot down, when out of use.

L

LADDER-PROP — A prop or leg that is pivoted to the pole of a curricle, and let down to support the vehicle when the horses are not harnessed to it. This prop is held up by a spring catch or leather strap when not in use.
LAMP BARREL — That part of the lamp which contains the candle.
LAMP FORK or PROP — A small iron fixture, which keeps the lamp barrel steady.

LAMP IRONS – Are what the lamps are fixed by to the body.
LAMP SPRING – A spiral wire, placed in the lamp barrel, which forces the candle to rise as it consumes.
LAMP STRAPS – Small straps, which buckle round the barrels.
LANDAU – A carriage built in the manner of a coach, but with the upper part of the body to open at pleasure.
LANDUALET – A chariot made the same as above.
LAYS – A strip of leather, which is sewed on the top of another that is broader, for the purpose of additional strength, or to confine a smaller buckle; also particular stripes in the lace, which are always of silk, called silk lays.
LEADING HARNESS, or parts thereof – Are what belong to the fore horses, when more than the ordinary number are used, commonly called leaders.
LIGHTS – The windows of the body, such as door, side, front, or back lights.
LINCH PIN – A small iron pin, which goes through the axletree point, and secures the nut to keep the wheel on.
LINING – Covering the wood work on the inside of the body with cloth, &c., or repairing any part that is worn.
LOCKING PLATES – Short, thick iron plates, fixed to the sides of the perch, to preserve it from injury, by the wheel rubbing against it when the carriage is turning.
LOCKING STOP – A piece of timber fixed to the fore bed, to prevent the wheel striking at all against the perch.
LOOPS – See *body loops* or *running loops*.
LOOP SPRING – A short, leaf spring sometimes used in place of an ordinary body loop, to assist the braces in providing a more comfortable suspension.
LUGGAGE BOOT – A boot with a loose cover, convenient to carry luggage.
LUGGAGE IRONS – The iron frames, of which those boots are made.
LUGG PLATE – An iron plate, with a part branching from the side, to unite or hang two things by.

M

MAIN BRACES – The strong leathers, by which the body hangs.
MANTLE – A painted ornament, in form of a curtain, in which the arms, crest, or cyphers are placed.
MARTINGALE – A temporary addition to the bridle, placed so as to prevent the horse throwing his head back, sometimes used as an ornament.
MIDDLE PILLAR, or PARTITION PIECE – That which divides the front windows into two.
MIDDLE RAILS – The middle framing of the body.
MORTOISE – A square hole, made in one timber, to receive the end of another, called a tennon, for the framings to be fastened by.

N

NAVE – The center or stock of the wheels, in which all the spokes are fixed, and through which the axletree arms go.
NECK COLLAR – Same as *round collar*.
NECK PLATES – Thin iron plates, fixed on the slats or wood work of chaise heads, which move by means thereof.
NECK or WITHER STRAP – A part of the harness, which crosses the withers of a horse, and supports the breast collar.
NET – A convenience placed across the roof, on the inside of a coach or chariot.
NEWMARKET STRAP – A part of the harness, which buckles together the housing and collar.
NOSE BAND – A leather, which crosses the nose of the horse, and buckles to the cheek of the bridle.
NOSE PLATE – A short iron plate, fixed across the chops or nose of the futchells to keep them fast, and on which the pole rests.
NUNTERS – Are short timbers, framed across the beds, or transoms of the carriage, to strengthen them.
NUTS – Square pieces of iron, which are screwed on the bottom of the bolt.

O

OCTAGON or OVAL LIGHT — The small window at the back of the body.

OILED COVERS — Are temporary covers for the body of a carriage, to preserve the paint from the injury of the road-dirt, or boughs, while traveling: oil-skin covers are frequently used, and are so made that the doors may open and shut with the cover on; every part of the body, except the windows and bottom, is covered; it is made to fit to the exact form of the body, and looped on to small plated buttons, so as to be taken off occasionally; they are made of common oil linen, lined with a soft baize, and bound with a worsted tape.

OIL SKIN — Linen dressed with oil, used as covers for hammer cloths, &c.

OIL SKIN PATENT — Woollen cloth, prepared in a peculiar manner, for the same use as the linen, but is more durable.

P

PAD CLOTH — A cloth usually bound with lace, and put under the pad or housing on the horse's back.

PANNEL BARS — Same as *bottom bars*.

PANNELS — Are what fills the framing of the body, and are called door, side, quarter, or back pannels.

PARTITION PIECE — Same as *middle pillar*.

PASTING LACE — A narrow lace, which is nailed and pasted over the nailed edges of the cloth.

PERCH — The long or main timber of a carriage, which unites the hind and fore end together.

PERCH BOLT — A strong round iron pin, on which the fore carriage turns.

PERCH BOLT HOLE — The hole in the timber through which this pin passes.

PERCH BOLT KEY, or COTTERELL — Is a thin piece of iron, fixed through the eye of the perch bolt, to keep it from rising.

PERCH BOLT NUT — An iron screw, fixed on the perch bolt, for the purpose of additional security.

PERCH CARRIAGE — The carriage made with a perch.

PERCH HOOP — The hoop that unites the other timbers to the perch.

PICKING OUT — The painting with various colours the mouldings, &c.

PINNING — The nailing with small headed iron nails, called pins, used only to the leather or lining.

PIPE BOX — See *axletree box*.

PLATED — The strengthening the timber with iron plates; or, covering the furniture of either carriage or harness, superficially with silver or other metal.

POINT STRAPS — Small straps, which buckle down the points of the main braces.

POLE — The long leaver, by which the carriage is conducted.

POLE HOOK — An iron fitting on the end of the pole, intended either for draught, or for ornament only.

POLE PIECES — Strong leather straps, which fasten the horses to the pole end.

POLE PIN — A round iron pin, which passes through the futchell ends and pole, to keep it from coming forward.

POLE PIN CAP — A leather, which secures the pole pin.

POLE RING — A ring fixed on the pole end, with loops for the pole pieces to be fastened to.

POLE STAPLE — A staple drove [sic] into the back end of the pole, with which it is fastened by a gib.

PORTSMOUTH BIT — A bit made of a peculiar form, for hard-mouthed horses.

PRIVATE LOCKS — Those fixed in the standing pillars, by which the doors are occasionally locked up.

PROPS — The iron fixtures, on which the joints of chaise or landau heads are fixed.

PUMP or PLOW HANDLES — The long projecting timbers, on the hind part of the carriage, on which the footboard is placed.

Q

QUARTERS — The sides of a coach, divided by the middle rails into four parts; in a chariot, only into two: the sides within the body are also called quarters.

R

RABBET — An edge of the timber sunk below the surface, for others to be lapped in.
RAISED HIND or FORE END — Is when the budget or footboard is raised on blocks, for the ornament of the carriage.
RAISER — A small pillar or block, for any other thing to rest on.
REIN RINGS and HOOK — Are conveniences for the reins to run in, or be hung by.
RIMS — Narrow stripes of leather, of various sorts, which are buckled to the bridle to manage the horse by.
ROCKERS — The flat pieces of timber fixed within the bottom side, on which the bottom boards are nailed, for the purpose of sinking the bottom, to give more height within the body.
ROLLERS — See *glass* and *splinter bar rollers*.
ROOF RAILS — The top framing of a coach or chariot body, on which the roof is fixed.
ROSES — Round ornaments for a horse's head, mostly made up of silk or worsted ribbons; also a small trimming, through which the hand holders are fixed.
ROUND COLLAR — The type of collar that goes round a horse's neck (also called a *neck collar*), as distinguished from a breast collar.
ROUND ROBBINS — Broad rims fixed to the ends of the axletree bed, to cover the back of the fore wheel, and for preventing dirt from falling in to injure the arms of the axletree.
RUNNING LOOPS — Leather loops, which slide on the reins to keep the points down.

S

SAFE BRACES — Braces, which are placed so as to support the body, if by accident, its other supporters should break.
SALISBURY BOX or BOOT — A coach box of a peculiar form, imitating those originally made to the Salisbury stages.
SCREWING A BOLT — Mending the thread of it, when injured by rust, or a bruise.
SCREWING UP THE BOLT — Is the tightening the nuts to keep the work firm.
SCROLL — An ornament, carved at the end of the timber.
SCROLL SPRINGS or SCROLL LOOPS — Are springs and loops, when bent round in the form of a scroll.
SCUTCHEONS — Small plates, fixed between the leather, and the shoulders of the territts, etc.
SEAMING LACE — A round lace, which is sewed in the corners, and round the edges of the linings.
SEAT BOARDS — The boards, nailed to the seat rails, on which are placed the cushions.
SEAT BOX — A box, which slides under the seat of the body.
SEAT FALL — A piece of cloth, nailed on the edge of the seat, trimmed with lace, and placed for ornament and also to cover the vacant space.
SEAT-IRONS — Strong irons made in the form of a T, with loops at the end for the cradle to be fixed to, on which the coachman's seat is placed.
SEAT RAILS — The cross framing, on which the seat boards are nailed.
SEAT ROLLS — A strip of cloth, nailed along the front of the seat, and stuffed in form of a roll, to keep the cushions in their place.
SHACKLE — A square iron loop, which is hung on the top of the springs, for the braces to hang by.
SHAFTS — The long timbers, in which the horse is placed, to a two-wheeled chaise.
SHAFT TUG — Part of a chaise harness, in which the shafts of a one-horse chaise are hung.
SHAM JOINT — A metal joint, made in imitation of a real joint, applied to a carriage body with a fixed head, for ornamentation only.
SHUTTER STRING — A tape nailed on the shutter, by which it is pulled up or down.
SLATT — The wooden ribs of a chaise or landau head.
SLEEPING CUSHIONS — See *squabs*.
SLIDING SEAT — A seat, which occasionally moves higher or lower, to accommodate ladies in their head dress; also a small seat that draws out to accommodate a third person to sit on.

SOCKET — An iron ferrule on the end of a wooden part, either for strength, or the attachment of some other part.
SPLASHING LEATHER — Same as *dashing leather*.
SPLINTER BAR — The fore bar, which the horses are fastened to, and draw by.
SPLINTER BAR ROLLS, or ROLLER BOLTS — Are strong bolts, with large round flat heads, and thick rollers, round which the traces are fastened.
SPLINTER BAR SOCKETS — Iron ferrules, for the splinter bar ends.
SPOKES — The timbers, which support the rim of the wheel from the center.
SPRING — A part made of either metal or wood, used to support the body on the carriage, and provide an easier movement to the occupants.
SPRING BACK PLATE — The outside, or main plate of a spring.
SPRING BARS, BEDS, or TRANSOMS — The timbers, on which the springs are placed.
SPRING CORDING — The purpose of cording a set of springs is to prevent danger and delay, if by accident a plate should break, and also to strengthen them when required to be heavy loaded; to carriages that have heavy imperials, and much luggage in the body, it is very necessary, which is done by placing a thin piece of ash, or a length of cord along the back, and afterwards twisting a small, but strong, cord round, and fastening it well at the top.
SPRING CURTAIN — A silk cutain, which draws down over the lights or windows, and instantly rises on pulling the trigger, by means of a concealed spring.
SPRING GUT PLATE — The inside plate of the spring.
SPRING HOOP — The hoop which confines the plates.

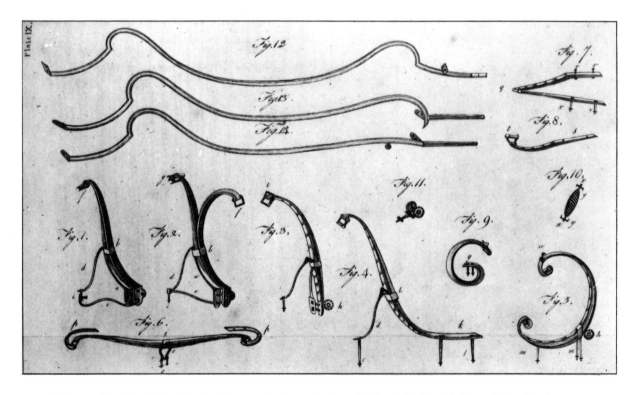

Springs and perches. (From William Felton: *Treatise on Carriages,* 1796, vol. 1, plate 9.) Figure 1, S-spring, for COACHES and CHARIOTS; 2, double spring, for COACHES and CHARIOTS; 3, GIG or CURRICLE hind spring; 4, long-tail PHAETON spring; 5, scroll spring, for various carriages, such as PHAETONS, COACHES, etc.; 6, grasshopper, or double elbow spring, for light WHISKIES or CHAIRS; 7, single elbow spring, for PHAETON or GIG foreends; 8, loop spring; 9, French-horn spring, for the fore part of a CURRICLE or GIG; 10, worm, or spiral spring, used between the double of a main brace; 11, spring jack; fixed to the bottom of a spring, it receives the brace, and is used to take up the brace; 12, double-bow crane for CRANE-NECK CARRIAGE, 13, half-bow crane for CRANE-NECK CARRIAGE; 14, common crane for CRANE-NECK CARRIAGE.

SPRING PLATE — One of the members of a spring.
SPRING STAY — The irons which support the springs.
SQUABS, or SLEEPING CUSHIONS — Soft thin cushions, hung on the inside of the body, for the shoulders and head to lean against.
STANDARD — The principal part of the coach box, or the perpendicular framings in other parts, such as the fore and hind standards.
STANDARD PLATES or IRONS — The iron work, which fixes the standards in their place.
STANDING PILLAR — An upright part of the framing of the body, which supports the roof, on which the doors hang, and shut against.
STAYS — The iron work, which supports or strenghtens any separate article, such as the horn bar stay, the spring stay, &c.
STEP PIECE BODY — The name of a peculiar formed chaise body.
STEP PLATES — Thin iron plates, for the joints of the steps to wear on, and to preserve the timber.
STEP STOPS — Small iron fixtures, against which the folding steps rest, when let down.
STRAKE — The short pieces of iron, with which the ordinary wheel is shod or rung.
STRAKE NAILS — Long strong nails, with which the strakes are fastened to the wheel.
SULKY — The name of a chariot, which can hold only one person.
SURCINGLE — A leather strap and buckle, sewed to a chaise saddle, the same as a belly band to a housing.
SWAY BAR — A compassed timber, fixed on the futchell, which keeps the fore carriage steady.
SWAY BAR PLATE — A plate screwed on the sway bar to strengthen it.
SWORD CASE — The same as a boodge.

T

TANDUM [sic] — The manner of driving two horses in a team.
TERRITTS — The harness furniture, through which the reins are conducted.
THIMBLE HOOKS AND EYES — Are the iron work, on which the shafts for one-horse phaetons are hung.
THROAT BAND or THROAT LATCH — A strap which buckles on each side of the bridle, placed under the throat.
THROAT BAND DEE — A D fixed on the throat band, to contract the bearing reins.
THUMB NUT or SCREW — A nut with lugs, to be screwed on with the finger and thumb.
TRACE — That part of the harness, by which the horse draws.
TRANSOMS — The timbers of the carriage, which are framed across the perch, on which the springs are fixed.
TREAD — Part of a step or flat place, reserved for the foot to be placed on, when getting in.
TRIMMING — The covering with lace, cloth, leather, &c., the inside or outside of a carriage.
TRUNK FASTENERS — Small iron screws with square heads, by which the trunk is kept steady.
TRUNK STRAPS — Straps, by which the trunk is fastened.
TUB BOTTOM BODY — A body, with a roundish formed bottom.
TUG PLATE — A plate, fixed on the shafts, in which the tugs of a one-horse harness is placed.
TUGS — Part of the harness, which supports the bearings, such as collar or breeching tugs, &c.
TYRE — The iron which rims the wheels.

U

UNDER CARRIAGE — The fore carriage, which conducts the other.

V

VALLENS [sic] — The top rows of broad lace, to the inside of a coach or chariot body, and the front strips of leather, used to the head of a one-horse chaise, &c.
VARNISHING — The covering with a glutinous transparent liquid, which gives lustre to, and preserves the paint.

VENETIAN BLIND — A blind, for the purpose of letting in the air, and shading from the sun, which serves also as a shutter when closed.

VIS-A-VIS — A small body, of a coach form, meant only to contain two passengers, fronting each other.

W

WEBB LACE — A thick coarse kind of lace, mostly used for footman holders.

WELL — A strong box, conveniently placed at the bottom of the body, to carry luggage.

WELTING — Is the sewing a narrow strip of leather over the corner seams of that part which covers the upper part of a body, or boot of a carriage, and which forms a round moulding, and keeps out the wet.

WHEEL FORE END — Is when the front of an upper carriage, has a whole or half circular plate, place horizontally, for the more steady bearing, when the carriage locks or turns.

WHEEL IRONS — Strong irons, which hook or bolt on the end of the splinter bar sockets, and go on to the end of the fore axletree arm, between the wheel stock and nut, in order to stay and strengthen the splinter bar, and assist the coachman in mounting.

WHEEL-PIECE — A casing on the horizontal half-wheel plate, placed there for no other purpose than to ornament the iron.

WHEEL PLATE — The circular iron flat plate, on the fore end of the carriage.

WHISKEY — A lighter sort of a one-horse chaise than usual.

WINGS — The extended timbers of a carriage; also what is fixed to the sides of a chaise or phaeton body for the elbows to rest on.

WINKER — A broad leather on each side of the bridle, which prevents the horse from seeing any way but before him.

WITHER STRAP — A part of the harness, which goes round the withers of the horse to hold up the collar.

WOODCOCK EYE — A small iron instrument, fixed to the end of a trace, which hooks on the splinter bar end for drawing by.

WORM SPRING — A narrow steel plate, twisted round in a spiral form, fixed in the double of the main brace, to assist it in giving ease.

Selected Bibliography

Adams, William Bridges. (1837) *English Pleasure Carriages:* London, Charles Knight & Co.
Akers, Dwight. (1938) *Drivers Up - The Story of American Harness Racing:* New York, G.P. Putnam's Sons.
Anderson, M.B. (1877) *A Treatise on Carriage Trimming:* Chicago, J.E. Packard & Co.
Arlot, M. (1873) *Complete Guide for Coach Painters:* Philadelphia, H.C. Baird.
Arnold, James. (1969) *The Farm Wagons of England and Wales:* London, John Baker.
Ashford, W.G. (1893) *Whips and Whip-Making:* Walsall, England, T. Kirby & Son.
Austin, K.A. (1967) *The Lights of Cobb & Co:* Adelaide, Australia, Rigby Limited.
Back, Joe. (1959) *Horses, Hitches and Rocky Trails:* Denver, Sage Books.
Baines, Frederick Ebenezer. (1895) *On the Track of the Mail-Coach:* London, R. Bentley & Son.
Banning, W. and G.H. Banning. (1930) *Six Horses:* New York, Century Co.
Beach, Belle. (1912) *Riding and Driving for Women:* New York, Charles Scribner's Sons.
Beaufort, The Duke of. (1889) *Driving:* London, Longmans, Green & Co.
Beebe, Lucius, and Charles Clegg. (1949) *U.S. West - The Saga of Wells Fargo:* New York, E.P. Dutton & Co.
Belloc, Hilaire. (1926) *The Highway and its Vehicles:* London, The Studio, Ltd.
Berkebile, Don H. (1959) *Conestoga Wagons in Braddock's Campaign, 1775:* Paper 9, U.S. National Museum Bulletin 218; Washington, Smithsonian Institution.
Bird, Anthony. (1969) *Roads and Vehicles:* London, Longmans, Green & Co., Ltd.
Bishop, John Leander. (1868) *A History of American Manufactures from 1608 to 1860:* 3 volumes, Philadelphia, Edward Young & Co.
Blew, William C. A. (1894) *Brighton and its Coaches - A History of the London and Brighton Road:* London, J.C. Nimmo.
Bloodgood, Lida Fleitmann. (1959) *The Saddle of Queens; the Story of the Side-Saddle:* London, J.A. Allen.
———— and Piero Santini. (1964) *The Horseman's Dictionary:* New York, Dutton.
Bourn, Daniel. (1763) *A Treatise upon Wheel-Carriages:* London, S. Crowder.
Bradley, Tom. (1889) *The Old Coaching Days in Yorkshire:* Leeds, England, Yorkshire Conservative Newspaper Co.
Burgess, James W. (1881) *A Practical Treatise on Coach-Building:* London, Crosby Lockwood & Co.
Carlisle, Lilian Baker. (1956) *The Carriages at Shelburne Museum:* Shelburne, Vermont, The Shelburne Museum.
Carter, William Giles Harding. (1918) *Horses, Saddles and Bridles:* Baltimore, The Lord Baltimore Press.
Clark, G.W. (1870) *The Carriage Makers' and Painter's Guide:* Buffalo, New York, Haas & Kelley.
Clear, Charles R. (1955) *John Palmer, Mail Coach Pioneer:* London, Blandford Press.
Coleman, J. Winston, Jr. (1935) *Stage-Coach Days in the Bluegrass:* Louisville, Ky., The Standard Press.
Collins, Herbert R. (1965) *Red Cross Ambulance of 1898:* Paper 50, U.S. National Museum Bulletin 241; Washington, Smithsonian Institution.
Collins, Ivan L. (1953) *Horse Power Days:* Stanford, Calif., Satnford University Press.
Conkling, R.P., and M.B. Conkling. (1947) *The Butterfield Overland Mail:* 3 volumes; Glendale, Calif., A.H. Clark Co.
Cook, John. (1817-1818) *Cursory Remarks on the Subject of Wheel Carriages:* 2 volumes; London, R.S. Kirby.

Corbett, Edward. (1890) *An Old Coachman's Chatter:* London, Richard Bentley & Son.
Cross, Thomas. (1861) *The Autobiography of a Stage-Coachman:* London, Hurst and Blackett.
Daly, H.W. (1916) *Manuel of Pack Transportation:* Washington, Government Printing Office.
Damase, Jacques. (1968) *Carriages:* New York, G.P. Putnam's Sons.
Deacon, William. (1808) *Remarks on Wheel Carriages, etc.:* London, W. Wilson.
Dollar, Jno. A.W., and Albert Wheatley. (1898) *A Handbook of Horseshoeing:* Edinburgh, Scotland, David Douglas.
Dunbar, Seymour. (1915) *A History of Travel in America:* Indianapolis, The Bobbs-Merrill Co.
Durrenberger, Joseph Austin. (1931) *Turnpikes:* Valdosta, Ga., Southern Stationary & Printing Co.
Dwyer, Francis. (1869) *On Seats and Saddles:* Philadelphia, J.B. Lippincott & Co.
Earle, Alice Morse. (1900) *Stage-Coach and Tavern Days:* New York, The Macmillan Co.
Edgeworth, Richard Lovell. (1817) *An Eassay on the Construction of Roads and Carriages:* London, R. Hunter.
Edwards, E. Hartley. (1972) *Saddlery:* London, J.A. Allen & Co.
Eggenhofer, Nick. (1961) *Wagons, Mules and Men:* New York, Hastings House Publishers.
Erskine, Albert Russel. (1924) *History of the Studebaker Corporation:* Chicago, Donnely and Sons.
Farr, William and George A. Thrupp. (1888) *Coach Trimming:* London, Chapman and Hall.
Faudel-Phillips, H. (1943) *The Driving Book:* Waltham Cross, Herts, England; published by the author.
Felton, William. (1796) *A Treatise on Carriages:* 2 volumes; London, printed for the author.
Fitzgerald, William N. (1881) *The Carriage Trimmer's Manuel and Guide Book and Illustrated Technical Dictionary:* New York, n.p.
————. (1875) *The Harness Makers' Illustrated Manuel:* New York.
Forbes, Allan and Ralph M. Eastman. (1953-1954) *Taverns and Stagecoaches of New England:* 2 volumes; Boston, The Rand Press.
Fox, Charles Philip. (1953) *Circus Parades:* Watkins Glen, New York, Century House.
Fry, Joseph Storrs. (1820) *An Eassay on the Construction of Wheel-Carriages:* London, J. & A. Arch.
Gardner, Franklin B. (1879) *The American Method of Carriage Painting:* New York, Valentine & Co.
————. (1871) *The Carriage Painters' Illustrated Manuel:* New York, S.R. Wells.
Gilbey, Walter. (1903) *Early Carriages and Roads:* London, Vinton & Co.
————. (1901) *Riding and Driving Horses:* London, Vinton & Co.
Giraud, Byng. (1891) *Stable Building and Stable Fitting:* London, Batsford.
Gregg, Josiah. (1844) *Commerce of the Prairies:* New York, H.G. Langley.
Guiet, A. (1894) *Carrosserie, Harnais, Velocipedes et Accessories:* Paris, Imprimerie Nationale.
Hampton, Jehiel B. (1886) *A Pocket Manuel for the Practical Mechanic, or the Carriage Maker's Guide:* Indianapolis, F.H. Smith.
Harper, C.G. (1903) *Stage-Coach and Mail in Days of Yore:* 2 volumes; London, Chapman and Hall.
Harris, Stanley. (1885) *The Coaching Age:* London, R. Bentley & Son.
————. (1882) *Old Coaching Days:* London, R. Bentley & Son.
Hasluck, Paul N. (1905) *Harness Making:* London, Lockwood & Co.
Hedges, E.W. (1893) *The World's Carriage Building Center, Cincinnati, Ohio, U.S.A.:* Cincinnati, Robert Clarke & Co.
Heergeist, C.A. (1916) *Vehicle Trimming:* Philadelphia, Ware Bros.
Hillick, M.C. (1898) *Practical Carriage and Wagon Painting:* Chicago, Press of the Western Painter.
Holmstrom, J.G. (1904) *Modern Blacksmithing and Horseshoeing:* Chicago, Drake.
Hope, Charles Evelyn Graham. (1965) *The Pony Owner's Encyclopedia:* London, Pelham Books.
Hornung, Clarence P. (1959) *Wheels Across America:* New York, A.S. Barnes.
Houghton, George W.W. (1889) *Lessons in Lettering and Monogram-Making:* New York, The Hub.
————. (1892) *The Hubs Vocabulary of Vehicles:* New York, Trade News Publishing Co.
Howlett, Edwin. (1894) *Driving Lessons:* New York, R.H. Russell & Son.
Hub Publishing Co. (1873) *Carriage Drafts:* New York, Hub Publishing Co.
————. (1876) edited by George W.W. Houghton. *Draft-Book of Centennial Carriages:* New York, Hub Publishing Co.
Hulbert, Archer B. (1902-1905) *Historic Highways of America:* 16 volumes; Cleveland, The A.H. C Clark Co.

Hungerford, Edward. (1949) *Wells Fargo, Advancing the American Frontier:* New York, Bonanza Books.

Inman, Col. Henry. (1897) *The Old Santa Fe Trail, The Story of a Great Highway:* New York, The Macmillan Co.

Jacob, Joseph. (1773) *Observations on the Structure and Draught of Wheel-Carriages:* London, E. & C. Dilly.

Jenkins, J. Geraint. (1961) *The English Farm Wagon:* Lingfield, Surrey, England, The Oakwood Press.

Kirkman, Marshall M. (1895) *Classical Portfolio of Primitive Carriers:* Chicago, The World Railway Publishing Co.

Lee, Charles E. (1962) *The Horse Bus as a Vehicle:* London, British Transport Commission.

Lee, Wilson H. (1887) *American Carriage Directory:* New Haven, Conn., Price and Lee Co., published annually beginning in 1887.

Longstreet, Stephen. (1952) *A Century on Wheels - The Story of Studebaker:* New York, H. Holt & Co.

Lungwitz, Anton. (1902) *The Complete Guide to Blacksmithing, Horseshoeing, Carriage and Wagon Building and Painting:* Chicago, M.A. Donohue Co.

Lynch, James. (1806) *An Essay on the Construction and Properties of Wheel Carriages:* Dublin, n.p.

Malet, Harold E. (1876) *Annals of the Road:* London, Longmans, Green & Co.

Marlowe, George Francis. (1945) *Coaching Roads of Old New England:* New York, Macmillan.

Masury, John W. (1871) *The Carriage Painters' Companion:* New York, W.J. Read.

Maudslay, Athol. (1888) *Highways and Horses:* London, Chapman & Hall.

McAdam, John Loudon. (1821) *Remarks on the Present System of Road Making:* London, H. Bryer.

McCausland, Hugh. (1948) *The English Carriage:* London, The Batchworth Press.

Meyer, Balthaser H. (1917) *History of Transportation in the United States before 1860:* Washington, D.C., Gibson Brothers.

Mitchell, Edwin V. (1937) *Horse and Buggy Age in New England:* New York, Coward-McCann.

Mitman, Carl W. (1935) *An Outline of Highway Travel, Especially in America:* Washington, D.C., Government Printing Office.

Moody, Ralph. (1967) *Stagecoach West:* New York, Thomas Y. Crowell Co.

Moore, H.C. (1902) *Omnibuses and Cabs:* London, Chapman and Hall, Ltd.

Omwake, John. (1930) *Conestoga Six-Horse Bell Teams of Eastern Pennsylvania:* Cincinnati, The Ebbet and Richardson Co.

Osbahr, Alan. (1962) *Felton's Carriages:* London, Hugh Evelyn, Ltd.

Parnell, Henry. (1838) *A Treatise on Roads:* London, Longmans.

Partington, Charles Frederick. (1825) *The Coach-Makers' and Wheelwrights' Complete Guide:* London, Sherwood, Gilbert and Piper.

Philipson, John. (1897) *The Art and Craft of Coachbuilding:* London, Geo. Bell & Sons.

-----. (1882) *Harness, As it Has Been, As it Is, and As it Should Be:* London, Edward Stanford.

-----. (ed.) (1890) *Reports on Carriages in the Paris Exhibition, 1889:* Newcastle-upon-Tyne, J. K Kemp & Co.

Reck, Franklin M. (1938) *The Romance of American Transportation:* New York, Thomas Y. Crowell Co.

Reid, James. (1933) *The Evolution of Horse-Drawn Vehicles:* n.p., Institute of British Carriage and Automobile Manufacturers.

Rich, George E. (1907) *Artistic Horseshoeing:* Akron, Ohio, The Commercial Printing Co.

Richardson, Milton Thomas. (1888-1891) *Practical Blacksmithing:* 4 volumes; New York, M.T. Richardson.

-----. (1892) *Practical Carriage Building:* New York, M.T. Richardson.

-----. (1889) *The Practical Horseshoer:* New York, M.T. Richardson.

Riker, Ben. (1948) *Pony Wagon Town:* Indianapolis, Bobbs-Merrill.

Rittenhouse, Jack D. (1948) *American Horse-Drawn Vehicles:* Los Angeles, Floyd Clymer.

-----. (1961) *Carriage Hundred:* Houston, Stagecoach Press.

Rogers, Fairman. (1899) *A Manuel of Coaching:* New York & London, Lippincott.

Root, Frank A. and William E. Connelley. (1901) *The Overland Stage to California:* Topeka, Kansas, published by the authors.
Rose, Albert C. (1953) *Historic American Highways:* Washington, D.C., American Association of State Highway Officials.
Roubo, Andre Jacob. (1771) *L'Art du Menuisier - Carrossier:* Academie des Sciences; Descriptions des arts et metiers, volume 17, Paris.
Schriber, Fritz. (1891) *The Complete Carriage and Wagon Painter:* New York, M.T. Riachardson.
Searight, Thomas B. (1894) *The Old Pike:* Uniontown, Penna., published by the author.
Self, Margaret Cabell. (1963) *The Horseman's Encyclopedia:* New York, Barnes.
Settle, Raymond W. and Mary Lund. (1966) *War Drums and Wagon Wheels — The Story of Russell, Majors and Waddell:* Lincoln, Nebr., University of Nebraska Press.
Shumway, George, Edward Durell and Howard C. Frey. (1964) *Conestoga Wagon, 1750-1850:* York, Pa., Trimmer Printing, Inc.
Skeat, Walter W. (1893) *An Etymological Dictionary of the English Language:* New York, Macmillan & Co.
Society for Promoting Christian Knowledge. (1853) *The Book of Carriages:* London, R. Clay.
Spafford, Horatio Gates. (1815) *Some Cursory Observations on the Ordinary Construction of Wheel-Carriages:* Albany, E. & E. Hosford.
Stratton, Ezra M. (1878) *The World on Wheels:* New York, published by the author.
Straus, Ralph. (1912) *Carriages and Coaches:* London, Martin Secker.
Sturt, George. (1958) *The Wheelwright's Shop:* Cambridge, England, Cambridge University Press.
Summerhays, Reginald Sherriff. (1952) *Encyclopaedia for Horsemen:* London and New York, F. Warne.
Taplin, William. (1812) *The Gentlemen's Stable Directory:* Philadelphia, James Webster.
Tarr, Laszlo. (1969) *The History of the Carriage:* New York, Arco Publishing Co., Inc.
Taylor, Louis. (1966) *Bits, Their History, Use and Misuse:* New York, Harper and Row.
Thompson, G.B. (1958) *Primitive Land Transport of Ulster:* Belfast, Ireland, Belfast Museum and Art Gallery.
Throm, Edward L. (ed.) (1952) *Popular Mechanics' Picture History of American Transportation:* New York, Simon and Shuster.
Thrupp, G.A. (1877) *The History of Coaches:* London, Kerby & Endean.
Timmis, Reginald Symonds (1965) *Driving and Harness:* London, J.A. Allen.
Trew, Cecil G. (1951) *The Accoutrements of the Riding Horse:* London, Seeley Service & Co.
Tristram, William Outram. (1888) *Coaching Days and Coaching Ways:* London, Macmillan & Co.
Tunis, Edwin. (1955) *Wheels, A Pictorial History:* Cleveland and New York, World Publishing Co.
Underhill, Francis T. (1897) *Driving for Pleasure:* New York, A. Appleton & Co.
U.S. Army. (1882) *Specifications for Means of Transportation, Paulins, Stoves and Ranges, and Lamps and Fixtures for Use in the United States Army:* Washington, D.C., Government Printing Office.
U.S. War Department. (1917) *Manuel for Farriers, Horseshoers, saddlers and Wagoners or Teamsters:* Washington, D.C., Government Printing Office.
Vestal, Stanley. (1939) *The Old Santa Fe Trail:* Boston, Houghton Mifflin Co.
Vince, John. (1970) *Discovering Carts and Wagons:* Tring, Hertfordshire, England, Shire Publications.
Walker, Henry Pickering. (1966) *The Wagonmasters:* Norman, Oklahoma, University of Oklahoma Press.
Wall, Margaret V. (1954) *The Carriage House of the Suffolk Museum:* Stony Brook, Long Island, N.Y., The Suffolk Museum.
Ware, Francis M. (1903) *Driving:* New York, Doubleday Page & Co.
————. (1903) *First-Hand Bits of Stable Lore:* Boston, Little, Brown.
Ware, I.D. (1875) *The Coach-Makers' Illustrated Hand-Book:* Philadelphia, I.D. Ware.
————. (1876) *Directory of Carriage and Wagon Manufacturers of the United States and Canada:* Philadelphia, I.D. Ware.
Watney, Marylian. (1961) *The Elegant Carriage:* London, Allen & Co.
Webley, A. (1763) *The Nobleman and Gentleman's Director and Assistant in the True Choice of Their Wheel-Carriages:* London, A. Webley.
White, Major General Geoffrey H.A. (1943) *Single and Pair Horse Driving:* London, National Horse Association.
Wilson, Violet A. (1922) *The Coaching Era:* New York, E.P. Dutton & Co.
Winther, Oscar Osburn. (1945) *Via Western Express and Stagecoach:* Stanford, Calif., Stanford University.

PERIODICALS

Of the many trade journals published for the use of the carriage trade, the following are given because they are fairly representative of the industry. *The Hub,* and *The Carriage Monthly* are the most important and the most readily available, both having enjoyed a long period of publication.

The American Blacksmith. New York, 1901-1924, became *The American Garage and Auto Dealer,* 1924-1927.
The American Coach-Makers' Illustrated Monthly Magazine. New York, 1855-1857.
The Blacksmith and Wheelwright. New York, 1880-1932.
The Carriage Builders' and Harness Makers' Art Journal. London, 1859-1862.
The Carriage Journal. Staten Island, New York, 1963 to date. (This is not a trade publication, but one intended for modern-day carriage enthusiasts.)
Coach-Makers' International Journal. Philadelphia, 1865-1873; its title changed to *The Carriage Monthly,* 1873-1915, using the numbering of the above journal; its title changed to *The Vehicle Monthly,* 1916-1921; its title changed to *Motor Vehicle Monthly,* 1921-1930.
The Coach Painter. Newark, New Jersey, 1880-1884.
Le Guide du Carrossier. Paris, 1855-1913.
Harness and Carriage Journal. New York, 1857-1883, merged into *Coach, Harness and Saddlery,* (1882-1886) in 1883, the latter assuming the numbering of the above journal in 1884.
The Hub began in 1869 and ran until 1871, then began a new format of *The Hub* in 1871, absorbing *The New York Coach-Maker's Magazine* (see next entry below), assuming its numbering, replacing also the older *Hub,* and running until 1919. The title then changed to *Automotive Manufacturer,* 1919-1927.
The New York Coach-Maker's Magazine. 1858-1871.
Varnish. Philadelphia, 1888-1904.

THE LIBRARY
ST. MARY'S COLLEGE OF MARYLAND
ST. MARY'S CITY, MARYLAND 20686

088735

THE LIBRARY
ST. MARY'S COLLEGE OF MARYLAND
ST. MARY'S CITY, MARYLAND 20686